# Microarrays

# Integrated Analytical Systems

Series Editor:
Radislav A. Potyailo
GE Global Research Center
Niskayana, NY

For other titles published in this series, go to
www.springer.com/series/7427

Kilian Dill • Robin Hui Liu • Piotr Grodzinski
Editors

# Microarrays

## Preparation, Microfluidics, Detection Methods, and Biological Applications

Springer

*Editors*
Dr. Kilian Dill
CombiMatrix Corporation, Inc.
6500 Harbour Heights Parkway
Suite 301, Mukilteo, WA 98275
USA
redwoodranch@yahoo.com

Dr. Robin Hui Liu
Osmetech Molecular Diagnostics
757 S. Raymond Ave.
Pasadena, CA 91105
USA
Robin.Liu@osmetech.com

Dr. Piotr Grodzinski
National Institutes of Health
National Cancer Institute
31 Center Drive
Bethesda, MD 20892
USA
grodzinp@mail.nih.gov

ISBN: 978-0-387-72716-5 e-ISBN: 978-0-387-72719-6
DOI: 10.1007/978-0-387-72719-6

Library of Congress Control Number: 2008937468



Printed on acid-free paper

springer.com

# Series Preface

In my career I've found that "thinking outside the box" works better if I know what's "inside the box."

Dave Grusin, composer and jazz musician

Different people think in different time frames: scientists think in decades, engineers think in years, and investors think in quarters.

Stan Williams, Director of Quantum Science Research,
Hewlett Packard Laboratories

Everything can be made smaller, never mind physics;
Everything can be made more efficient, never mind thermodynamics;
Everything will be more expensive, never mind common sense.

Tomas Hirschfeld, pioneer of industrial spectroscopy

## Integrated Analytical Systems

Series Editor: Dr. Radislav A. Potyrailo, GE Global Research, Niskayuna, NY

The book series *Integrated Analytical Systems* offers the most recent advances in all key aspects of development and application of modern instrumentation for chemical and biological analysis. The key development aspects include: (i) innovations in sample introduction through micro- and nanofluidic designs, (ii) new types and methods of fabrication of physical transducers and ion detectors, (iii) materials for sensors that became available due to the breakthroughs in biology, combinatorial materials science, and nanotechnology, and (iv) innovative data processing and mining methodologies that provide dramatically reduced rates of false alarms.

A multidisciplinary effort is required to design and build instruments with previously unavailable capabilities for demanding new applications. Instruments with more sensitivity are required today to analyze ultratrace levels of environmental pollutants, pathogens in water, and low vapor pressure energetic materials in air. Sensor systems with faster response times are desired to monitor transient in vivo events and bedside patients. More selective instruments are sought to analyze specific proteins in vitro and analyze ambient urban or battlefield air.

For these and many other applications, new analytical instrumentation is urgently needed. This book series is intended to be a primary source of both fundamental and practical information on where analytical instrumentation technologies are now and where they are headed in the future.

Looking back over peer-reviewed technical articles from several decades ago, one notices that the overwhelming majority of publications on chemical analysis has been related to chemical and biological sensors and has originated from departments of chemistry in universities and divisions of life sciences of governmental laboratories. Since then, the number of disciplines has dramatically increased because of the ever-expanding needs for miniaturization (e.g., for in vivo cell analysis, embedding into soldier uniforms), lower power consumption (e.g., harvested power), and the ability to operate in complex environments (e.g., whole blood, industrial water or battlefield air) for more selective, sensitive, and rapid determination of chemical and biological species. Compact analytical systems that have a sensor as one of the system components are becoming more important than individual sensors. Thus, in addition to traditional sensor approaches, a variety of new themes has been introduced to achieve an attractive goal of analyzing chemical and biological species on the micro- and nanoscale.

# Preface

The area of microarrays enjoyed unprecedented growth in the last two decades the. The early demonstration of the concept in the seminal work of Fodor et al. led to the formulation of new standards for gene expression studies and molecular diagnostics methodologies. The highly parallel microarray concept evolved from the initial area of genomic studies to proteomics and single-cell studies benefiting a multitude of fields from fundamental systems biology to practical diagnostic tests.

In the mid 1980s, dot-blots and slot-blots were utilized in the earliest array experiments. These low-density and early-stage microarrays were used to assess the identity of the material present as well as the concentration of the constituents. Later, they evolved into modern DNA and protein microarrays mostly produced on membranes. In the late 1980s and early 1990s, DNA microarrays were developed further and were either spotted or synthesized via photolithography on solid surfaces. Affymetrix developed photolithography synthesis method into a mature technology and relied upon a step-by-step base synthesis with “photomasks” and light-labile protecting groups on oligonucleotide synthetic blocks. With time, Affymetrix became one of the largest suppliers of commercial DNA microarrays. In parallel, spotting of presynthesized oligonucleotides on the surface, in predetermined two-dimensional patterns allowed for an inexpensive approach to building custom-designed, “home-made” arrays in many research labs.

Nowadays, DNA microarrays have become a widespread tool used in life sciences, drug screening, and diagnostic applications. With the abundant availability of gene targets and combinatorial chemistry/biology libraries, researchers leveraged the ability to study the effects of diseases, environmental factors, drugs, and other treatments on thousands of genes at once using microarrays. DNA microarrays are used in pharmacogenomic studies that include gene expression profiling, the measurement and analysis of regulated genes under various conditions, and genotyping, the detection of polymorphisms or mutations in a gene sequence. DNA microarrays are also useful in molecular diagnostics, which includes genetic screening (e.g., detection of mutations or inherited disorders), identification of pathogens and resistance in infections, and molecular oncology (cancer diagnosis). Protein microarrays, on the other hand, consist of antibodies, proteins, or protein fragments and are used to screen and assess patterns of interaction with samples containing distinct proteins or classes of proteins. Similar to DNA systems developed earlier, protein microarrays find their use in the identification of diagnostic targets and drug screening. Finally, cell arrays allow for the immobilization of single cells on the solid surface, while maintaining their viability and can be used in ion-channel studies and drug screening experiments as well as in monitored tissue growth.

In this book, not only do we discuss the use of microarrays in DNA studies, but also include peptide arrays, protein arrays, combinatorial chemistry arrays, cell-based arrays, and glycoarrays, to name a few. The book is organized around several features of the microarray field: the biological material studied on the array, the detection methodology, and the application of the array toward specific study or diagnosis of the disease. This organization of the chapters demonstrates the advancement of the field in many different facets and shows the implementation of new technological advances into microarray systems and the subsequent expansion of possible utilization of these systems.

Undoubtedly, we will witness further progress in microarrays due to the introduction of microtechnology, nanotechnology, and modern molecular biology into the field. This book will prove a useful source of current information for researchers in the field of microarrays and for those who are just entering the field of microarray research.

We wish to thank all of the contributing authors for their enthusiasm for the project and their commitment to provide high-quality manuscripts. We are also thankful to our families and coworkers for their patience and support during the course of completing this project.

Kilian Dill
Robin Hui Liu
Piotr Grodzinski

# Contents

# Contributors

**Gregory T. Carroll**
Department of Chemistry, Columbia University, New York, NY

**Grace Y. J. Chen**
Departments of Chemistry and Biological Sciences, National University of Singapore, Singapore 117473

**Bill Coty**
Osmetech Molecular Diagnostics, Pasadena, CA

**David Danley**
CombiMatrix Corp., 6500 Harbour Heights Parkway, Mukilteo, WA 98275

**Anil K. Deisingh**
Department of Chemical and Biomolecular Engineering, Clemson University, Clemson, SC

**Kilian Dill**
Independent Consultant, P. O. Box 1442, Sultan, WA

**H. Sho Fuji**
CombiMatrix Corp., 6500 Harbour Heights Parkway, Mukilteo, WA 98275

**Andrey Ghindilis**
Sharp Labs of America, 5750 NW Pacific Rim Bld, Camas, WA

**Piotr Grodzinski**
Nanotechnology in Cancer, National Cancer Institute, Bethesda, MD; National Institutes of Health, National Cancer Institute, 31 Center Drive, Bethesda MD 20892, USA

**Anthony Guiseppi-Elie**
Department of Chemical and Biomolecular Engineering, Clemson University, Clemson, SC

**Amy E. Herr**
Department of Bioengineering and Biological Sciences, University of California-Berkeley, Berkeley, CA 94720

**Arul Jayaraman**
Department of Chemical Engineering, Texas A & M University, College Station, TX

**Kevin R. King**
Center for Engineering in Medicine, Mass. General Hospital & Harvard Medical School, Boston, MA

**Jeffrey T. Koberstein**
Department of Chemical Engineering, Columbia University, New York, NY

**Robin H. Liu**
Osmetech Molecular, Diagnostics, 757 S. Raymond Ave, Pasadena, CA; CombiMatrix Corporation, Inc, 6500 Harbour Heights Parkway, Suite 301, Mukilteo, WA 98275, USA

**Michael Lodes**
CombiMatrix Corp, 6500 Harbour Heights Parkway, Mukilteo, WA

**Karl Maurer**
CombiMatrix Corp., 6500 Harbour Heights Parkway, Mukilteo, WA

**Andrew McShea**
CombiMatrix Corp., 6500 Harbour Heights Parkway, Mukilteo, WA

**Kevin Moeller**
Department of Chemistry, Washington University, St. Louis, MO

**Uwe R. Müller**
Neoclone, Inc., Madison, WI
and
BANYAN BIOMARKERS, 12085 Research Drive, Alachua, FL 32615

**Sebastian J. Osterfeld**
Departments of Materials Science and Engineering, Stanford University, Stanford, CA

**Resmi C. Panicker**
Departments of Chemistry and Biological Sciences, National University of Singapore, Singapore 117473, Republic of Singapore

**Roger G. Perkins**
Z-Tech Corporation, 3900 NCTR Road, Jefferson, AR

**Michael R. Reed**
Osmetech Molecular Diagnostics, 757 S. Raymond Ave, Pasadena, CA

**Leming Shi**
Center for Toxicoinformatics, US FDA, 3900 NCTR Rd., HFT-020, Jefferson, AR

**Dominic Suciu**
CombiMatrix Corp., 6500 Harbour Heights Parkway, Mukilteo, WA

**Hongyan Sun**
Departments of Chemistry and Biological Sciences, National University of Singapore, Singapore 117473, Republic of Singapore

**Mervyn Thomas**
Emphron Informatics Inc., Queensland 4066, Australia

**Weida Tong**
Center for Toxicoinformatics, US FDA, 3900 NCTR Rd., HFT-020, Jefferson, AR

**Nicholas J. Turro**
Department of Chemistry, Columbia University, New York, NY

**Denong Wang**
Carbohydrate Microarray Laboratory, Stanford University School of Medicine, Stanford University, Stanford, CA 94305

**Joseph Wang**
Department of Nanoengineering, University of California San Diego, San Diego, CA

**Shan X. Wang**
Departments of Material Science, Engineering & Electrical Engineering, Stanford University, Stanford, CA

**Shao Q. Yao**
Departments of Chemistry and Biological Sciences, National University of Singapore, Singapore 117473, Republic of Singapore

**Martin L. Yarmush**
Center for Engineering in Medicine, Mass. General Hospital & Harvard Medical School, Boston, MA;
Center for Bioelectronics and Biosensors, Arizona State University, Tempe, AZ 85387

# Part I
# Overview and New Detection Method

# Chapter 1
# The Current Status of DNA Microarrays

**Leming Shi, Roger G. Perkins, and Weida Tong**

**Abstract** DNA microarray technology that allows simultaneous assay of thousands of genes in a single experiment has steadily advanced to become a mainstream method used in research, and has reached a stage that envisions its use in medical applications and personalized medicine. Many different strategies have been developed for manufacturing DNA microarrays. In this chapter, we discuss the manufacturing characteristics of seven microarray platforms that were used in a recently completed large study by the MicroArray Quality Control (MAQC) consortium, which evaluated the concordance of results across these platforms. The platforms can be grouped into three categories: (1) in situ synthesis of oligonucleotide probes on microarrays (Affymetrix GeneChip® arrays based on photolithography synthesis and Agilent's arrays based on inkjet synthesis); (2) spotting of presynthesized oligonucleotide probes on microarrays (GE Healthcare's CodeLink system, Applied Biosystems' Genome Survey Microarrays, and the custom microarrays printed with Operon's oligonucleotide set); and (3) deposition of presynthesized oligonucleotide probes on bead-based microarrays (Illumina's BeadChip microarrays). We conclude this chapter with our views on the challenges and opportunities toward acceptance of DNA microarray data in clinical and regulatory settings.

**Keywords** *DNA microarray, fabrication, gene expression, microarray experimental process, reproducibility, data analysis, clinical applications, regulatory decision-making*

**List of Abbreviations** MAQC: MicroArray Quality Control; CV: Coefficient of variation; DNA: Deoxyribonucleic acid; ERCC: External RNA Controls Consortium; FC: Fold change; FDA: U.S. Food and Drug Administration; GAPDH: Glyseraldehyde-3-phosphate dehydrogenase; GO: Gene ontology; HCA: Hierarchical cluster analysis; MM: Mismatch; NT: Nucleotide; PCA: Principal

---

L. Shi(✉) and W. Tong
National Center for Toxicological Research, U.S. Food and Drug Administration, 3900 NCTR Road, Jefferson, Arkansas 72079, U.S.A. e-mail: Leming.Shi@fda.hhs.gov.

R.G. Perkins
Z-Tech Corporation, 3900 NCTR Road, Jefferson, Arkansas 72079, U.S.A.

K. Dill et al. (eds.), *Microarrays: Preparation, Microfluidics, Detection Methods, and Biological Applications*,

component analysis; PLIER: Probe logarithmic intensity error; QC: Quality control; qPCR: Quantitative polymerase chain reaction; PM: Perfect match; RIN: RNA integrity number; RNA: Ribonucleic acid; SAM: Significance analysis of microarrays; SD: Standard deviation; SNR: Signal-to-noise ratio; VGDS: Voluntary Genomic Data Submission; WG: Working group

## 1.1 DNA Microarrays: A Revolutionary Tool for Gene Expression Analysis

DNA microarrays exploit the preferential binding of complementary nucleic acid sequences to simultaneously measure expression levels of thousands of genes. In most applications, profiles of gene expression between two or more biological sample groups (e.g., normal versus disease, or control versus drug treatment) are comparatively evaluated. Each gene found to be significantly differentially expressed between samples constitutes a hypothesis for investigating underlying biological mechanisms distinguishing the samples. Patterns of differentially expressed genes may provide biomarkers or fingerprints of, for example, disease or toxicity, that when adequately validated may have utility in such clinical applications as diagnostics, prognostics, and treatment selection. Since their introduction just over a decade ago, microarrays have become ubiquitous in biological and medical research and drug discovery. The scientific community's excitement about microarrays may be unprecedented for a new technology, as evidenced by both the rapid development of many new commercial experimental platforms and an extensive literature documenting their use in highly diverse applications [1–3].

DNA microarrays detect RNAs that may or may not be translated into active proteins. With tens of thousands of distinct probes on an array, each microarray experiment can accomplish the equivalent number of genetic tests in parallel. DNA microarrays, together with protein microarrays, cell microarrays, tissue microarrays, and SNP microarrays offer the promise of providing a detailed, global, and objective survey of the biological system under study. Hope runs high that these high-throughput, parallel assay systems will enable holistic monitoring of the progression of diseases or the temporal response to external stimuli such as drugs or toxicants.

The vexing problem of variation in drug efficacy and toxicity may eventually be solved at least in part by these technologies, launching a new era of personalized medicine matching the right drug at the right dose to the right patient based on genetic profiles and information from SNPs, mRNA expression, protein expression, and microfluidics-based assay systems. The major genetic risk factors for many key human diseases such as diabetes, cancer, heart disease, autism, hypertension, bipolar illness, asthma, Alzheimer's, osteoporosis, and many others may be identified in the next few years with array technologies playing a major role.

Similarly, prospective studies to elucidate the interaction of genetic and environmental risk factors will be possible. Future arrays are envisioned to profile an individual's genetic risk factors as well as adverse drug sensitivities and metabolic characteristics. Ideally, a blockbuster drug would need to have an indication and be effective for a wide diverse patient population. However, as indicated above, differences in the genetic profiles of individual patients determine the differences in drug responses. Therefore, it makes economical and ethical sense to identify the subsets of patients that will most likely respond to individual treatments. Similarly, microarray-based toxicogenomics could enable identification of sensitive populations that are susceptible to adverse responses to drugs or xenotoxicants.

The FDA published the "*Guidance for Industry: Pharmacogenomic Data Submissions*" in order to foster progress in applying pharmacogenomic data in drug development and medical diagnostics (http://www.fda.gov/cder/genomics/). Although most microarray data currently submitted to the U.S. FDA is on a voluntary basis within a program called the Voluntary Genomic Data Submission (VGDS) [4], it is anticipated that such data will eventually become a routine part of new drug or medical device applications to the FDA. The FDA is currently at an early stage of developing guidance on pharmacogenomic codevelopment that addresses the usage of pharmacogenomic data from diagnostic products in guiding therapeutic treatment selection. Although many challenges remain, efforts leading toward individual, personalized treatment selection based on genetic profile are well underway with FDA involvement.

Gene expression profiling with DNA microarrays involves many experimental and analysis steps that can be categorized, as depicted in Fig. 1.1, as sample collection, RNA extraction, cDNA/cRNA synthesis, labeling with fluorescent dye, hybridization, image acquisition and quantification, data analysis, and biological interpretation. As in other experimental systems, each step of a microarray experiment has limitations in measurement precision and in consistent control of numerous sources of variability. Errors or lack of quality controls in each step can undermine quality and reproducibility as a whole and, concomitantly, the outcome and conclusions of a study.

Factors affecting the accuracy and reproducibility can be conveniently grouped into four categories, technical, instrumental, computational, and interpretive. An error in any of these categories could conceivably render data useless and nonreproducible in an otherwise well-performed microarray experiment. Low concordance in some reported cross-platform and cross-lab studies may be attributed to the lack of intralab reproducibility. When data from one lab or platform are unreliable, conclusions regarding comparability of cross-lab or cross-platform studies should be carefully considered. We argue that a careful evaluation of the technical, instrumental, computational, and interpretive factors used in individual microarray studies will enable the determination of the fidelity of those microarray studies. This will, in turn, facilitate determining whether microarray data from any given study are suitable for comparison with other similar studies deemed reliable and

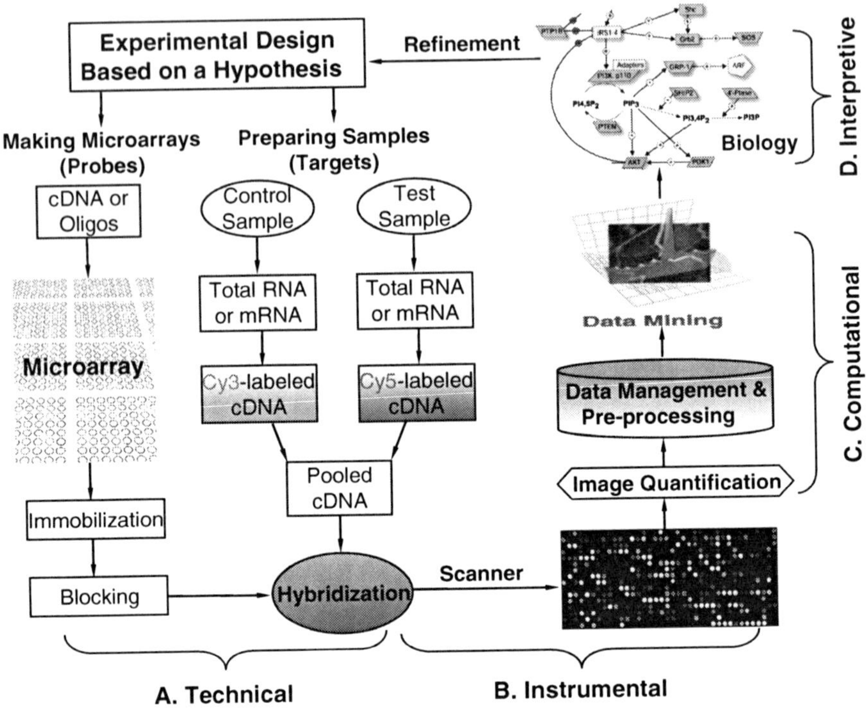


**Fig. 1.1** The microarray experimental process. Adapted from Shi et al. (2004)

of good fidelity. Finally, we urge the microarray community to establish a set of objective criteria for microarray QA/QC so that scientific or regulatory decisions are not made on unreliable data.

Microarray studies are complex experiments having many options for carrying out the many steps depicted in Fig. 1.1. As such, it is unrealistic for regulatory agencies such as the FDA to regulate how the individual microarray steps are conducted. Rather, microarray vendors and investigators should make every possible effort to optimize and standardize microarray operating procedures and establish QA/QC criteria that pertain to each individual step. Otherwise, the quality of data and findings of a complex experimental process can be in doubt. For a regulatory agency, the quality of the final data generated from a microarray study and the availability of appropriate quality assessment measures are more important than regulating individual steps [1].

DNA microarrays were identified by the FDA's Critical Path Initiative (http://www.fda.gov/oc/initiatives/criticalpath/) as a methodology holding "vast potential" for advancing medical product development and personalized medicine through the identification of biomarkers. The Critical Path Opportunities released by the FDA on March 16, 2006 identified 76 areas in priority fields such as genomics, proteomics, and bioinformatics; "biomarker qualification" and "standards for microarray and proteomics-based identification of biomarkers" were highlighted as the top two opportunities.

Excitement about prospects runs high, however, a gap remains between the technology level in use today and levels required for demonstrable improvements in product development and regulatory decision-making. A recent concern pertaining to microarrays is that lists of gene differentiating samples (e.g., disease versus control) cannot be reliably reproduced with similar or identical study designs and/or with different platforms or laboratories [5, 6]. Correspondingly, the reliability and utility of classification models derived from microarray studies for diagnostics purposes or prediction of patient outcomes, for example, has been challenged in recent literature [7–9]. A recent survey of publications examining prediction of cancer outcomes using microarrays revealed that serious flaws in the statistical analysis of microarray data are common [10].

The level of concern across many areas of microarray application tempered the excitement about microarrays, and motivated the FDA to increase its involvement in the standardization of this technology that could benefit the FDA Critical Path Initiative. On February 11, 2005, scientists at the FDA's National Center for Toxicological Research (NCTR), Jefferson, Arkansas, formally launched the MicroArray Quality Control (MAQC) project (http://edkb.fda.gov/MAQC/). The project's mission was to assess reliability, performance, and quality of microarray data, as well as to begin the process of developing standards for microarray use and data analysis [11].

Phase I of the MAQC project (MAQC-I, from February 11, 2005 to September 8, 2006) focused on assessing technical reliability of microarray technology for the identification of differentially expressed genes between a pair of well-established reference RNA samples. MAQC-I involved 137 scientists from 51 organizations, including the six FDA centers, other government agencies (the U.S. Environmental Protection Agency, the National Institutes of Health, and the National Institute of Standards and Technology), manufacturers of microarray platforms and RNA samples, microarray service providers, academic laboratories, and other stakeholders. The MAQC participants donated their time and resources for the completion of MAQC-I. Phase II of the MAQC project (MAQC-II) was officially launched on September 21, 2006 and focuses on the development and validation of predictive models or classifiers, especially those that hold potential for clinical and preclinical (toxicogenomic) applications.

There exist numerous microarray experimental systems (platforms) that utilize markedly different technical features and fabrication techniques, and new, more advanced, and more capable platforms seem to be constantly emerging [12–14]. In this chapter, we discuss the manufacturing characteristics of seven microarray platforms that took part in the MAQC phase I project. These platforms, summarized in Table 1.1, can be organized in three categories: (1) in situ synthesis of oligonucleotide probes on microarrays (Affymetrix GeneChip® arrays based on photolithography synthesis and Agilent's arrays based on inkjet synthesis); (2) spotting of presynthesized oligonucleotide probes on microarrays (GE Healthcare's CodeLink system, Applied Biosystems' Genome Survey

**Table 1.1** Microarray Gene Expression Platforms Included in the MAQC Main Study

| Type | Manufacturer | Code | Protocol | Platform | Number of Probes | Probe Length | Detection |
|---|---|---|---|---|---|---|---|
| In situ synthesis | Affymetrix | AFX | One-Color Microarray | HG-U133 Plus 2.0 GeneChip® | 54,675* | 25 | Fluorescence |
| In situ synthesis | Agilent | AGL | Two-Color Microarray | Whole Human Genome Oligo Microarray, G4112A | 43,931 | 60 | Fluorescence |
| In situ synthesis | Agilent | AG1 | One-Color Microarray | Whole Human Genome Oligo Microarray, G4112A | 43,931 | 60 | Fluorescence |
| Pre-synthesized oligos | Applied Biosystems | ABI | One-Color Microarray | Human Genome Survey Microarray v2.0 | 32,878 | 60 | Chimiluminescence |
| Pre-synthesized oligos | GE Healthcare | GEH | One-Color Microarray | CodeLink™ Human Whole Genome | 54,359 | 30 | Fluorescence |
| Pre-synthesized oligos | NCI_Operon | NCI | Two-Color Microarray | Operon Human Oligo Set v3 | 37,632 | 39–70 | Fluorescence |
| BeadChip microarrays | Illumina | ILM | One-Color Microarray | Human-6 BeadChip, 48 K v1.0 | 47,293 | 50 | Fluorescence |

*Indicates the number of probesets, each of which usually contains 11 pairs of probes, half of which are perfect matches and the other half are mismatches.

Microarrays, and the custom microarrays printed with Operon's oligonucleotide set); and (3) deposition of presynthesized oligonucleotide probes on microsphere-based microarrays (Illumina's BeadChip microarrays).

Researchers should bear in mind that different types of microarray platforms will produce different types of errors and allow different types of quality control. Platforms that synthesize oligonucleotides in situ reduce errors that would otherwise result from the hydrolyzing the oligonucleotide from its synthetic support and reattaching it to the microarray. However, the in situ approach precludes the ability to confirm sequence synthesis and the ability to purify the oligo.

## 1.2 In Situ Synthesis of Oligonucleotide Probes on Microarrays

### *1.2.1 Affymetrix GeneChip Platform: Photolithographic Synthesis*

The Affymetrix GeneChip microarrays are the most widely used microarray system for gene expression profiling. This platform comprises high-density, standardized microarrays and reagents, and instrumentation, together with tools for managing, processing, and analyzing the array data. Affymetrix manufactures its GeneChip microarrays using photolithography. The robotic manufacturing process integrates semiconductor fabrication, solid-phase chemical synthesis, combinatorial chemistry, and molecular biology [15, 16].

During microarray fabrication, 5 in. × 5 in. quartz wafers are coated with a light-sensitive chemical that prevents coupling between the wafer and the first nucleotide of the DNA probe being created. Computer-controlled lithographic masks are next successively placed on wafer locations according to a program that specifies whether light should be transmitted or blocked on specific locations of the wafer surface (Fig. 1.2). The surface is then successively flooded with solution containing adenine, thymine, cytosine, or guanine, causing coupling to occur at wafer positions unprotected by illumination. Because each coupled nucleotide carries a light-sensitive protecting group, the masking and coupling can be consecutively repeated until the probe at each position reaches its design length.

The Affymetrix GeneChip oligos are normally 25 nucleotides in length, and currently available arrays are typically manufactured at a density of more than 1.3 million unique features per array (for the U133 Plus2.0). Each quartz wafer can be subdivided into tens to more than 100 distinct centimeter-square, independent arrays, depending on the requirements of the experiment and the number of probes needed for each array. For example, the semiconductor fabrication technology enables array feature size to be reduced and the advances in optics make it possible to scan the arrays at high resolution, allowing more probes per unit area, providing

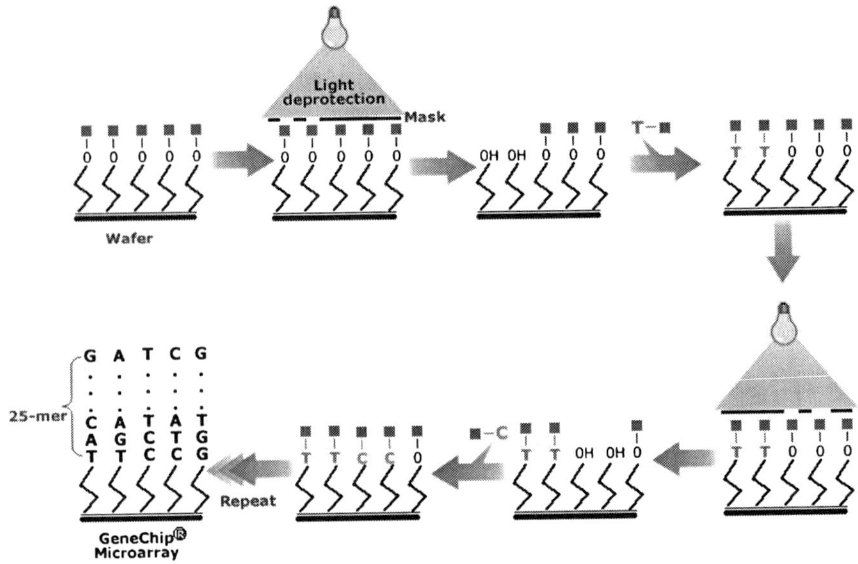


**Fig. 1.2** Affymetrix uses a unique combination of photolithography and combinatorial chemistry to manufacture its GeneChip® microarrays. Adapted from Affymetrix, Inc

a corresponding increase of expression information. Sampling control hybridizations on arrays from every wafer provides a measure of batch quality.

Unique to Affymetrix GeneChip microarrays is the concurrent use of multiple probes, called a probeset, that enables a direct means of control of false signals from nonspecific sequences or cross-hybridization, for example. The high density generally allows some 22 probes (or 11 pairs of probes) to be used for each transcript's expression measurement. Of these, half are "perfect match (PM)" and the remaining half are "mismatches (MM)" to the corresponding PM probes. An MM probe contains a single mismatch located exactly in the middle (i.e., the 13th base) of the 25-base PM probe sequence. A PM probe provides the entire fluorescence measurement for the target sample binding to it, whereas the paired MM probe provides the means for estimating nonspecific fluorescence in the measurement. Many algorithms (e.g., MAS 5.0, dCHIP, RMA, PLIER, and other variants) are available to use probe-level data to derive probeset-level data that correlate with transcript abundance. It should be noted, however, the relative merits of such "summary" methods remain in evaluation and debate [17].

Probe design is paramount for all microarrays, and constitutes an ongoing effort aimed at continual improvement. At any point in time, the best possible probes for a target species are limited by the extant genome sequence information. Multiple probes are designed to uniquely represent each transcript or DNA sequence to be interrogated. The design process requires that millions of raw sequences and SNPs from multiple databases in the public domain be screened. Probes are designed to yield efficient hybridization, desired intensity, and relative concentration-dependence, as well as to minimize cross-hybridization. The probe sequence descriptions and annotations are openly available at the NetAffx Analysis Center (http://www.affymetrix.com/analysis/index.affx), which provides researchers the ability to examine results in the context of array design and annotation information.

The Affymetrix technology offers a system platform with the flexibility to study expression at the gene, exon, or whole-genome level. Arrays for gene-level expression profiling are available for human, mouse, rat, and other major model organisms. For the "standard" gene expression analysis, probes are designed to be 3′-end biased. Alternative splicing and biomarker research can be conducted at the exon level by using the newly released GeneChip exon microarrays, which can profile some million exons representing essentially all coding genes on a single microarray. GeneChip tiling microarrays enable researchers to look in areas of the genome that were previously unexplored for understanding novel transcripts and protein/DNA interactions.

### 1.2.2 Agilent Platform: Inkjet Synthesis

Agilent's microarray technology is based on Hewlett-Packard's noncontact inkjet printing technology [18]. Nucleotides are printed on glass slides in a manner analogous to printing a document with an inkjet printer on paper, where a computer controls location, color, and amount of ink from reservoirs. Glass wafers are coated with a hydrophobic surface with exposed hydroxyl groups that will strongly bond with both the glass and the nucleic acids to be printed. Probe oligos are synthesized by printing one base at a time using standard phosphoramidite chemistry. In phosphoramidite synthesis reactions, the reactive sites on the nucleotides are blocked with chemical groups that can be selectively removed, enabling base chains to be built in a highly controlled manner. After printing the first base, the trityl group that protects the 5′ hydroxyl group on the nucleotide is removed and oxidized to activate it, allowing it to react with the 3′ group on the next nucleotide. Excess reagent is washed away between each step to preclude residual contamination. The overall process is depicted in Fig. 1.3.

Nucleotide printing is by de-tritylation, oxidation, and washing, and is repeated 60 times in order to synthesize oligo probes having 60 bases. After printing the last base, the microarray undergoes a final unprotect step, which is followed by scribing, dicing, and quality control testing. After manufacture, arrays are diced to a standard size of $25 \times 75$ mm that accommodates commercial scanners. Sample microarrays from the beginning, middle, and end of each printing run are selected for quality control testing; none of the microarrays from that print run is packaged and shipped to customers unless the sample microarrays pass quality control. Agilent uses the described noncontact in situ synthesis process to produce both "off-the-shelf" commercial microarrays, as well as custom oligonucleotide-based microarrays.

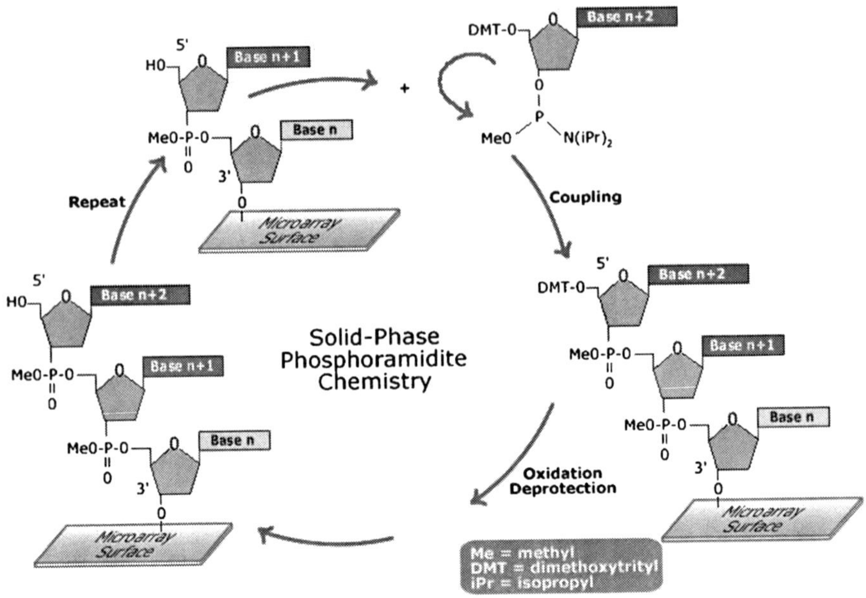


**Fig. 1.3** The general cycle of oligo synthesis via phosphoramidite chemistry used by Agilent. Adapted from Agilent Technologies, Inc

Agilent also produces deposition microarrays that, in contrast to synthesized oligo microarrays, are printed on the microarray surface using DNA molecules prepared prior to printing. These cDNA microarrays are printed with clone PCR products from Incyte Genomics' LifeSeq clone sets. They are purified and run on gels to check for clone size and absence of multiple bands, thus verifying PCR products' identity and ensuring no cross-contamination. DNA concentrations are then measured, and clones not having concentrations requisite for optimal gene expression performance are flagged. The ratio data between spotted Agilent cDNA microarrays and Agilent's 60-mer oligo microarrays have been found to be highly consistent [18], leading Agilent to discontinue its offering of cDNA microarrays in 2004.

Agilent's microarrays have traditionally used a two-color design where each of two samples under study is each distinctively labeled with two dyes (Cy3 and Cy5) that are cohybridized on the same microarray in a single hybridization. This experimental scheme is the same as the early spotted microarrays developed by Stanford University [19, 20]. To correct inevitable dye-biases [21], Agilent recommends a dye-swap experimental design for two-color experiments, where the hybridization is repeated after reversely labeling the samples, with the two results subsequently averaged. However, such an experimental design becomes cumbersome when there are multiple sample groups unless a common reference RNA is used. To allow more flexible and tractable experimental designs, Agilent recently adopted a one-color experimental design that was demonstrated in the MAQC Phase I [11], where the compared results between the two-color and one-color designs showed high concordance [22]. With the dual-mode, that is, one or two color, Agilent technology provides the user the flexibility to choose the most cost-effective approach.

Improvements leading to increased scanning resolution and corresponding feature density have recently allowed Agilent to offer multiplexed arrays on the standard 25×75 mm slides that contain up to 244,000 unique spots. The multiplexed microarrays can be provided in a variety of formats. Four-plex microarrays provide four microarrays on a single slide, each having 44,000 probes. An eight-plex format offers eight microarrays, each having 15,000 probes. These arrays can be particularly cost-effective for targeted profiling of a large number of samples.

MammaPrint, the first microarray-based gene expression diagnostic assay approved by the U.S. FDA on February 6, 2007 (http://www.fda.gov/bbs/topics/NEWS/2007/NEW01555.html), is based on the Agilent two-color platform design. The MammaPrint assay has recently been implemented in an eight-plex format with a few hundred genes printed in triplicates [23].

## 1.3 Spotting of Presynthesized Oligonucleotide Probes on Microarrays

### *1.3.1 GE Healthcare's CodeLink Platform*

The CodeLink™ microarray platform [24] utilizes a silanized glass slide to achieve coverage with long-chain alkyl groups. A lightly cross-linked polymer film is created by photocoupling a prepolymer of acrylamide. The prepolymer contains an activated ester providing attachment sites for C6-amino-oligonucleotides. Then, 5′-amine-terminated oligonucleotides are deposited onto the polymer by piezoelectric dispensing robots. A fluorescein-derivative dye is dispensed along with the oligonucleotides, rendering every feature element of the slide capable of being scanned and inspected. Placing slides in a humidified environment causes the attachment of the oligonucleotide to the polymer to occur, and for which nonspecific binding or attachment due to the exocyclic amine groups has been shown to be negligible when 5′-hydroxyl-terminated oligonucleotides are used. Additional sites are then blocked and slides are washed, rinsed, and dried, after which an integrated, proprietary, polypropylene hybridization chamber is attached.

The design of CodeLink activated slides is based on the covalent immobilization of amine-modified DNA. They are coated with a novel three-dimensional surface layer composed of a long-chain, hydrophilic polymer containing amine-reactive groups that is covalently cross-linked to itself as well as the slide surface. The cross-linked polymer together with the means of attachment in a gel matrix orients the immobilized DNA away from the surface of the slide, thus enhancing interaction between the probe and the target. Additionally, the hydrophilic nature of the polymer provides stabilization of the immobilized DNA, thus lowering background. CodeLink activated slides are compatible with most available arraying and scanning instruments.

CodeLink slides are manufacured by SurModics and were formerly sold as 3D-Link™ Activated Slides. The Human Whole Genome Bioarray gives comprehensive coverage of the human genome with some ~57 000 transcripts and ESTs, including about 45,000 well-characterized human gene and transcript targets. GE Healthcare decided to discontinue its CodeLink product in April, 2007.

### *1.3.2 Applied Biosystems' Genome Survey Microarrays*

The microarrays manufactured by Applied Biosystems use 60-mer oligonucleotides (oligos) as probes [25]. In general, oligos of this length are considered to provide a better balancing of sensitivity and specificity. The probes on the Applied Biosystems microarrays are built using standard phosphoramidite chemistry and solid-phase synthesis. The probes are deposited and covalently bound at the 3′ end onto the microarray's derivatized nylon substrate that is, in turn, bound to a glass slide.

Digoxigenin-UTP labeled cRNA is generated and amplified from $1\,\mu g$ of total RNA from each sample using Applied Biosystems Chemiluminescent RT-IVT Labeling Kit. The RT labeling process begins with reverse transcription of poly(A) mRNA using an oligo (dT) primer. Transcription of the mRNA into cDNA incorporates a nucleotide modified with digoxigenin (dUTP DIG). Extraneous mRNA is then removed from the mixture. After purification, the DIG-labeled cDNA is ready for hybridization. DIG-labeled cDNA or cRNA is incubated with the microarray in a hybridization chamber for 16 hours at 55°C. After washing to remove unhybridized DIG-labeled molecules, an alkaline phosphatase–antibody conjugate is added to bind to the DIG-labeled target. The addition of substrate and a chemiluminescence enhancer initiates the chemiluminescent reaction. The presence of enhancer further strengthens the chemiluminescent signal. The microarray is then ready for data collection in the Applied Biosystems 1700 Chemiluminescent Microarray Analyzer.

For quality control purposes, a 24-mer oligo, internal control probe (a 24-mer oligo) is cospotted on the microarray along with the unique 60-mer probe (Fig. 1.4). Then, an internal control target oligo, complementary to the internal control probe and prelabeled with a fluorescent dye oligo, is included in the hybridization mixture. The combination control–target control provides the 1700 Microarray Analyzer in fluorescence mode the means to pinpoint each feature site on the array with absolute certainty, independent of the chemiluminescence signal measuring gene expression. The internal control probes serve to monitor manufacturing quality, provide measures of feature-to-feature and array-to-array variations useful for normalization, and provide a means for finding specific spots or defining grids within the array.

The Applied Biosystems 1700 Chemiluminescent Analyzer uses a CCD camera for acquisition of chemiluminescent and fluorescent signals. Custom software translates the intensities to gene expression values, and provides quality control capabilities. The Applied Biosystems microarrays are in closed systems that are incompatible with other microarray and associated detection systems.

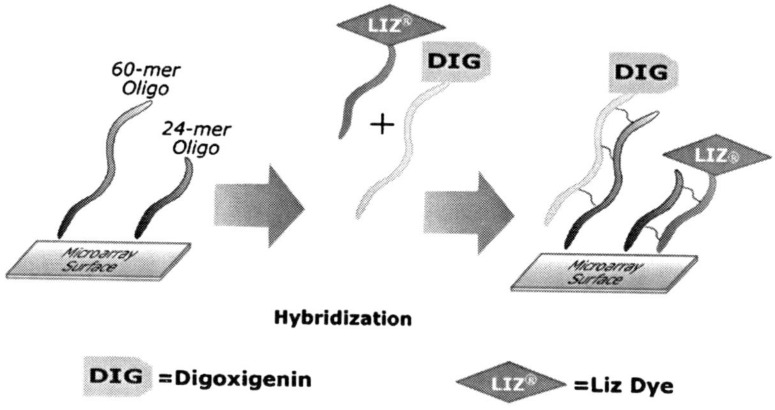


**Fig. 1.4** Applied Biosystems' microarrays use fluorescence to image the microarray features. A 24-mer oligo labeled with the fluorescent LIZ® dye is present during hybridization. This oligonucleotide is complementary to one that is co-deposited at every feature during microarray manufacture. The fluorescent signal, which has a close spatial correlation with the chemiluminescent signal, locates all features on the microarray, even in the absence of gene expression products. The transcripts are quantitated by chemiluminescence. Adapted from Applied Biosystems

### *1.3.3 Custom Microarrays Printed with Operon Oligonucleotides*

The NCI microarray core facility prints customized oligo microarrays based on oligo sets manufactured by Operon, Inc. [26] (Alameda, CA). Their V3 oligoset contains over 30,000 oligonucleotides that are about 70 bases in length. The oligonucleotides are resuspended in 150 mM sodium phosphate, 4% sodium sulfate at pH 8.5. The final concentration of the oligonucleotides is 30 uM. The oligonucleotides have amine groups at their 5′ ends that can react with the activated epoxy slides obtained from Schott (www.schott.com). The epoxy activated slides are Nexterion® Slide E and can react with NH2–, SH–, OH– or other nucleophilic groups.

The oligonucleotides are printed with a GeneMachines® instrument (San Carlos, CA) containing 48 Telechem International SMP 2.5 pins (Sunnyvale, CA) in a 50% humidified atmosphere. After printing, the slides are incubated in the humidified conditions overnight to allow covalent attachment of the oligonucleotides, after which the unused activated sites must be destroyed or inactivated. The slides are first washed to remove unbound oligonucleotides and buffer salts. Slides are then rinsed once for five minutes at room temperature with 0.1% Triton X-100 and twice for two minutes at room temperature with 1 mM HCl. Next, they are washed for ten minutes at room temperature with 100 mM KCl and finally for one minute with deionized water. Immediately after the last wash, the slides are blocked by incubating in Nexterion Blocking Reagent E at 50°C for one hour (at least 100 mls/ 5 slides). The slides are then rinsed once for two minutes in de-ionized water at room temperature followed by a one-minute wash in 95% ethanol. The slides are immediately dried by centrifugation at 200 X g for three minutes. The slides can be used at this point for hybridization to labeled targets or stored at room temperature in a dry atmosphere in the dark. The shelf life of the printed slides should be six months or more.

For use in the MAQC project, CapitalBio Corporation (Beijing, China) generated microarray data from its customized printing of Operon oligo set V2.1 [27]. Both two-color and one-color modes were used and comparable results were obtained [22].

## 1.4 Bead-Based Microarrays: Illumina's BeadChip Microarrays

The manufacturing process of Illumina's microarrays is substantially different than those of other vendors. The process relies on random self-assembly of microspheres into ordered microwells [28]. Semiconductor technology is used to build substrates containing millions of wells in highly ordered, predefined patterns. Presynthesized oligonucleotide probes are immobilized onto microspheres to produce the array elements. These microspheres are quantitatively pooled and introduced to the etched microwell substrates where they spontaneously assemble into the wells. The resultant Human-6 Expression BeadChip has more than 48,000 different microspheres with unique sequences; the HumanRef-8 Expression BeadChip has more than 24,000.

Each microsphere contains hundreds of thousands of copies of covalently attached oligonucleotide probes. The spheres are assembled into more than 1.6 million pits, each measuring 3 μm in diameter. The configuration provides an average 30-fold measurement redundancy for each represented sequence on the array, increasing the precision of each gene's measurement. After the microspheres are assembled on the array, a hybridization-based procedure is used to map gene positions [28].

The oligos covalently attached to beads contain a 29-base address concatenated to a 50-base gene-specific probe (Fig. 1.5). The address is used to map and decode the microarray, whereas the gene-specific probe is used to quantify expression levels of corresponding transcripts. Only a minimal amount of total RNA (50–100 ng) is required for the single-round transcription (IVT) reaction.

Data for design of the HumanRef-8 Expression BeadChip came from the Curated RefSeq Database of NCBI [29], supplemented by sequences of less well-characterized genes from other sources. An empirical screening process was used to select the best probe for each gene. Multiple probes were made for each gene (three for RefSeq Database genes and two for genes from other sources). These probes were then hybridized to a panel of RNA samples representing 26 human tissue types. The data were analyzed with a proprietary algorithm to select the best probe for each gene on the basis of its responsiveness and specificity across the tissue types.

The Human-6 and HumanRef-8 Expression BeadChips can be considered as "arrays of arrays." That is, multiple arrays exist on each slide, six in the case of the Human-6 product and eight for HumanRef-8. The arrays on each slide are

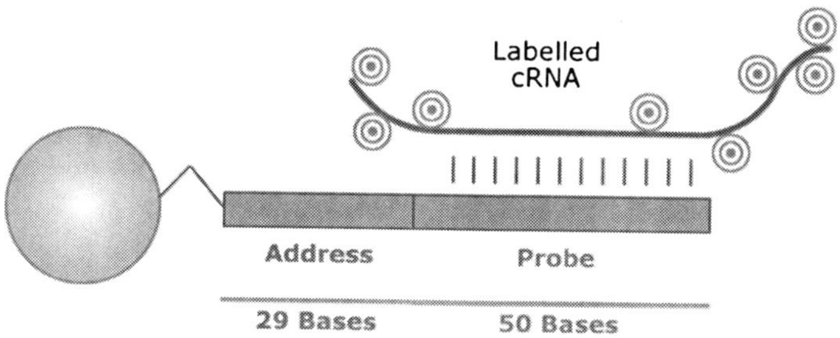


**Fig. 1.5** Gene-specific probes are attached to beads that are then assembled into the arrays. For simplicity, this diagram shows only one oligomer attached to the bead; actual beads have hundreds of thousands of copies of the same sequence attached. Adapted from Illumina, Inc

separated by a seal that enables each array to be hybridized to a different sample. In other words, either six or eight samples can be simultaneously done, which would require six or eight regular microarrays. The resulting parallel conduct of all steps after hybridization potentially reduces handling, time, and cost. Human-6 and HumanRef-8 Expression BeadChips are scanned on the Illumina BeadArray™ reader, a submicron resolution scanner with an adapter tray to simultaneous scan three BeadChips (18 or 24 arrays).

Every array on each Human-6 Expression BeadChip also includes more than 1000 microspheres as controls; more than 775 are included for the HumanRef-8 Expression BeadChip. The controls enable monitoring of such parameters as sample quality, labeling reaction success, hybridization stringency, and signal generation. Additionally, background intensity is estimated by using over 1000 "negative" probes that are designed not to bind to transcripts of the target species.

## 1.5 Concordance of Results from Different Microarray Platforms

Microarrays are indeed remarkable in their ability to generate and test so many hypotheses in parallel. Hundreds or thousands of genes within a genome are identified as those most likely to be associated with the biological phenomena being studied. These genes constitute the so-called differentially expressed (or significant) gene list, the determination of which is a fundamental goal of a microarray experiment. When viewed as a generator of hypotheses, there is little argument that microarray data provide an immense resource for biological interpretation of disease and toxicological mechanisms, for example. Clues are provided regarding biological mechanisms, mechanistic steps within mechanisms, pathways, transcription factors, genetic variability, and so forth, and also provide insights regarding whether and which genes act in concert.

When viewed as a hypothesis tester, in contrast, microarrays present new and vexing problems for the valid statistical interpretation of results. The reliability and validity of interpreted results and conclusions will be judged according to the normal considerations of sensitivity, specificity, and reproducibility, as they should be. Particularly in the area of experiment reproducibility, a paramount scientific principle, microarrays have come under criticism in the past few years [6–8, 30–32] because of poor concordance of significant genes from different platforms and/or laboratories. Shi et al. [11, 33], however, concluded that much of the published criticism was unwarranted because the genes for the compared lists were selected by statistical methods inappropriate for judging microarray experiment reproducibility.

As confirmed in the MAQC study [11, 34], lists of genes selected based on ranking significance by magnitude of differential expression are highly concordant and reproducible. Selecting genes by ranking significance using *P*-values from simple *t*-tests, in contrast, assures poor concordance and reproducibility, especially for short lists of genes. The inappropriateness of stringent, simple statistical gene selection methods is a direct consequence of the huge number of hypotheses being tested in parallel, the usually small number of samples or replicates, and the considerable uncontrolled experimental and biological variability. Selecting genes with small *P*-values results in a short list of true positives that is unlikely to be reproducible. In addition, a long list of highly biologically relevant false negative genes is likely to be lost in the noise when selecting a short list of significant genes with a stringent *P*-value.

The recognition that microarray data require statistical approaches beyond simple *t*-tests has resulted in a huge number of new or revitalized methods. In fact, the number of methods and in many cases the method complexities may seem daunting to many biological researchers. Many methods recognize that using only a specific gene's variance obtained from a small number of replicates risks a false negative, and that the error structure across all array genes is relevant to each specific gene. Such methods, in effect, reduce the denominator in the *t*-test to achieve more realistic error estimates. Significance analysis of microarrays (SAM) [35] is one such popular method, but there are many variations to accomplish the same goal [36].

The many methods for analyzing microarray data have so far not undergone adequate scientific vetting of their capabilities, implications, and limitations [37], with the consequent pressing need to critically evaluate currently available methods with relevant and objective criteria. For example, reproducibility has seldom been, but in the future should be, used as a critical criterion to judge the performance of data analysis procedures. In addition, several differential gene expression profiling studies have demonstrated that the relative expression measures (i.e., difference in transcript abundance between sample types) are more consistent than the absolute gene expression levels. The MAQC dataset is expected to be widely utilized by the community in order to promote and reach consensus on the appropriate analysis methods for microarray data. The MAQC project mapped the probe sequences from Affymetrix and other platforms to the RefSeq [29] and AceView [38] databases. The numbers of RefSeq NMs or genes targeted by the platforms are comparable (Table 1.2).

**Table 1.2** Summary of Probe-Sequence Mapping per Gene Expression Platform

| Platform[a] | Number of Probe Sequences Analyzed[b] | Number of Probes that Met Mapping Criteria[c] (Percent of All Probes, %) | | Number of RefSeq NM Accessions Mapped to Probes[e] | Number of Entrez Gene IDs Mapped to Probes via | | Number of Genes, Not in Entrez, Mapped to Probes via AceView[d,f,g] |
|---|---|---|---|---|---|---|---|
| | | RefSeq | AceView[d] | | RefSeq[e] | AceView[d,f] | |
| ABI | 32,878 | 18,547 (56.4) | 25,566 (77.8) | 21,963 | 16,763 | 18,676 | 3,267 |
| AFX | 54,675 | 24,694 (45.2 ) | 44,693 (81.7) | 21,318 | 15,965 | 18,911 | 10,129 |
| AG1 | 41,000 | 22,677 (55.3) | 32,024 (78.1) | 21,890 | 16,493 | 18,051 | 4,055 |
| GEH | 53,423 | 16,881 (31.6) | 43,540 (81.5) | 20,230 | 15,429 | 16,984 | 18,408 |
| ILM | 47,282 | 20,140 (42.6) | 31,229 (66.0) | 22,161 | 16,990 | 18,797 | 8,666 |
| NCI | 35,235 | 21,555 (61.2) | 29,396 (83.4) | 20,987 | 15,899 | 17,641 | 1,411 |
| *Union* | 264,493 | 125,216 (47.3) | 206,448 (78.1) | 23,971 | 18,114 | 21,662 | 32,025 |
| *Intersection* | | | | 15,615 | **12,091** | 13,327 | 9 |

[a]For the AFX platform, the length of each individual probe is given. For the QGN platform, the length of the intended target is given.

[b]The number of probes for which mapping was attempted may slightly differ from the number of probes arrayed (Table 1.1) because of the removal of control probes and replicate spots. For the AFX platform, the number of probe sets is given.

[c]Probes were mapped as described in the Methods section. An exact sequence match was required and probes that match more than one gene were excluded. For the AFX platform, there were generally 11 perfect-match probes per probeset, and each probe was mapped individually. An exact match of 80% of the probes in a probe set was required for the probe set to qualify as a perfect match.

[d]AceView is a transcriptome database that combines RefSeq, GenBank, and dbEST entries [38]. For the details on the AceView mapping, please refer to the supplementary materials online at ftp://ftp.ncbi.nlm.nih.gov/repository/acedb/MAQC/MaqcMapping2AceViewTranscripts.zip.

[e]The numbers in these columns illustrate the source of the common set of 12,091 genes represented on the six high-density microarray platforms that have an overlap of 906 genes with the TAQ platform. The data do not fully reflect the coverage of each platform because the degree to which RefSeq and non-RefSeq sequences are emphasized during probe design and selection differs among the platforms.

[f]The number of Entrez genes specifically assayed through any of their alternative transcript variants is given in these columns. Probes with a few gaps or mismatches were permitted, but at the same time, probes with even a minor risk of cross-hybridization to another gene (with up to 30% mismatches) were ignored.

[g]Genes, not yet in Entrez, are supported by cDNAs in GenBank, and are described in AceView. The sum of genes in Entrez (via AceView) and genes not in Entrez that mapped to probes (via AceView) is the total number of genes in the AceView database that are matched by each platform under the mapping criteria chosen for this study.

Contrary to the high-visibility claims of microarray irreproducibility, the MAQC project observed good intraplatform as well as high interplatform concordance in terms of genes identified as differentially expressed. The MAQC gene expression data came from four titration pools from two distinct, commercially available reference RNA samples. Data were produced at multiple test sites using a variety of microarray-based and alternative technology platforms. The resulting rich dataset has over 1300 microarray hybridizations and additional measurements for over 1000 genes from alternative technologies such as qPCR.

In general, the platforms described above in Sections 1.2 to 1.4, despite markedly different manufacturing approaches, often gave highly comparable results in terms of differential gene expression. In other words, the differential gene expression patterns observed represented the same biology despite differences in platform technology. It is important to note that similarly good concordances of results were observed from a rat toxicogenomics dataset [34], validating the extensibility of the major findings from data generated from the reference RNA samples. Findings of the MAQC project were published in six research papers in *Nature Biotechnology*, September 8, 2006; all papers are freely available at http://www.nature.com/nbt/focus/maqc/index.html. Data are available through GEO (series accession number: GSE5350), ArrayExpress (accession number: E-TABM-132), ArrayTrack (http://www.fda.gov/nctr/science/centers/toxicoinformatics/ArrayTrack/), and the MAQC Website (http://www.fda.gov/nctr/science/centers/toxicoinformatics/maqc/).

Another important goal of the MAQC study was to estimate the best performance achievable with microarray technology under consistent experimental conditions, so that end users will be able to judge whether the quality of their microarray data is comparable to the achievable performance of the platform. Having such a standard of comparison provides the possibility of identifying and correcting procedural failures of a laboratory or operator before precious study samples are profiled. The commercial availability of the two reference RNA samples coupled with the large reference datasets also enables objective evaluations of new array products, reagents, or protocols.

The high level of comparability of microarrays to qRT-PCR (TaqMan assays) results from the MAQC project [11] is shown in Fig. 1.6. Similar correlation values were observed between the microarray platforms and other alternative platforms, StaRT-PCR, and QuantiGene [39].

## 1.6 Conclusions

Demonstrating reproducible technical performance is mandatory for the acceptance of microarray-based gene expression data in clinical and regulatory environments. The microarray technology itself has been generically criticized whenever findings were found irreproducible. As with any other analytical instruments, technical proficiency of the laboratory performing microarray studies is expected to lead to reproducible results. As stated in the FDA's concept paper on "*Recommendations*

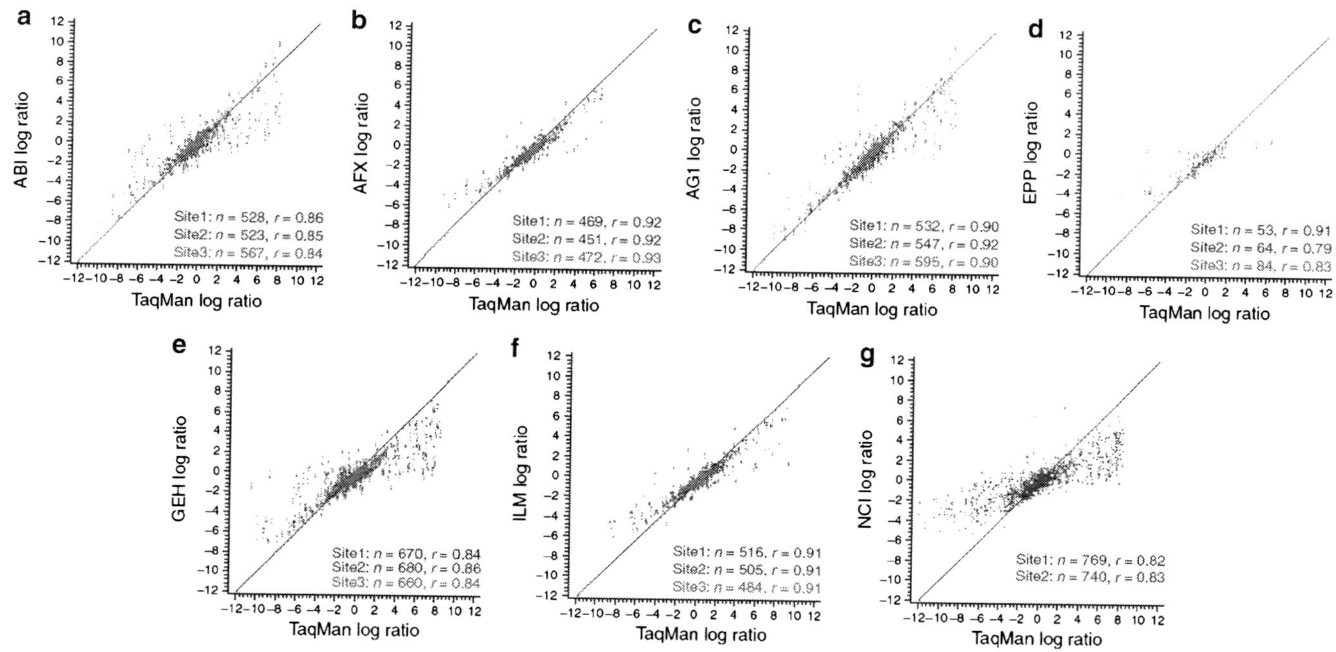


**Fig. 1.6** Comparability of microarray platforms with qRT-PCR (TaqMan®) in detecting differential gene expression. The scatterplots compare the log ratio differential expression values (using A versus B replicates) from each microarray platform relative to values obtained by TaqMan assays. Each point represents a gene that was measured on both the microarray and TaqMan assays. The spot coloring indicates whether the data were generated in test site 1 (black), test site 2 (blue), or test site 3 (red) for the microarray platform. Only genes that were generally detected in sample type A replicates and sample type B replicates were used in the comparisons. The exact number of probes analyzed for each test site and its correlation to TaqMan assays are listed in the bottom right corner of each plot. (Reproduced with permission from L. Shi et al., 2006.)

*for the Generation and Submission of Genomic Data*," the many steps involved in a microarray study must each be done correctly in order to achieve adequate quality in a study (http://www.fda.gov/cder/genomics/conceptpaper_20061107.pdf). Investigators should take proper measures to ensure that each step meets the quality control and quality assurance criteria, because an error in any step can render the entire study doubtful.

Although many metrics have been proposed for microarray quality control, there remains a current lack of consensus regarding the relative importance among them. More problematically, there is little information regarding the acceptable thresholds for each quality control metric. The establishment of quality control thresholds will depend on the availability of large collections of historical microarray data and those from international standardization efforts such as the ERCC (External RNA Controls Consortium) [40] and the MAQC [11, 34]. The community's willingness to share relevant data is critical. Ensuring good quality data is the essential first step toward reproducible and reliable microarray results in terms of the lists of differentially expressed genes, the associated biological interpretations from them, and in the ability to use the genes for prediction of clinical outcomes.

The microarray community needs a diligent and concerted effort to assess microarray data analysis methods and then reach a consensus on those that are most appropriate generally, and those for selecting significant genes, particularly. The vetting of methods needs to account for what methods are suitable for what applications and should consider their understandability by and availability to the scientific community. Depending on the application, such metrics as sensitivity and specificity will be important, but the scientific principle of reproducibility cannot be circumvented in the assessment of methods. The too common practice in the past of selecting differentially expressed genes with a simple $t$-test [41] has been recently shown to be too simplistic [11, 33, 34], inasmuch as such practice predestines nonreproducible gene lists, irrespective of the platforms, sample pairs, or normalization methods. The MAQC study recommended the use of fold-change ranking plus a nonstringent $P$ cutoff as a baseline practice in order to attain reproducible lists of differentially expressed genes. The fold-change criterion enhances reproducibility, and the $P$ criterion enables balancing and tradeoffs between sensitivity and specificity. It is important to note that, as would be expected, nonreproducible lists of differentially expressed genes result in disparate enriched gene ontology terms or pathways, and correspondingly cause differing biological interpretations [34]. Therefore, the reproducibility of gene lists is needed to ensure the reproducibility of corresponding biological interpretations.

The criticisms of microarrays have had a demonstrably sobering effect on the initial excitement and concomitant race to be among early microarray publishers across numerous areas of specialization. The phase in which microarray technology now finds itself is more vested in quality in experiment design and performance, and in a search for real outcomes with medical impacts that can be validated. Although the range of applications is in no way diminished in number, the deluge of publications should slow as peer reviewers become more insistent on validation of conclusions. The generalized criticisms of microarrays, we believe, will ultimately be shown to have been unwarranted. In the end, general criticisms will narrow to specific criticisms pertaining to adequate quality control, appropriate statistical inference, or some combination of these.

**Acknowledgments** We thank participants of the MicroArray Quality Control (MAQC) project for their dedication in generating and analyzing the large datasets. Many MAQC participants contributed to the sometimes heated discussions on the choice of methods for identifying differentially expressed genes. The common conclusions and recommendations evolved from this extended discourse. We invite the microarray community, in particular, statisticians, to participate in the debate in order to reach consensus on microarray data analysis.

## References

1. Shi L, Tong W, Goodsaid F, Frueh FW, Fang H, Han T, Fuscoe JC, Casciano DA (2004) Expert Rev Mol Diagn 4: 761
2. Shi L, Goodsaid FM, Frueh FW, Tong W (2007) In: Resch-Genger U (Ed) Standardization in Fluorometry: State-of-the-Art and Future Challenges. Springer: New York, in press
3. Yauk C, Berndt L (2007) Enviromental and Molecular Mutagenesis
4. Frueh FW (2006) Nat Biotechnol 24: 1105
5. Marshall E (2004) Science 306: 630
6. Tan PK, Downey TJ, Spitznagel EL, Jr., Xu P, Fu D, Dimitrov DS, Lempicki RA, Raaka BM, Cam MC (2003) Nucleic Acids Res 31: 5676
7. Michiels S, Koscielny S, Hill C (2005) Lancet 365: 488
8. Ioannidis JP (2005) Lancet 365: 454
9. Ein-Dor L, Zuk O, Domany E (2006) Proc Natl Acad Sci USA 103: 5923
10. Dupuy A, Simon RM (2007) J Natl Cancer Inst 99: 147
11. Shi L, Reid LH, Jones WD, Shippy R, Warrington JA, Baker SC, Collins J, Longueville Fd, Kawasaki ES, Lee KY, Luo Y, Sun Y, Willey JC, Setterquist RA, Fischer GM, Dix DJ, Dragan YP, Goodsaid FM, Herman D, Jensen RV, Johnson CD, Lobenhofer EK, Puri RK, Scherf U, Thierry-Mieg J, Tong W, Wang C, Wilson M, Zhang L, Amur S, Bao W, Barbacioru CC, Bergstrom-Lucas A, Bertholet V, Boysen C, Bromley B, Brown D, Brunner A, Canales R, Cao XM, Cebula TA, Chen JJ, Cheng J, Chu T-M, Chudin E, Corson J, Corton JC, Croner LJ, Davison TS, Delenstarr G, Deng X, Dorris D, Eklund AC, Fan X-h, Fang H, Frueh FW, Fulmer-Smentek S, Fuscoe JC, Gallagher K, Ge W, Guo L, Guo X, Hager J, Haje PK, Han J, Han T, Harbottle HC, Harris SC, Hatchwell E, Hauser CA, Hester S, Hong H, Hurban P, Jackson SA, Ji H, Knight CR, Kuo WP, LeClerc JE, Levy S, Li Q-Z, Liggett W, Liu C, Liu Y, Lombardi MJ, Ma Y, Magnuson SR, Maqsodi B, McDaniel T, Mei N, Myklebost O, Ning B, Novoradovskaya N, Orr MS, Osborn TW, Papallo A, Patterson TA, Perkins R, Peters EH, Peterson R, Philips KL, et al. (2006) Nat Biotechnol 24: 1151
12. Beaucage SL (2001) Curr Med Chem 8: 1213
13. Barbulovic-Nad I, Lucente M, Sun Y, Zhang M, Wheeler AR, Bussmann M (2006) Crit Rev Biotechnol 26: 237
14. Kronick MN (2004) Expert Rev Proteomics 1: 19
15. Fodor SP, Read JL, Pirrung MC, Stryer L, Lu AT, Solas D (1991) Science 251: 767
16. Lockhart DJ, Dong H, Byrne MC, Follettie MT, Gallo MV, Chee MS, Mittmann M, Wang C, Kobayashi M, Horton H, Brown EL (1996) Nat Biotechnol 14: 1675
17. Irizarry RA, Wu Z, Jaffee HA (2006) Bioinformatics 22: 789
18. Hughes TR, Mao M, Jones AR, Burchard J, Marton MJ, Shannon KW, Lefkowitz SM, Ziman M, Schelter JM, Meyer MR, Kobayashi S, Davis C, Dai H, He YD, Stephaniants SB, Cavet G, Walker WL, West A, Coffey E, Shoemaker DD, Stoughton R, Blanchard AP, Friend SH, Linsley PS (2001) Nat Biotechnol 19: 342

19. Schena M, Shalon D, Davis RW, Brown PO (1995) Science 270: 467
20. Schena M, Shalon D, Heller R, Chai A, Brown PO, Davis RW (1996) Proc Natl Acad Sci USA 93: 10614
21. Shi L, Tong W, Su Z, Han T, Han J, Puri RK, Fang H, Frueh FW, Goodsaid FM, Guo L, Branham WS, Chen JJ, Xu ZA, Harris SC, Hong H, Xie Q, Perkins RG, Fuscoe JC (2005) BMC Bioinformatics 6 Suppl 2: S11
22. Patterson TA, Lobenhofer EK, Fulmer-Smentek SB, Collins J, Chu T-M, Bao W, Fang H, Kawasaki ES, Hager J, Tikhonova I, Walker SJ, Zhang L, Hurban P, Longueville Fd, Fuscoe JC, Tong W, Shi L, Wolfinger RD (2006) Nat Biotechnol 24: 1140
23. Glas AM, Floore A, Delahaye LJ, Witteveen AT, Pover RC, Bakx N, Lahti-Domenici JS, Bruinsma TJ, Warmoes MO, Bernards R, Wessels LF, Van't Veer LJ (2006) BMC Genomics 7: 278
24. Ramakrishnan R, Dorris D, Lublinsky A, Nguyen A, Domanus M, Prokhorova A, Gieser L, Touma E, Lockner R, Tata M, Zhu X, Patterson M, Shippy R, Sendera TJ, Mazumder A (2002) Nucleic Acids Res 30: e30
25. Wang Y, Barbacioru C, Hyland F, Xiao W, Hunkapiller KL, Blake J, Chan F, Gonzalez C, Zhang L, Samaha RR (2006) BMC Genomics 7: 59
26. Petersen D, Chandramouli GV, Geoghegan J, Hilburn J, Paarlberg J, Kim CH, Munroe D, Gangi L, Han J, Puri R, Staudt L, Weinstein J, Barrett JC, Green J, Kawasaki ES (2005) BMC Genomics 6: 63
27. Guo Y, Guo H, Zhang L, Xie H, Zhao X, Wang F, Li Z, Wang Y, Ma S, Tao J, Wang W, Zhou Y, Yang W, Cheng J (2005) J Virol 79: 14392
28. Gunderson KL, Kruglyak S, Graige MS, Garcia F, Kermani BG, Zhao C, Che D, Dickinson T, Wickham E, Bierle J, Doucet D, Milewski M, Yang R, Siegmund C, Haas J, Zhou L, Oliphant A, Fan JB, Barnard S, Chee MS (2004) Genome Res 14: 870
29. Pruitt KD, Tatusova T, Maglott DR (2005) Nucleic Acids Res 33: D501
30. Miller RM, Callahan LM, Casaceli C, Chen L, Kiser GL, Chui B, Kaysser-Kranich TM, Sendera TJ, Palaniappan C, Federoff HJ (2004) J Neurosci 24: 7445
31. Miklos GL, Maleszka R (2004) Nat Biotechnol 22: 615
32. Frantz S (2005) Nat Rev Drug Discov 4: 362
33. Shi L, Tong W, Fang H, Scherf U, Han J, Puri RK, Frueh FW, Goodsaid FM, Guo L, Su Z, Han T, Fuscoe JC, Xu ZA, Patterson TA, Hong H, Xie Q, Perkins RG, Chen JJ, Casciano DA (2005) BMC Bioinformatics 6 Suppl 2: S12
34. Guo L, Lobenhofer EK, Wang C, Shippy R, Harris SC, Zhang L, Mei N, Chen T, Herman D, Goodsaid FM, Hurban P, Philips KL, Xu J, Deng X, Sun YA, Tong W, Dragan YP, Shi L (2006) Nat Biotechnol 24: 1162
35. Tusher VG, Tibshirani R, Chu G (2001) Proc Natl Acad Sci U S A 98: 5116
36. Allison DB, Cui X, Page GP, Sabripour M (2006) Nat Rev Genet 7: 55
37. Eisenstein M (2006) Nature 442: 1067
38. Thierry-Mieg D, Thierry-Mieg J (2006) Genome Biol 7 Suppl 1: S12 1
39. Canales RD, Luo Y, Willey JC, Austermiller B, Barbacioru CC, Boysen C, Hunkapiller K, Jensen RV, Knight C, Lee KY, Ma Y, Maqsodi B, Papallo A, Peters EH, Poulter K, Ruppel PL, Samaha R, Shi L, Yang W, Zhang L, Goodsaid FM (2006) Nat Biotechnol 24: in press
40. Baker SC, Bauer SR, Beyer RP, Brenton JD, Bromley B, Burrill J, Causton H, Conley MP, Elespuru R, Fero M, Foy C, Fuscoe J, Gao X, Gerhold DL, Gilles P, Goodsaid F, Guo X, Hackett J, Hockett RD, Ikonomi P, Irizarry RA, Kawasaki ES, Kaysser-Kranich T, Kerr K, Kiser G, Koch WH, Lee KY, Liu C, Liu ZL, Lucas A, Manohar CF, Miyada G, Modrusan Z, Parkes H, Puri RK, Reid L, Ryder TB, Salit M, Samaha RR, Scherf U, Sendera TJ, Setterquist RA, Shi L, Shippy R, Soriano JV, Wagar EA, Warrington JA, Williams M, Wilmer F, Wilson M, Wolber PK, Wu X, Zadro R (2005) Nat Methods 2: 731
41. Chen JJ, Wang SJ, Tsai CA, Lin CJ (2007) Pharmacogenomics J 7:212

# Chapter 2
# Electrochemical Detection on Microarrays

**Kilian Dill and Andrey Ghindilis**

## 2.1 Introduction

Microarray detection methods have long been based upon optical methods: visible detection, fluorescence, luminescence, and surface plasmon resonance (1–6). In the case of the visible detection method an enzyme produces a substrate that forms an insoluble precipitate at the spot site, which becomes visible to the naked eye. Surface plasmon resonance also uses visible light, but in this case the energy (and reflected angle) of the light absorption on the metal surface is altered; the researcher finds the absorption minimum using an array detector. This method utilizes more expensive equipment in the process of detecting the material on the chip/slide (4).

More sensitive methods such as fluorescence and luminescence have also been employed for the detection of material on the chip/slide surface. The fluorescence-based system requires a laser, so the beam can be rastered over the chip area. Or conversely, the chip can be illuminated with light and then a signal detected with a CCD camera using the appropriate filters (3). Similarly, luminescence detection utilizes a CCD camera after an enzyme (such as alkaline phosphatase) has generated a product that luminesces (1).

Both these methods require expensive optics and equipment that may cost \$60,000 to \$500,000. Moreover, the footprint of the larger and more expensive systems requires a benchtop area on one side of the room.

A less expensive, mobile, and more sensitive platform for microarray detection purposes is highly desirable. To this endeavor, many researchers in the field have instituted programs for electrochemical detection on microarrays (6–20). In order to do this, one must have active electrodes on the chip or glass surface. Thus many of the spotted slides out in the marketplace contain surface areas where the spots are not electroactive and do not qualify for electrochemical detection.

---

K. Dill and A. Ghindilis
CombiMatrix Corporation, Inc.
6500 Harbour Heights Parkway, Suite 301
Mukilteo, WA 98275

K. Dill et al. (eds.), *Microarrays: Preparation, Microfluidics, Detection Methods, and Biological Applications*,

Even in the case of electrochemical detection, there are a host of methods that can be employed. They range from inductive measurements to simple redox tags, and to compounds that are wedged into the grooves of a DNA duplex to enzyme amplification (6–20). Even the latter may be performed by various methods: product redox shuttle (between two electrodes) or a method whereby the product is reduced/oxidized at the single electrode. In the former case the current output is monitored while voltage to the electrodes is changed in sign (cycled). In the former case, the signal is stored on a capacitor on the chip and is accessed every so often (usually once). The charge on the capacitor is then converted to amperage and hence the amount of enzyme at the surface of the electrode. Electrochemical methods of detection are discussed in detail below.

## 2.2 Electrochemical Detection

There are numerous advantages of electrochemical detection (ECD) over conventional fluorescence.

1. *Superior Performance*. In optical detection schemes, the signal-to-noise ratio is often limited by the amount of stray light (often from the incident beam) that gets into the detector channel. With ECD there is no incident background. The only background that exists comes from the inherent background currents in the measurement systems and the capacitive charging currents at the chip. Both of these are relatively small, thus a much higher signal-to-noise ratio is achieved in the ECD mode. As a result greater sensitivity (perhaps several orders of magnitude) can be achieved.
2. *Fewer Components*. For ECD there is no need for optical components such as the light source, mirrors, filters, detectors, their support mechanics, or the movement mechanics for chip scanning. Consequently, the system is simpler and less expensive. Also, the fewer component requirements enable smaller footprints and weight.
3. *Portability*. Because the instrument is small and relatively light, and the power can be supplied via batteries, ECD systems can potentially be portable. This characteristic alone will enable significant applications in the IVD marketplace where small, inexpensive, and portable systems are necessary.

ECD systems for microarray analysis are portable and low cost, and coupled with their superior performance will revolutionize the use of microarrays not only for IVD applications but also for conventional R&D studies. One can imagine a situation where array readers can be purchased for a few hundred dollars (as opposed to the hundreds of thousands for the top-of-the-line optical scanners) and can be taken anywhere.

We discuss four ECD methods that have been described in the literature for microarray applications (6–20). All of these methods have a spatial array of electrodes, which are hardwired to be individually addressable. Scanning is performed by measuring an electrical property, independently, at each electrode, in either a serial or parallel manner.

### 2.2.1 Capacitance Measurements

Capacitance is determined by charged ions at the surface of a layer covering the electrode (20). These counter-ions and water dipoles lie at the surface of the bound material. There is also a more remote layer that consists of diffuse counter-ions known as the Guoy–Chapman layer (20). The inner layer, tightly captured, changes the nature of the electrode capacitance. This surface transductance results in a capacitance change.

There are numerous traditional and new methods for detecting capacitance change at the electrode surface. These may include chronoamperometry and impedance spectroscopy. Both are difficult to measure and require multiple electrodes in any given electrical cell. Additional sample conditions will play a great role, which may include salts found within the sample.

In the case of DNA analysis, the capture probe must be placed upon the chip surface. This in itself changes the capacitance of the electrode and introduces a sodium ion countercharge (to offset the negatively charged phosphate groups). Duplex formation (hybridization with the complementary strand) introduces additional charges to the outer layer (more sodium ions). See Fig. 2.1 for details.

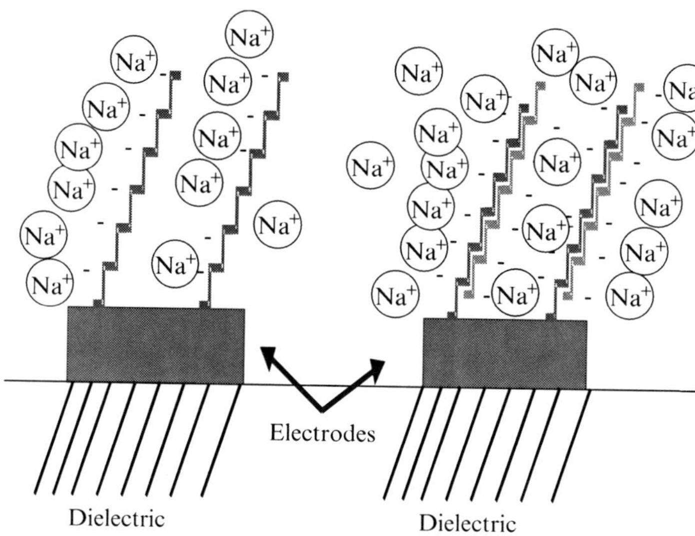


**Fig. 2.1** Detection of DNA via capacitance measurements. Hybridization allows the accumulation of charge near the electrode due to the phosphate groups on each strand. One strand of DNA is chemically attached to the electrode. The complementary strand is then hybridized to the stationary strand

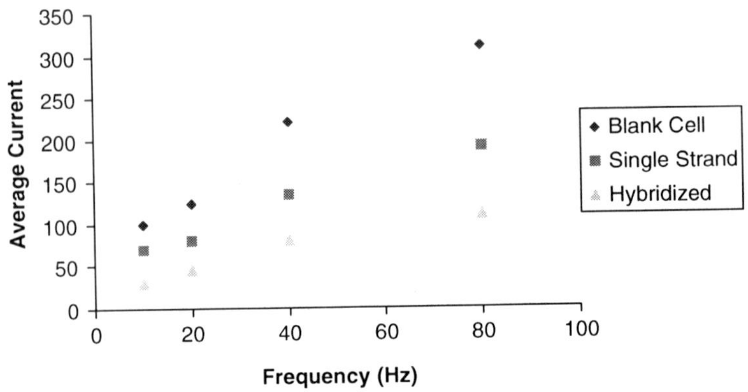


**Fig. 2.2** Results from a capacitance measurement based upon single-strand and hybridized DNA. Data was obtained from Table 2.1

**Table 2.1** Frequency Measurement of Average Current Flowing Through the Cell

| F(Hz) | Cell | Single Strand | Hybridized |
|---|---|---|---|
| 10 | 100 | 70 | 30 |
| 20 | 125 | 80 | 45 |
| 40 | 220 | 135 | 80 |
| 80 | 310 | 190 | 110 |

How can this be measured in a simpler manner? If the voltage applied to the capacitive cell is AC in nature, then the ion double layer moves with the frequency of the applied voltage. From a static system to a dynamic system the movement of ions can be mathematically described as

$$I_{avg} = I_{DC} + C(\Delta V / 2) f.$$

Then the current on average can be described as based upon the DC current followed by a cell constant and the frequency of the applied AC voltage; ΔV/2 is periodic, the pulse applied to the electrode to allow the capacitor to fully charge and discharge (20).

Data provided above are graphed in Fig. 2.2. Note that upon duplex formation, the average current is decreased as the frequency applied increases. Of course a blank measurement is also provided and this is based upon the cell containing all components but the attached oligomer.

## 2.3 Faradaic Current Measurements Through DNA

DNA is a chain comprised of organic and inorganic molecules, which can act to inhibit current flow, and it is well known that duplex DNA is a poorer insulator than single-stranded DNA. By introducing a redox couple, such as ferricyanide/ferrocyanide and measuring the current at different electrodes with attached DNA,

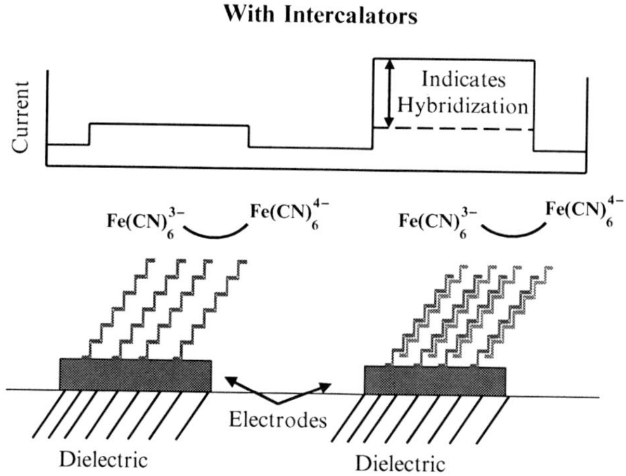


**Fig. 2.3** Detection of DNA via ferrocyanide/ferricyanide couple. Differences can be observed between single-stranded and double-stranded DNA in the presence of ferrocyanide/ferricyanide. However, they are heavily dependent on solution conditions with the potential to oxidize the sample. A method that does show reproducible results is when intercalators such as methylene blue are present. These mediators can be oxidized by the ferricyanide and the electron is shuttled through the DNA base stacked network to the electrode. The choice of an appropriate intercalator is crucial

one can determine the existence of a duplex by noting an increase in current at a particular applied voltage (Fig. 2.3). Similar to the case above, this system is very sensitive to interference by solution components, so events occurring at the chip surface may be difficult to differentiate.

The system described in Fig. 2.3 shows the best results when coupled to an intercalator or groove binder, which themselves aid in mediating electron transfer. Intercalators bind to the DNA duplex but not single-stranded DNA. These intercalators or groove binders may or may not be sequence-specific. The intercalators may be an organic molecule, such as methylene blue or Hoechst 33258, or transition metal complexes composed of cobalt. The existence of the intercalator makes the duplex even more conductive resulting in greater current at electrodes where hybridization has occurred. This method has had some commercial success (9, 18). Some of these intercalators are themselves redox active and may act as signal generators when the voltage is ramped up, so as to induce a current flow. The main drawbacks are specificity and lack of signal amplification as the data are collected on a single acquisition cycle. Additionally, the intercalators bind to all DNA duplexes formed, but some may prefer certain nucleotides in a specific order.

### *2.3.1 Direct Oxidation of DNA (or Mediated Oxidation)*

DNA can be electrochemically oxidized directly but the process is slow and the signal weak. There are also methods that utilize metals such as osmium or ruthenium, which oxidize the sample as the reading occurs (9, 18). Guanine is oxidized using a $Ru(bpy)_3$ complex. The oxidized $Ru(bpy)_3$ is reduced at the electrode. The signal generated would be amplified to a small degree based upon the number of guanine residues present. By design, the capture probe would contain a limited (or no) G in the sequence making single-stranded DNA and hybridized, double-stranded DNA distinguishable by this electrochemical technique. An alternative nucleotide is used, so as not to interfere.

## 2.4 Signal Transduction Using Various Labeless Systems

A new concept has evolved called labeless detection, somewhat incorrectly as one does need a redox species to provide the signal and that redox species does need to be attached (bound) to the DNA duplex. In most cases it is hybridized to the duplex extension or binds to one of the grooves of the duplex. It is not thought to be part of the sample modification as the amplicon contains no external modifiers (11, 9, 18).

We discuss two cases in some detail. Case one is used by a company called Osmetech (www.osmetech.com; the technology is discussed in a later chapter). The redox signal generator is a modified piece of DNA (containing ferrocene molecules) that is complementary in sequence to a captured amplicon which has a long overhang from the capture probe. The ferrocene moiety readily undergoes a transition from the 2+/3+ iron state when the voltage is ramped higher. It is the amperage that is monitored and can only exist if a duplex has formed with the correct complementary pairs.

Another method utilizes a redox dye that intercalates into a DNA duplex. The redox dye then undergoes an oxidation/reduction that is monitored. The most obvious dye is Hoescht 33258 used by Toshiba (11). It only binds to DNA duplexes and not single-strand DNA. Thus, if that electrode site does not contain a hybridized species, which can bind the redox active dye, no signal is observed. Excess dye is readily washed away, so background signal is minimal. In both systems, the temperature dependence of the signal is used to measure SNPs. The melting temperatures are different enough to afford a signal difference.

### 2.4.1 *Redox Enzyme Mediated Measurements*

A method which we feel is the most suitable for array applications is one that utilizes redox enzymes as the labels (9, 18) or enzymes that create products that are redox active. There are several approaches to this method. One uses a three-electrode system to each electrode and lets the enzyme (alkaline phosphatase) convert a redox inactive substrate to a product that can be oxidized and, conversely, reduced (called redox cycling). Thus with a three-electrode system, the enzyme substrate can then be oxidized/reduced in a ping-pong-like manner by switching voltages. The greater the concentration of enzyme that is present, the greater the amperometric signal as more product can be oxidized/reduced. The downside to this and many other three-electrode systems is that one focuses on a single electrode at one time and the number of leads is large. Hence, the number of electrodes per array will remain low until the bottleneck can be released (time present per a single electrode).

Another method is to use an electrode to monitor the output of a redox enzyme at the surface of the electrode. The redox enzyme functions to provide the signal indicating hybridization as well as to provide amplification comparable to optical methods (Fig. 2.4). Although this methodology has not been utilized for microarrays, it has been used for years in colorimetric enzyme-linked immunosorbent assays (ELISAs). In addition, redox enzymes have been used in very commercially successful glucose meters. This approach is simple, commercially successful, and technically superior to the methods discussed above. The signal is amplified by the action of the enzyme, and only minor modification to conventional gene expression protocols is necessary.

CombiMatrix has developed a commercial system that is based on this approach and unique semiconductor-based microelectrode arrays. The CombiMatrix system has the ability to address each electrode individually and measure the signal present at that electrode site. Fig. 2.5a shows a micrograph of one section of a chip that has roughly 1170 electrodes per sq. cm. Fig. 2.5b shows a photograph of the newer version 12 K chip (17,778 electrodes/sq. cm). On both chips, each electrode is individually addressable and can have unique oligomers synthesized at each site. Analysis of hydridization can be accomplished by fluorescence methods or by ECD.

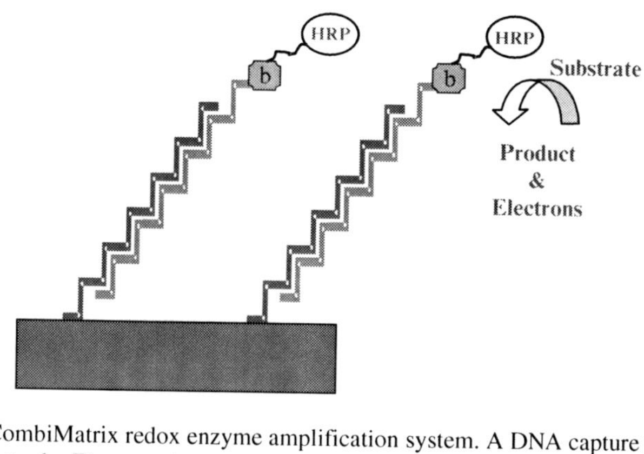


**Fig. 2.4** The CombiMatrix redox enzyme amplification system. A DNA capture probe is synthesized at the electrode. The complementary target is a PCR product containing a biotin molecule that may be attached at the end of the sequence or to bases within the sequence. Streptavidin-labeled horseradish peroxidase is then added to the sample, and HRP binds to biotin on the DNA strand. Addition of substrate allows HRP to produce a product and a current at the electrode

**Fig. 2.5a** A white light photograph of a portion of the 902 chip. The electrode density is > 1000/sq. cm

a

**Fig. 2.5b** Photograph of the commercial 12 K chip

b

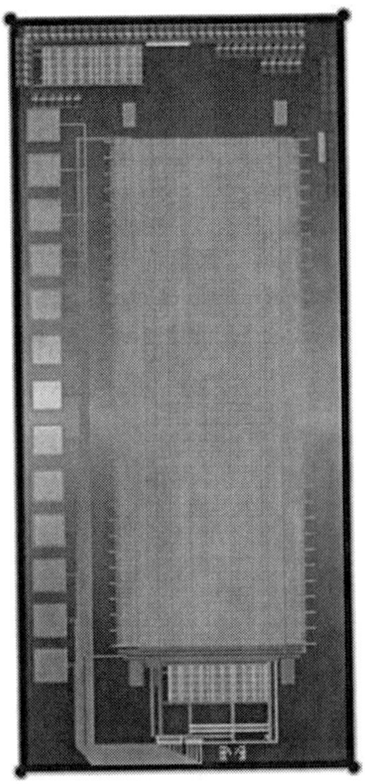

In the CombiMatrix signal amplification method, the attached reporter group is usually horseradish peroxidase. The enzyme oxidizes substrate to product in the presence of hydrogen peroxide. The product in return is reduced at the electrode under the appropriate conditions. Because the enzyme continues to create product at an extremely fast pace, we have an amplified signal that can be detected as current at the electrode. Fig. 2.6 depicts the ECD output of a gene expression

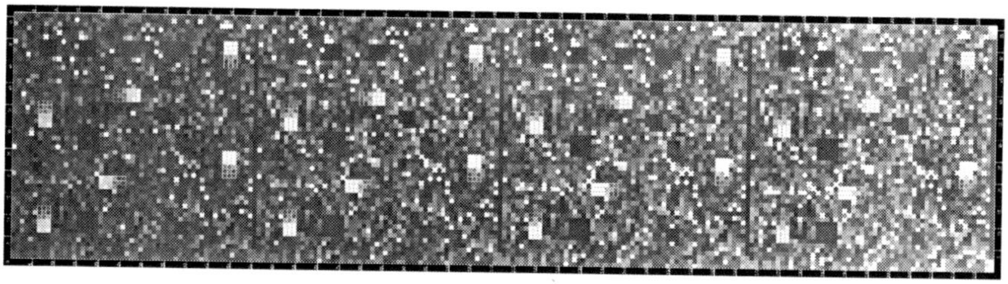

Fig. 2.6 ECD output of a lambda spike experiment (0.375 pM, 0.75 pM, 1.5 pM, 3 pM, 6 pM, 12 pM) into a complex sample of biotinylated cRNA from a leukemic cell line. The boxed areas in yellow are where the various lambda spike-in DNA should bind. Other probes on the chip are complementary to specific sets of genes, which are either expressed or not expressed in this particular cell line. Twenty-four repeats of each concentration range were measured

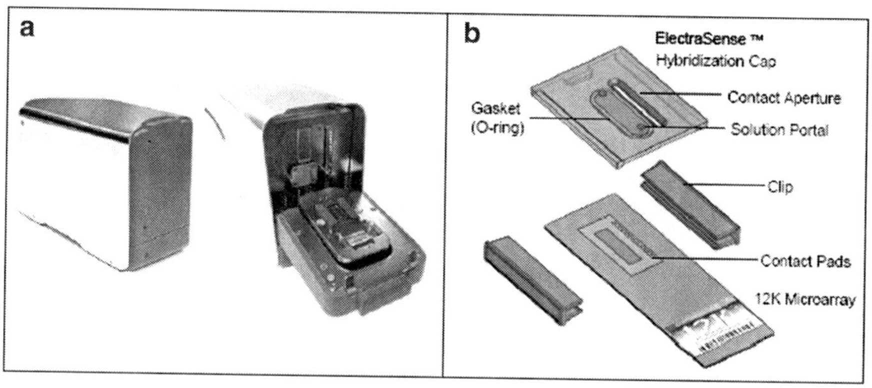


Fig. 2.7 ElectraSense™ Reader (a) and ElectraSense™ 12 K microarray with hybridization cap (b)

experiment for a leukemic cell line containing lambda spike-in controls. The data were generated from a 12 K chip where given lambda capture probes were synthesized in specific regions of the chip (as indicated in the figure). Limit of detection was determined to be 0.750 pM of biotinylated lambda cRNA. Twenty-four replicates were taken for each lambda concentration with a given sequence.

Shown in Fig. 2.7 is the ElectraSense enzyme amplified electrochemical unit used in the CombiMatrix studies. The electroactive microarray is contained within a chamber that contains the appropriate buffers and enzyme substrate. It is then placed within a holder and contact is made with the reader through several pins. Each electrode is addressed individually and a measurement is taken via the charge buildup at a capacitor associated with each electrode. That charge release results in a nanoamp current flow.

In summary, the scope of this chapter is to provide some insights into new electrochemical methods of DNA and immunoassay detection that have the potential to revolutionize microarray application and hasten their use in IVD markets.

## References

1. Cheek BJ, Steel AB, Torres MP, et al., Chemiluminescence detection for hybridization assays on the flow-thru chip, a three dimensional microchannel biochip, Anal. Chem. 73: 5777–5783, 2001.
2. Fodor SPA, DNA Sequencing-massively parallel genomics, Science 277: 393–403, 1977.
3. Epstein JR, Brian I, Walt DR, Fluorescence-based nucelic acid detection and microarrays, Anal. Chim. Acta 469: 3–36, 2002.
4. McDonnell JM, Surface plasmon resonance: Towards an understanding of mechanisms of biological recognition, Curr. Opin. Chem. Biol. 5: 572–577, 2001.
5. Storhoff JJ, Elghanian R, Mucic RC, et al., One-pot colorometric differentiation of polynucleotides with single base imperfections using gold nanoparticle probes, J. Am. Chem. Soc. 120: 1959–1964, 1988.
6. Dill K, Montgomery DD, Ghindilis AL, et al., Immunoassays and sequence specific DNA detection on a microchip using enzyme amplified electrochemical detection, J. Biochem. Biophys. Meth. 59: 181–187, 2004.
7. Drummond TG, Hill MG, and Barton JK, Electrochemical DNA sensors, Nature Biotechnol. 21: 1192–1198, 2003.
8. Wang J, Survey and summary: From DNA biosensors to gene chips, Nucleic Acids Res. 28: 3011–3016, 2000.
9. Gau JJ, Lan EH, Dunn B, et al., A MEMS based amperometric detector for E, coli bacteria using self assembled monlayers, Biosens. Bioelectron. 16: 745–755, 2001.
10. Hashimoto K and Ishimori Y, Preliminary evaluation of electrochemical PNA array for detection of single base mismatch mutations, Lab on a Chip 1: 61–63, 2001.
11. J.S. Rossier and H.H. Girault, Enzyme linked immunosorbent assay on a microchip with electrochemical detection, Lab on a Chip 1 (2001): 153–157.
12. Kelly SO, Boon EM, Barton HK, et al., Single-base mismatch detection based on charge transduction through DNA, Nucleic Acid Res. 27: 4830–4837, 1999.
13. Yu CJ, Wan Y, Yowanto H, et al., Electronic detection of single-base mismatches in DNA with ferrocene-modified probes, J. Am. Chem. Soc. 123: 11155–11161, 2001.
14. Yang IV and Thorp HH, Kinetics of metal-mediated one-electron oxidation of guanadine in polymeric DNA and in oligonucleotides containing trinucleotide repeat sequences, Inorg. Chem. 39: 4969–4976, 2000.
15. Aguilar ZP and Fritsch I, Immobilized enzymed-linked DNA-hybridization assay with electrochemical detection for *cryptosporidium parvum* hsp70 mrna, Anal. Chem. 75: 3890–3897, 2003.
16. Dill K, Montgomery DD, Ghindilis AL, et al., Immunoassays based on electrochemical detection using microelectrode arrays, Biosens. Bioelectron. 20: 736–742, 2004.
17. Drummond TG, Hill MG, Barton JK, Electrochemical DNA sensors, Nature Biotechnol. 21: 1192–1199, 2003.
18. Kerman K, Kobayashi M, Tamiya E, Recent trends in electrochemical DNA biosensor technology, Meas. Sci. Technol. 15: R1–R11, 204.
19. Dill K, Schwarzkopf K, Ghindilis A, Multiplexed analyte and oligonucleotide detection on microarrays using several enzymes in conjunction with electrochemical detection, Lab Chips 6: 1052–1055, 2006.
20. Guiducci C, Stagni C, Zuccheri G, Boglioli A, Benini L, Samori A, Ricco B, A biosensor for direct detection of DNA seqeunces based on capacitance measurements, 32nd European Solid-State Device Research Conference, Firenze, Italy, 2002.

# Part II
# Fluidic Manipulation and Microarrays

# Chapter 3
# Fully Integrated Microfluidic Device for Direct Sample-to-Answer Genetic Analysis

**Robin H. Liu and Piotr Grodzinski**

**Abstract** Integration of microfluidics technology with DNA microarrays enables building complete sample-to-answer systems that are useful in many applications such as clinic diagnostics. In this chapter, a fully integrated microfluidic device [1] that consists of microfluidic mixers, valves, pumps, channels, chambers, heaters, and a DNA microarray sensor to perform DNA analysis of complex biological sample solutions is present. This device can perform on-chip sample preparation (including magnetic bead-based cell capture, cell preconcentration and purification, and cell lysis) of complex biological sample solutions (such as whole blood), polymerase chain reaction, DNA hybridization, and electrochemical detection. A few novel microfluidic techniques were developed and employed. A micromixing technique based on a cavitation microstreaming principle was implemented to enhance target cell capture from whole blood samples using immunomagnetic beads. This technique was also employed to accelerate DNA hybridization reaction. Thermally actuated paraffin-based microvalves were developed to regulate flows. Electrochemical pumps and thermopneumatic pumps were integrated on the chip to provide pumping of liquid solutions. The device is completely self-contained: no external pressure sources, fluid storage, mechanical pumps, or valves are necessary for fluid manipulation, thus eliminating possible sample contamination and simplifying device operation. Pathogenic bacteria detection from ~mL whole blood samples and single-nucleotide polymorphism analysis directly from diluted blood were demonstrated. The device provides a cost-effective solution to direct sample-to-answer genetic analysis, and thus has a potential impact in the fields of point-of-care genetic analysis, environmental testing, and biological warfare agent detection.

---

R.H. Liu
Osmetech Molecular Diagnostics, Robin.Liu@osmetech.com

P. Grodzinski
National Cancer Institute, grodzinp@mail.nih.gov

K. Dill et al. (eds.), *Microarrays: Preparation, Microfluidics, Detection Methods, and Biological Applications*,

## 3.1 Introduction

Current developments of DNA microarrays have been focused on different microarray platforms, development and optimization of detection techniques, and the use of microarrays in various applications. For most conventional microarrays, purified DNA/RNA or homogeneous samples are used as input samples, whereas the labor-intensive and time-consuming front-end sample preparation steps to prepare such purified samples are carried out off-chip using traditional benchtop methods. With the need of developing practical clinical and environmental analysis approaches that require processing of samples as complex and heterogeneous as whole blood or contaminated environmental fluids, the importance of building complete sample-to-answer systems emerged. Rapid developments in back-end microarray platforms have shifted the bottleneck, impeding further progress in rapid analysis devices, to front-end sample preparation where the "real" samples are used.

Microfluidics technology is a powerful technology that can allow all functions including sample preparation, mixing steps, chemical reactions, and detection to be performed in a miniature fluidic device. The miniaturization of biological assays to the chip level carries several advantages. In general, on-chip assays use reduced volumes of reagents (2–3 orders of magnitude as compared to traditional bench approaches) and allow for reducing cost per reaction and improving reaction kinetics [2–4].

On-chip reactions are performed in miniature channels or chambers that can be distributed on the device wafer at high density. This high population of identical reaction paths allows for the development of highly parallel analytical systems with high system throughput [5, 16]. Furthermore, integration of several assay functions on a single chip leads to assay automation and elimination of operator involvement as a variable [7]. The microfluidic lab-chip device with capabilities of on-chip sample processing and detection provides a cost-effective solution to direct sample-to-answer biological analysis. Such devices will be increasingly important for rapid diagnostic applications in hospitals and in-field biothreat detection.

Microfluidics technology is descended from a field referred to as micro total analysis systems (μTAS), which first came about in the late 1980s. The benefits of miniaturizing various lab equipment down to the size of a chip have been pursued for many years; the first such innovations took place in as far back as the 1960s and 1970s with the development of microfabricated sensors for process monitoring and other applications [8]. This was the beginning of the microelectromechanical systems (MEMS) field, which has branched into many other industries such as bioMEMS, optics, and telecommunications [8].

Microfluidics technology involves movement and processing of samples and reagents inside the microchannels or microstructures of a chip. The early development of this field focused on microchip CE (capillary electrophoresis): on-chip DNA fragment separation using microfluidic channels that resulted in significantly reduced time of analysis as compared with traditional gel electrophoresis and capillary electrophoresis [9–11].

These early microchip CE devices used glass chips allowing for the definition of a nL volume sample plug and subsequent separation of DNA fragments in the presence of a strong electric field [9–11]. The applications of these microchip CE devices include nucleic acid analyses and protein assays for separating, sizing, quantifying, and identifying the content of a sample of DNA, RNA, or protein extracted from cells. These systems have been later expanded to 96-channel parallel microplate structures [5] and 384-lane capillary array electrophoresis microplates [6]. Moreover, most of the advanced integrated microfluidic work has been directed towards the integration of DNA amplification with CE [2, 12, 13].

Microfluidic techniques have been employed to enhance DNA microarrays. One inherent problem faced by a conventional microarray is its lengthy hybridization process. In general, DNA hybridization is a diffusion-limited process that relies on the diffusion of the target to the surface-bound probes in a sample solution at low Reynolds number. It has been recognized that mixing is important to achieve enhanced rates of mass transfer of target DNA in the hybridization process and various microfluidic techniques were devised to accelerate this process. They include electrokinetic methods to accelerate the transportation of DNA molecules [14–16], dynamic DNA hybridization using paramagnetic beads [17], rotation of the whole device [18], the use of a microporous three-dimensional biochip with the hybridization solution being pumped continuously through it [19], shuttling the sample in a microfluidic channel [20], shuttling sample plugs in a serpentine microtrench [21], chaotic micromixers [22, 23], and acoustic mixing [24].

In addition to hybridization enhancement, microfluidic techniques have also been employed to integrate front-end sample preparation (e.g., sample collection and pretreatment, DNA extraction, and amplification) with DNA microarrays. Kricka and Wilding have demonstrated physical filters relying on separation of biological cells by size [25]. Anderson et al. integrated monolithic genetic assay devices to carry out serial and parallel multistep molecular operations, including nucleic acid hybridization [26]. Liu and Grodzinski have developed plastic disposable chips for pathogen detection, performing PCR amplification, DNA hybridization, and a hybridization wash in a single device [27].

On-chip valving using phase change pluronics material was also implemented to facilitate a separation of different stages of the assay. Wilding, Kricka, and Fortina have also developed a prototype of an integrated semidisposable microchip analyzer for cell separation and isolation, PCR amplification, and amplicon detection which is described with preliminary results [28].

Our group has expanded the level of integration further to a complete self-contained biochip capable of magnetic, bead-based cell capture, cell preconcentration, purification, lysis, PCR amplification, DNA hybridization, and electrochemical detection of hybridization events [1]. The on-chip analysis starts with the preparation process of a whole blood sample, which includes magnetic bead-based target cell capture, cell preconcentration and purification, and cell lysis, followed by PCR amplification and electrochemical microarray-based detection. The device is completely self-contained and does not require external pressure sources, fluid storage, mechanical pumps, or valves. This integrated microfluidic device is desirable for applications valuing portable solutions such as point-of-care diagnostics, in-field environmental testing, and on-site forensics.

This chapter begins with a description of the design and fabrication of the self-contained integrated microfluidic device [1], followed by a discussion of the individual microfluidic components essential to this integrated device. Next, a demonstration of the performance of the device for integrated nucleic acid analysis, including pathogenic bacteria detection from ~mL whole blood samples and single nucleotide polymorphism analysis directly from blood samples, is presented.

## 3.2 Device Design and Fabrication

The microfluidic device (Fig. 3.1) consists of a plastic chip, a printed circuit board (PCB), and an eSensor® microarray chip [1]. The plastic chip includes a mixing unit for rare cell capture using immunomagnetic separation, a cell preconcentration/purification/lysis/PCR unit, and a DNA microarray chamber. The complexity of the chip design is minimized by using some of the chambers for more than one function. For example, the chamber to capture and preconcentrate target cells is also used for subsequent cell lysis and PCR. The PCB consists of embedded resistive heaters and control circuitry. The eSensor is a separate PCB substrate with 4 × 4 gold electrodes on which thiol-terminated DNA oligonucleotides are immobilized via self-assembly to detect electrochemical signals of hybridized target DNA [29, 30].

The operation of the microfluidic device is as follows. A biological sample (such as a blood sample) and a solution containing immunomagnetic capture beads are loaded in the sample storage chamber. Other solutions including a wash buffer, PCR reagents, and hybridization buffer are separately loaded in other storage chambers. The microfluidic device is then inserted into an instrument, which provides electrical power, PCR thermal cycling, DNA electrochemical signal readout, and magnetic elements for bead arrest. The PCR chamber of the plastic chip is sandwiched between a Peltier heating element (Melcor Corp., Trenton, NJ) and a permanent magnet.

The on-chip sample preparation starts with a mixing and incubation step in the sample storage chamber to ensure target cell capture from the blood using immunomagnetic capture beads. The sample mixture is then pumped through the PCR chamber, where target cell capture and preconcentration occur as the bead–bacteria conjugates are trapped by the magnet. The washing buffer is subsequently pumped through the PCR chamber to purify the captured cells. After the PCR reagents are transferred into the PCR chamber, all the normally open microvalves surrounding the chamber are closed, and thermal cell lysis and PCR are performed.

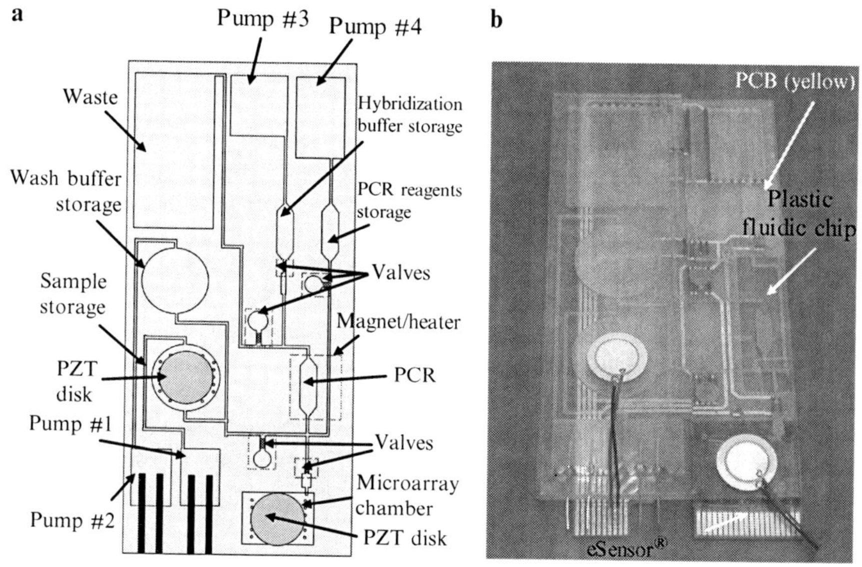


**Fig. 3.1** (A) Schematic of the plastic fluidic chip. Pumps 1–3 are electrochemical pumps, and Pump 4 is a thermopneumatic pump. (B) Photograph of the integrated device that consists of a plastic fluidic chip, a printed circuit board (PCB), and an eSensor microarray chip (reproduced with permission from [1], copyright 2004, American Chemical Society)

Once PCR is completed, the normally closed microvalves are opened, allowing the hybridization buffer and the PCR product to be pumped into the detection chamber, where acoustic mixing of the target and DNA hybridization reaction occurs. The electrochemical hybridization signals corresponding to the redox-reaction of the ferrocene-labeled signaling probes that hybridize with the target DNA bound to the immobilized probes are detected on the chip and recorded by the instrument.

The plastic chip measures 60 × 100 × 2 mm and has channels and chambers that range from 300 μm to 1.2 mm in depth and 1 to 5 mm in width. The plastic chip was machined in a polycarbonate (PC) substrate (1.5 mm thick) using conventional computer-controlled machining (Prolight 2500, Intelitek Inc., Manchester, NH), followed by sealing it with a thin PC cover layer (500 μm thick) using a solvent-assisted thermal bonding technique.

During the bonding process, acetone was first applied on one side of the thin cover layer. After 1 min, the cover layer with the gluey surface caused by the chemical reaction with acetone was bonded on the substrate plastic layer followed by a press of 1 ton force at 385 °F for 2 min in a hydraulic press (Carver, Inc., Wabash, IN). Platinum wires with 0.5 mm diameter were inserted into the electrochemical pumping chambers, which were then loaded with 20 μL of 5M NaCl solutions to form electrochemical pumps. Two piezoelectric disks (each 15 mm diameter, APC Inc., Mackeyville, PA) that were used to provide acoustic micromixing were bonded onto the external surfaces of the sample storage chamber and the microarray detection chamber, respectively, using a superglue (Duro, Loctite Corp., Avon, Ohio).

Fabrication of the paraffin-based microvalves in the device began with heating up the plastic chip using a hotplate with a temperature of 90 °C that is above the melting temperature of the paraffin. Solid paraffin (~10 mg) with a melting temperature $T_m$ of 70 °C was then placed into each of the paraffin access holes on the plastic chip. The paraffin was melted instantaneously and capillary force drove the molten paraffin into the channels. The chip was then removed from the hotplate. The paraffin solidified, resulting in an array of microvalves in the device. The paraffin access holes were subsequently sealed using an adhesive tape (Adhesive Research Inc., Glen Rock, PA). The plastic chip was then bonded with the PCB using a double-sided adhesive tape (Adhesive Research Inc., Glen Rock, PA). The eSensor microarray chip was attached to the detection cavity of the plastic chip using a double-sided adhesive tape.

## 3.3 Microfluidics

### 3.3.1 Fluidic Transport

The fluidic architecture takes advantage of the fluid gravity to remove the air bubbles from the system without the use of porous hydrophobic vents [26]. As in many other microfluidic devices, air plugs or bubbles trapped in the system are of great practical concern, because they often lead to difficulties in controlling the flow and hinder uniform mixing between fluids. Because all the chambers have dimensions of ≥500 μm and fluid volumes on the order of μL or mL were handled in the system, the Reynolds number for the flow is on the order of 10 or above [31]. The Reynolds number gives the ratio between inertial forces and viscous forces in a flow. The definition used here is $Re = (Q/A)D_h/v$, where $Q$ is the volumetric flow rate through the channel, $A$ is the cross-sectional area, $D_h$ is the hydraulic diameter of the channel (4$A$/wetted perimeter of the channel), and $v$ is the kinematic viscosity of the fluid.

When the Reynolds number is on the order of 10 or above, fluid gravity dominated surface forces when the chip was operated in a vertical position. Gas bubbles always migrated to the upper portion of the chamber due to buoyant forces whereas the liquid solution resided in the lower portion. For example, in the PCR chamber where fluids entered at the bottom of the chamber, all air bubbles trapped in the solution escaped towards the top of the chamber and subsequently traveled to the downstream waste chamber, leaving the PCR chamber bubble-free.

### 3.3.2 Micromixing

Rapid and homogeneous mixing is essential in our microfluidic device. First of all, the sample solution in the sample storage chamber needs to be mixed effectively with immunomagnetic capture beads in order to achieve efficient capture and binding of target cells. In the hybridization chamber, the hybridization buffer solution and the amplicon solution need to be mixed in order to obtain homogeneous solutions and achieve efficient and rapid DNA hybridization. Mixing in microfluidic systems is generally dominated by diffusion, because turbulence is not practically attainable in microscale channel flows or minisystems with small dimensions and thus small Reynolds (*Re*) numbers.

In most microfluidic systems, *Re* is typically below the critical value for transition to turbulence. Unfortunately, a pure diffusion-based mixing process can be very inefficient and often takes a long time. Mixing is particularly inefficient in solutions containing macromolecules that have diffusion coefficients one or two orders of magnitude lower than that of most liquid molecules. Thus, some mixing techniques must be employed to enhance micromixing.

A few interesting micromixing schemes, including inline micromixers that enhance mixing between two adjacent flow streams in a microchannel [23, 31–34] and chamber micromixers that utilize stirring mechanisms to mix the fluids in a microchamber [23, 35–34], have been developed.

One example of inline micromixers is a multistage multilayer lamination scheme developed by Branebjerg et al. [32]. The mixer divides and stacks two flow streams resulting in increased contact area and decreased diffusion length. Another inline micromixer concept was developed by Liu et al. [31], using a three-dimensional serpentine microchannel to create rapid stretching and folding of material lines associated with flowfield-induced chaotic advection. A two-dimensional chaotic mixer was developed by Stroock et al. [34]. Electrokinetic instability induced by fluctuating electric fields was also utilized to enhance mixing of electroosmotic channel flows [33].

Chamber micromixing is of particular interest in our applications. Examples of chamber micromixers include those of Moroney et al. [35] and Zhu et al. [36]. The former used ultrasonic lamb waves (4.7 MHz) traveling in a 4 μm thick composite membrane of silicon nitride and piezoelectric zinc oxide to induce convectional liquid flow in a chamber. The latter utilized loosely focused acoustic waves generated by an electrode-patterned piezoelectric film to enhance mixing in an open chamber. Microfluidic motion produced by loosely focused acoustic waves uses radio frequency (RF) sources with frequencies corresponding to thickness-mode resonance of the thin piezoelectric film. Both devices required a thin chamber membrane (with a thickness of a few microns) between the liquid solution and the piezoelectric film, which was fabricated by silicon (Si) micromachining. The piezoelectric films were driven in ultrasonic frequency range (~ MHz).

We have developed a novel chamber micromixing technique based on the principle of cavitation microstreaming [24, 37]. This technique uses air bubbles in a liquid medium as actuators. The bubble surface behaves as a vibrating membrane when the bubble undergoes vibration within a sound field. The behavior of bubbles in sound fields is determined largely by their resonance characteristics. For frequencies in the range considered here (~kHz), the radius of a bubble at resonant frequency $f$ (Hz) is given by the equation:

$$2\pi af = \sqrt{3\gamma P_o / \rho} \tag{3.1}$$

where $a$ is the bubble radius (cm), $\gamma$ is the ratio of specific heats for the gas, $P_o$ is the hydrostatic pressure (dynes/cm$^2$), and $\rho$ is the density of the liquid (g/cc). When the bubble undergoes vibration within a sound field, the frictional forces generated at the air/liquid interface induce a bulk fluid flow around the air bubble, called cavitation microstreaming or acoustic microstreaming [38].

It was found that cavitation microstreaming is orderly at low driving amplitudes when the insonation frequency drives the bubbles at their resonance frequency for pulsation and when the bubbles are situated on solid boundaries. The bubble-induced streaming is strongly dependent on frequency for a given bubble radius, and on bubble radius for a given frequency. Acoustic microstreaming arising around a single bubble excited close to resonance produces strong liquid circulation flow in the liquid chamber. This liquid circulation flow can be used to effectively enhance mixing.

As shown in Fig. 3.2, a set of air bubbles was trapped inside the solution using air pockets (500 μm in diameter and 500 μm in depth) in the cover layer of the chamber [24]. These bubbles were set into vibration by an acoustic field generated using an external piezoelectric transducer (PZT disk, 15 mm diameter, APC Inc., Mackeyville, PA) that was bonded on the external surface of the cover layer using a superglue (DuroTM, Loctite Corp., Avon, OH). Fluidic experiments showed that the time taken to achieve a complete mixing in a 50 μL chamber using cavitation

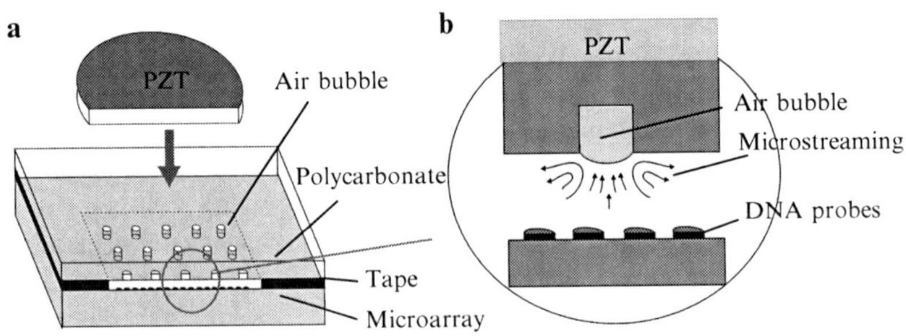


**Fig. 3.2** Schematic showing a number of air pockets in the top layer of the DNA biochip chamber: (a) overview; (b) sideview (reproduced with permission from [24], copyright 2003, American Chemical Society)

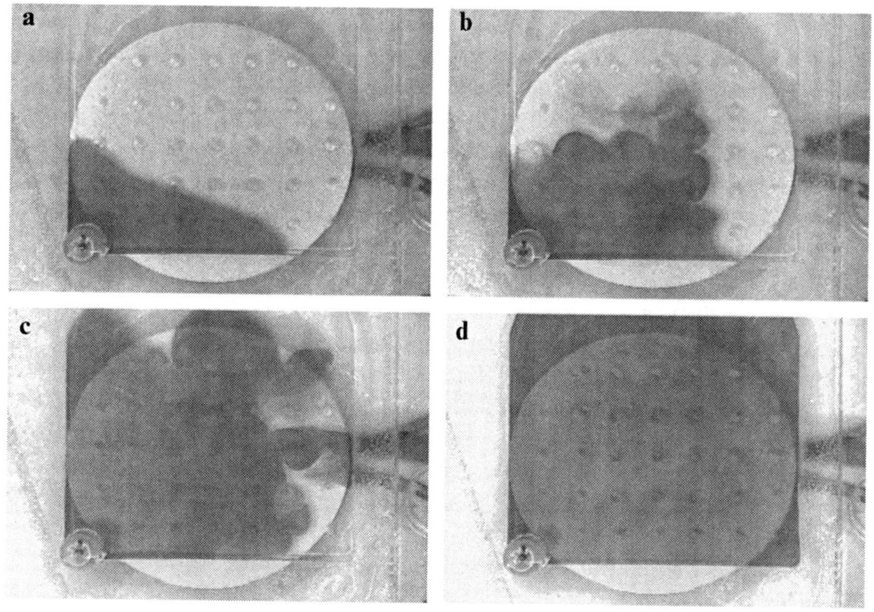

**Fig. 3.3** Photographs showing multibubble induced (7 × 5 top bubbles) cavitation microstreaming in a 16 × 16 × 0.2 mm chamber at (a) time 0; (b) 2 s; (c) 4 s; (d) 6 s. The PZT disk on the back side of the chamber was driven at 5 kHz and 40 $V_{pp}$ (reproduced with permission from [24], copyright 2003, American Chemical Society)

microstreaming was significantly reduced from hours (a pure diffusion-based mixing) to only 6 seconds, as shown in Fig. 3.3 [24]. The employment of cavitation microstreaming to enhance target cell capture and DNA hybridization is described in detail in Sections 3.4.1 and 3.4.3, respectively.

### *3.3.3 Microvalves*

The microvalve is another important component in the microfluidic device. Conventional microvalves can be divided into two major categories: passive microvalves (without actuation) and active microvalves (with actuation). Inasmuch as most passive microvalves are check valves that allow fluid flow only in one direction, they are not as diverse in their use as active valves that can open and close the fluid passage. Most conventional active microvalves couple a flexible diaphragm to an electromechanical actuator operating based on a thermopneumatic [39, 40], bimetallic [41, 42], shape-memory [43], electrostatic [44], piezoelectric [45], or electromagnetic principle [46].

For example, the thermopneumatic microvalve developed by Zdeblick et al. made use of resistive heating that caused boiling of a trapped liquid and resulted in a high pressure [39]. The pressure pushed a thin silicon (Si) membrane that modulated the "current" of fluid in the regulated channel. Jerman et al. demonstrated a normally closed thermal bimorph actuation scheme for a microvalve for gas flow control using an aluminum layer fabricated on a Si membrane [41]. Shoji et al. demonstrated the use of a separately fabricated stacked piezoelectric actuator as a means of closing a Si micromachined valve [45].

Although the active microvalves described above have shown good performance, most of them are fabricated using a multilayer Si process (multilayer Si stacking and bonding) that is a complicated fabrication process. As a result, the devices often suffer from high cost, poor reliability, and high power consumption (note that Si has a high Young's modulus, which makes the Si diaphragm difficult to deform). The displacement of conventional silicon or silicon nitride diaphragms is typically limited to tens of microns or less. Moreover, these active microvalves are made of Si and, thus, are too expensive to be used for single-use biomicrofluidic devices.

Other interesting active microvalves include pH-sensitive hydrogel valves [46, 47], and a pneumatically actuated PDMS valve [48]. Both devices were made using poly(dimethylsiloxane) (PDMS) microfabrication and used a thin PDMS diaphragm sandwiched between multilayers of PDMS. PDMS microvalve devices in general suffer from shortcomings of PDMS's physical properties, such as poor thermal conductivity (e.g., it is difficult to use PDMS to construct a DNA polymerase chain reaction device because PDMS expands and tends to degas when heated up) and high liquid/molecule absorption because of its porosity. The fabrication of the active microvalves described above involves multilayer construction, thus not only is the fabrication process of such microvalves complicated, but the integration into complex microfluidic systems has also proven to be nontrivial [8].

We have developed a novel microvalving technique that was implemented into the biochip to facilitate a sequential and multistage analysis [1]. In this valving mechanism, paraffin is used as an actuator material that undergoes solid–liquid phase transition in response to changes in temperature. Several one-shot valving schemes, including "close–open" and "open–close–open" valves (Fig. 3.4), are demonstrated. The "close–open" is a normally closed valve that can be opened once (one-shot valve) as shown in Fig. 3.4a,b [1].

The valve uses a bulk of paraffin, which is localized in a heating zone, to close the channel first. The paraffin can be melted using the heat generated by a heater (e.g., a resistive heater) underneath, when the channel needs to be opened. A pressure from the upstream channel section moves the molten paraffin downstream. The downstream channel is incorporated with a wide solidification section (~2 times wider than the upstream channel, 4~6 mm long), which is located 1 mm away from the heating zone and designed to trap paraffin after solidification and thus prevent the downstream channel from clogging.

During the fabrication of the valve, the channel device was first heated up to a temperature (i.e., 90°C), which is above the melting temperature of the paraffin,

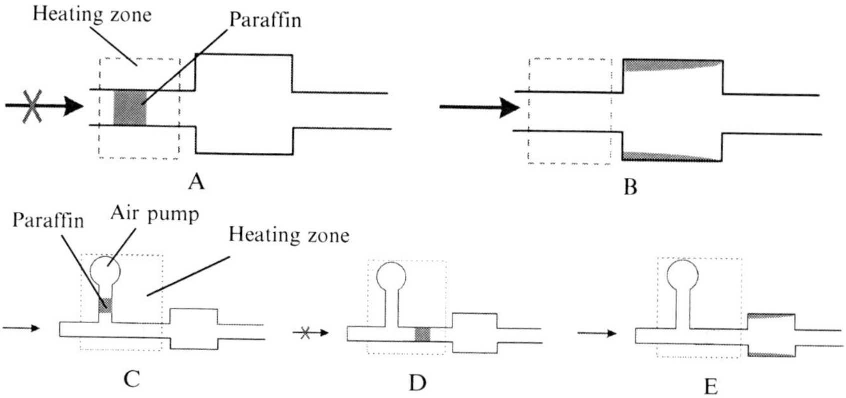


**Fig. 3.4** Schematic illustrations of a "close–open" paraffin microvalve design (a),(b) and an "open–close–open" microvalve design (c)–(e). The former has a block of paraffin that initially closes the channel (a). To open the channel, the paraffin is melted using the heater underneath and moved downstream by the pressure from the upstream channel. Once the molten paraffin moves out of the heating zone, it starts to solidify on the wall of a wide channel section resulting in an open channel (b). The latter is a normally open valve with a block of paraffin connected to an air pocket that acts as a thermally actuated air pump (c). When the heater is activated, the air in the pocket expands and pushes the molten paraffin into the regulated channel. If the heater is turned off immediately, the paraffin solidifies in the main flow channel, resulting in a closed channel (d). The channel can be reopened by reactivating the heater (e) (reproduced with permission from [1], copyright 2004, American Chemical Society)

using a hotplate. A small amount (20 mg) of solid paraffin (Signature Brands LLC, Ocala, FL) with a melting temperature $T_m$ of 70°C was then loaded into each of the paraffin access holes on the channel. The paraffin was melted instantaneously and capillary force drove the molten paraffin into the channel. The channel device was then immediately removed from the hotplate. The paraffin was solidified in the channel, resulting in normally closed microvalves in the channel device. The paraffin access holes were subsequently sealed using an adhesive tape (3 M, St. Paul, MN).

The fluidic experiments showed that no leakage occurred when the valves were in a "closed" position under normal flow conditions (e.g., pressure less than 20 psi). Maximum hold-up pressure was measured to be about 40 psi, above which we started to see some leakage at the interface of the paraffin and the channel wall. No dead fluid volume was observed in the valve because the resulting opened channel was a through channel where no fluid element was trapped. The response time of the valves was approximately 20 s.

It was found that the use of a wider solidification channel (i.e., 3 mm wide instead of 2 mm wide) resulted in an improved time response by 10% due to the increased space that allowed solidified paraffin to reside in and thus less time for the channel to be opened. It was also found that shortening the distance between the solidification section and the heating zone resulted in a slightly improved time response due to the shortened distance for the molten paraffin to travel prior to solidification. The actuation of a paraffin valve required 100~200 mW of power.

The "open–close–open" valve as shown in Fig. 3.4c–e is a normally open valve that can be closed and reopened, subsequently. The valve is integrated with a thermally actuated air pump that consists of an air pocket (with a diameter of 5 mm and a depth of 0.5 mm) attached with a heater (e.g., Peltier element). As shown in Fig. 3.4c, a small plug of paraffin is first placed in the channel branch connected with the air pump. The main flow channel is initially open. If the heater is turned on to a temperature above the paraffin melting temperature, the paraffin melts. The air trapped in the pocket is also heated up, resulting in increased pressure, which in turn pushes the molten paraffin into the regulated channel. If the heater is turned off immediately, the paraffin solidifies in the main flow channel, resulting in a closed channel (Fig. 3.4d). The channel can be reopened by turning on the heater to the temperature above the melting point of the paraffin until the paraffin moves out of the heating zone and solidifies downstream in a solidification section as discussed in previous paragraph (Fig. 3.4e).

Pressure drop measurements in the fluidic experiments showed that the maximum hold-up pressure was about 40 psi when the valve was in a closed mode. No leakage was observed when the channel was closed. The closing operation (including the temperature ramping time of the heater) took approximately 10 s. The closed channel was reopened after the heater was turned back on to heat up the channel and melt the paraffin, which then solidified in the downstream channel section. The reopening operation took about 12 s.

As shown in Fig. 3.5a, a plastic micro-PCR device that consists of a PCR chamber (5 mm in width, 16 mm in length, 0.5 mm in depth) and five paraffin-based microvalves was fabricated and tested for DNA amplification. A PCB substrate (not shown here) consisting of resistive heaters was attached to the plastic device to provide thermal actuation to the paraffin elements. Following the loading of the PCR reaction mixture into the reaction chamber through the channels regulated by the paraffin microvalves (2) and (3), the chamber was sealed by closing the valves to ensure no leakage. Note that the microvalves (4) and (5) were initially closed.

The PCR microdevice was then placed into the Peltier thermal cycler. During PCR thermal cycling, the micro-PCR chamber was sandwiched between the two Peltier elements, whereas all the paraffin microvalves were located at a distance of 10 mm away from the heating zone, in order to prevent the bivalve actuation during PCR thermal cycling. After PCR was completed, the microvalves (4) and (5) were opened to retrieve the reaction product from the chamber for off-chip electrophoresis analysis. The electrophoretic results, as shown in Fig. 3.5b, indicate that DNA was successfully amplified in this device. The PCR yield is similar to that of the control PCR performed in a conventional PCR tube (Molecular BioProducts, San Diego, CA) using a conventional DNA Engine™ Thermal Cycler (MJ Research Inc. South San Francisco, CA) under the same conditions, indicating that the paraffin is fully PCR compatible.

a

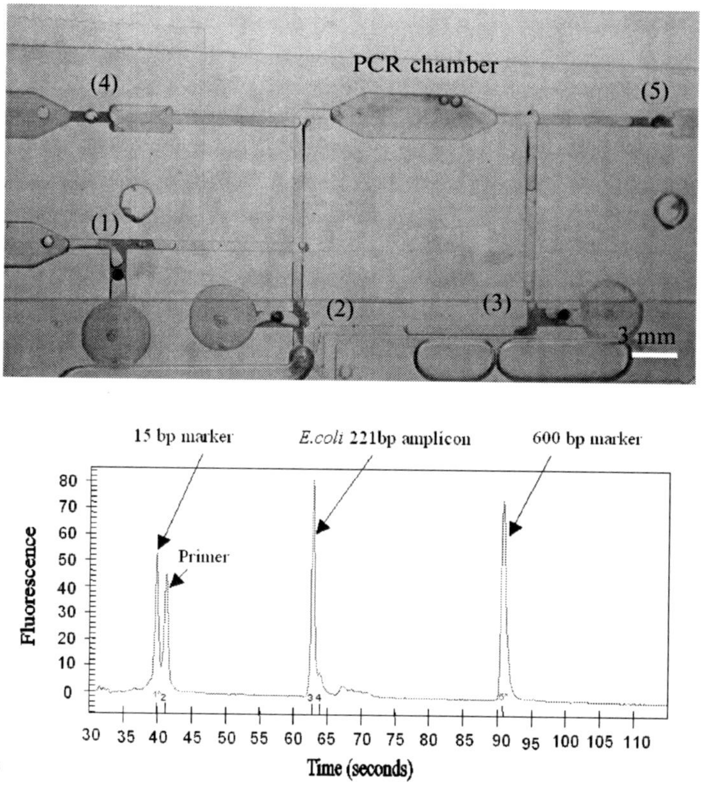


b

**Fig. 3.5** (a) A photograph of the polycarbonate micro-PCR device integrated with five paraffin-based microvalves: valves (1)–(3) are "open–close" valves, and (4) and (5) are "close–open" valves (note that the PCB substrate consisting of resistive heaters to actuate the valves is not shown here); (b) PCR results of the micro-PCR device. The *E. coli* K12-specific gene (221 bp fragment) was amplified and analyzed using Agilent BioAnalyzer DNA 500 (reproduced with permission from [1], copyright 2004, American Chemical Society)

The planar designs of the paraffin-based valves do not include a flexible diaphragm, and thus are simpler than traditional actuator/diaphragm designs that require multilayer structures. Although we have only demonstrated one-shot "close–open" and "open–close–open" valving schemes, other configurations can be easily achieved [49]. For example, a toggle valve that consists of a number of "open–close–open" segments operated in sequence can perform a number of "open–close" cycles as designated. Although the time response (~10 s) of the paraffin-based microvalves is relatively slow as compared to that of many conventional microvalves (~ms), the paraffin-based valves have proven to be practical and useful in our biochip device where rapid response is not critical. It is believed that the time response could be improved by employing a paraffin material with a lower melting temperature.

The fabrication process of the paraffin valves is compatible with many other material fabrication processes (such as Si, plastic, etc). The precise loading of paraffin material (melted volume on the order of pL) into the microchannel to form a microvalve can be achieved using a wax injector (Microdrop GmbH, Germany), leading to a simple manufacturing method to fabricate paraffin microvalves in complex microfluidic devices. This process is simpler than bulk processes (e.g., bulk etching of Si wafers), surface processes (e.g., thin-film processes), or chemical reactions (e.g., in situ polymerization [37, 50]) used in fabrication of conventional microvalves.

Because paraffin is a commonly used and inexpensive material, the paraffin-based microvalves are cost-effective and highly desirable for many single-use and disposable microfluidic applications. It is worth noting that the valving approach is not limited to paraffin and can be extended to many other materials that can undergo a phase transition from solid to liquid in response to changes in temperature [27].

### 3.3.4 Micropumps

In addition to microvalves, the micropump is another essential and important component in the integrated microfluidic device. Based on different pump (actuation) mechanisms, conventional micropumps can be classified into two major groups: membrane-actuated (mechanical) and nonmembrane-actuated pumps [8]. Membrane-actuated pumps can be further divided into different types: piezoelectric, electrostatic, and thermopneumatic, among others [48, 51, 52]. Most of these conventional pressure-driven membrane-actuated micropumps suffer from complicated designs, complicated fabrication, or high cost. Nonmembrane pumping principles include electrohydrodynamic, electro-osmotic, traveling wave, diffuser, bubble, surface wetting, rotary, and so on. Although much progress has been made, micropumps with the appropriate combination of cost, performance, and operating requirements are still not available for many practical applications.

The microfluidic device reported here requires integrated micropumps for transport of a wide range of sample volumes (μL–mL). In our device, two simple pressure-driven micropumping methods were employed: a thermopneumatic air pump (Pump 4 in Fig. 3.1a) for pumping of ~μL volumes, and electrochemical pumps (Pumps 1–3 in Fig. 3.1a) for ~mL volumes. The former made use of the air expansion in an air chamber, which was attached to a resistive heater in the PCB substrate, when heated up. The air expansion is a nearly linear function of temperature. The resulting air expansion pushed the solution from the storage chamber into the downstream channels and chambers. The latter relied on electrolysis of water between two platinum electrodes in a saline solution to generate gases when a DC current is applied [53]. The gas generated a pressure that in turn moved liquid solutions in the device (Fig.3.6).

Both pumping mechanisms do not require a membrane and/or check valves in their designs. As a result, their fabrication and operation are much simpler than most conventional micropumps. Flow experiments demonstrated that the thermopneumatic air pumps with air pockets of 50 μL internal volume could efficiently move up to 60 μL volume of fluids with a heater power consumption of less than 0.5 W. For pumping of ~mL solution volumes, the electrochemical pump is more efficient and consumes less power. A steady flowrate of up to 0.8 mL/min was achieved with a power consumption of <150 mW [1].

## 3.4 On-Chip Assays

### *3.4.1 Sample Preparation*

Front-end sample preparation represents the most time-consuming and labor-intensive procedure in DNA analysis. It also introduces one of the largest variables in subsequent analyses due to its complexity. The developments of back-end detection schemes, resulting in demonstrations of DNA microarray biochips [54, 55] and electrophoresis separation chips [9–11], have shifted the bottleneck, impeding further progress in rapid analysis devices, to front-end sample preparation where the "real" samples of bodily fluids are used.

Target cell isolation and preconcentration from the crude biological sample solution have been previously studied in the attempt to address the front-end stage

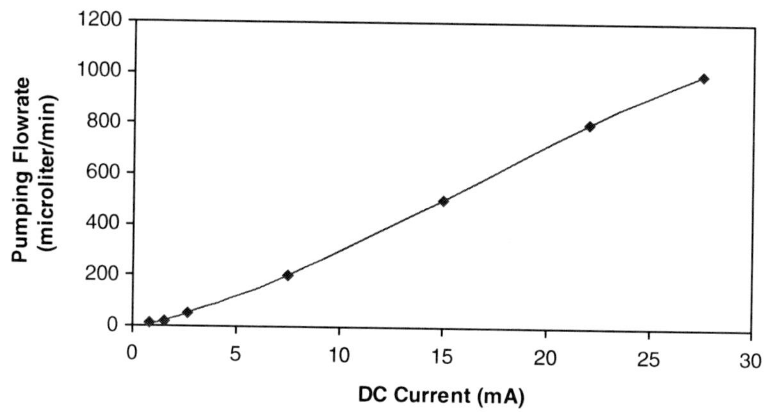


**Fig. 3.6** Measurement of the liquid pumping rate as a function of the applied DC current in an electrochemical pump that relies on the electrolysis of water between two platinum electrodes in a saline solution (20 μL 5 M NaCl) (reproduced with permission from [1], copyright 2004, American Chemical Society)

of sample preparation. Kricka and Wilding [4] have demonstrated physical filters relying on separation of biological cells by size. White blood cells were isolated from whole blood in silicon-glass 4.5 μL microchips containing a series of 3.5 μm feature-sized "weir-type" filters, formed by an etched silicon dam spanning the flow chamber. Genomic DNA targets, such as the dystrophin gene, were then directly amplified using PCR from cells isolated on the filters.

This dual-function microchip provides a means to simplify nucleic acid analyses by integrating in a single device two key steps in the analytical procedure, namely cell isolation and PCR [4, 56, 57]. Gascoyne et al. [58] developed dielectrophoretic (DEP) separation techniques that exploit the differential dielectric properties occurring among different biological cells. Cheng et al. [59] showed isolation of cervical carcinoma cells from blood. Ward and Grodzinski [60, 61] introduced immunomagnetic cell separation into microfluidic devices containing gradient producing soft magnetic ridges. These devices were used to separate *E. coli* and mammalian cells from blood samples. Saito et al. [62] worked on ultrasonic techniques to arrest euglena and paramecia cells in acoustic standing wave nodes.

In our integrated diagnostic device, an immunomagnetic separation technique has been applied to enrich rare target cells from a large-volume clinical sample such as blood. Considering that the blood sample is a highly heterogeneous cellular and genetic medium (e.g., 1 mL of whole blood contains $\sim 10^6$ white blood cells, $\sim 10^9$ red blood cells, and $10^7$ platelets, in addition to the serum components), a highly selective and sensitive target preconcentration posts a real challenge for on-chip integration of sample preparation steps.

The use of immunomagnetic separation techniques for this purpose has a number of advantages, including high selectivity, ease of implementation, and reduced assay complexity, as compared to other approaches [4, 26, 63, 64]. Moreover, the immunomagnetic approach allows for sample preparation and PCR amplification to be performed in the same chamber, which not only simplifies the device design, fabrication, and operation, but also eliminates sample loss due to unnecessary fluid transfer.

In the self-contained, integrated microfluidic system reported here, we have successfully demonstrated capturing $10^6$ *E.coli* cells (equivalent to 5 ng of genomic material) from 1 mL whole blood using the immunomagnetic bead separation technique [1]. Pathogenic bacteria detection from a whole blood sample was performed. *Escherichia coli* (*E. coli*) K12 cells inoculated in whole rabbit blood were used as a model for demonstration. Although this application may have no immediate practical use, it is easy to conceive a system where, for example, the rabbit blood is replaced by human blood and pathogenic or cancer cells are targeted instead of the *E. coli* K12 cells.

The *E. coli* / rabbit blood system was used because of the simplicity of performing the experiments as well as control assays in an ordinary laboratory setting (without BSL-2 or higher laboratory requirements). The input sample was 1 mL of whole citrated rabbit blood (Colorado Serum Company, Denver, CO) containing $10^3 \sim 10^6$ *E. coli* K12 cells. 10 μL of biotinylated polyclonal rabbit–anti-*E. coli* antibody (ViroStat, Portland, ME), and 20 μL of streptavidin labeled Dynabeads (total

$1.3 \times 10^7$ M-280 beads, Dynal Biotech, Inc., Lake Success, NY) were also added into the sample storage chamber. The bead–antibody–*E. coli* cell complexes were formed during a 20-min cavitation microstreaming-based mixing period.

In order to efficiently capture target cells (*E. coli* K12) from the whole blood sample using immunomagnetic capture beads, cavitation microstreaming was implemented in the sample storage chamber to achieve a homogeneous mixing for the bead–cell complex formation [37]. An experiment was designed and implemented to evaluate cell capture enhancement using a cavitation microstreaming technique. A 55 μL sheep blood solution suspended with 100 *E. coli* K12 cells that were already labeled with biotinylated polyclonal rabbit anti *E. coli* antibody (ViroStat, Portland, ME) was loaded into the mixing chamber followed by a separate load of 5 μL streptavidin-coated colloidal magnetic beads (Miltenyi Biotec, Inc., Auburn, CA).

During the mixing process, the bead–antibody–cell complexes would form through specific biotin–streptavidin interaction if the mixing were sufficient. The mixing chamber had a 4 × 4 array of air pockets (0.5 mm in diameter and 0.5 mm in depth) on the chamber wall [37]. Acoustic micromixing was then performed at 5 kHz and 10 $V_{pp}$ (square wave) for 30 min. Following mixing, the mixture solution was retrieved from the mixing chamber and passed through a miniMACS separation column (Miltenyi Biotec, Inc., Auburn, CA) where magnetically labeled bacteria were captured and subsequently plated overnight onto L-Broth Agar plates at 37°C.

The *E. coli* colonies formed on the plates were counted. Capture efficiency was calculated by dividing the number of colony-forming units found on the elution plate (captured bacteria) by the total number of colony-forming units (captured bacteria and escaped bacteria that were not captured by the beads). Capture efficiencies were compared to those using a standard protocol (i.e., vortexing in a microfuge tube) and those using no mixing enhancement except pure diffusion only. Calculated results of capture efficiencies are summarized in Fig. 3.7.

Results show that the capture efficiency of *E. coli* cells is comparable between cavitation microstreaming and a conventional vortex in a tube, indicating mixing efficiency is comparable between them. Both cavitation microstreaming and conventional vortex result in much higher capture efficiency than pure diffusion, suggesting that cavitation microstreaming significantly enhances cell capture efficiency during the sample preparation step.

Following the mixing, the blood sample mixture was pumped using an electrochemical pump (Pump 1 in Fig. 3.1a) at 0.1 mL/min into the PCR chamber, where the bead–antibody–cell complexes were collected by the magnet. The uncaptured samples were flowed into the waste chamber. Subsequently, 1 mL wash buffer was pumped at 0.1 mL/min through the PCR chamber using Pump 2 to wash the captured complexes. As a result of the cell preconcentration and purification steps, the purified bead–antibody–cell complexes were isolated from the blood sample solution and trapped in the PCR chamber. Cell thermal lysis followed by PCR was then performed in the PCR chamber to amplify the DNA of captured *E. coli* K12 cells.

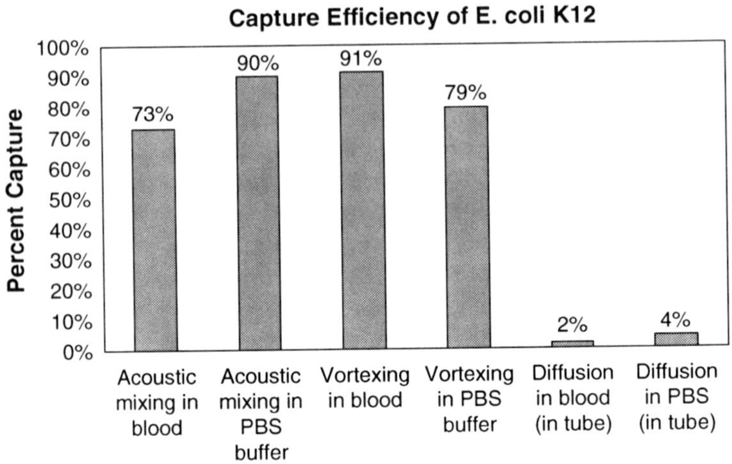


**Fig. 3.7** Comparison of normalized capture efficiencies of *E. coli* K12 cells from PBS or blood samples using acoustic conventional vortexing, or diffusion only (reproduced with permission from [37], copyright 2004, American Chemical Society)

## 3.4.2 *On-Chip DNA Amplification*

DNA amplification, such as polymerase chain reaction (PCR), is essential to most genetic analysis applications of integrated microfluidic biochips. Rapid operation, small sample volume, and parallel amplification of different DNA templates within the same chip are among the desired features of amplification in the microchip environment. The design and development of such chips is hindered by several challenges including: loss of sample to the chamber walls due to dramatic increase in surface-to-volume ratio, evaporation in small volume regime, and effective heat dissipation in order to achieve rapid thermal cycling. Micro-PCR devices have been successfully fabricated in glass [13, 65], silicon [25, 66], and plastic [67–69]. Silicon, due to its superior thermal conductivity (~10× of glass and ~700× of polymers), allows for fast temperature ramping and results in short on-chip processes [66].

Similarly, successful and fast amplification assays have been demonstrated in glass by many groups [65, 70]. Landers' group has used an IR heating scheme and a glass PCR device consisting of a 1.7 μL microchamber; amplification was achieved by 15 thermal cycles in 4 min [71]. In an integrated monolithic silicon–glass device, submicroliter (280 nL) volume was thermally cycled as fast as 30 s/cycle [72].

Using a real-time PCR device, Belgrader et al. reported PCR detection of *Erwinia*, a vegetative bacterium, in 7 min [66]. An integrated rapid PCR-detection system coupled with capillary electrophoresis analysis was presented by Khandurina et al. [73] and Lagally et al. [70] with amplification times in the 20 min range. Recently with increasing emphasis on disposable devices, the use of plastic and plastic fabrication methods have become very popular in microreactor development. Plastic chips are inexpensive, optically transparent, and biocompatible [74, 75].

Despite all the advantages, plastic possesses a major challenge to a designer of PCR microreactors due to its poor thermal conductivity and resulting difficulty to achieve rapid thermal cycling. We have demonstrated a successful DNA amplification in polycarbonate chips; 30 thermal cycles took 30 min, which considering the poor thermal conductivity characteristics of polymers is a significant achievement [76].

Despite all these demonstrations, the sensitivity aspect of micro-PCR chip assay has not been well studied. Most of the reports focused on achieving amplification per se, but the systematic evaluation of amplification yield and reaction sensitivity was usually not explored. The initial template concentration used to achieve fast and small volume amplification was usually high, ranging from 0.1 ng phage DNA [71] to hundred ng human genomic DNA [77].

The most sensitive micro-PCR assay was demonstrated by Lagally et al. [72]. A single molecule DNA template could be amplified in a glass integrated microfluidic device. Also, in an integrated glass sandwich structure, Mathies and coworkers [70] reported amplification with a starting template concentration as low as 5–6 copies was achieved in their nanoliter-volume glass microchambers.

For the "real" sample (containing target cells, rather than purified DNA) analysis, the most sensitive silicon microstructure that could perform rapid real-time PCR analysis from a sample containing low concentration of target (*Erwinia*) cells, was reported by Belgrader and colleagues [66]. A positive amplicon signal was detected in less than 35 cycles (17 s per cycle time) with the starting template concentration as low as 5 cells. In a flat polypropylene tube, Northrup and his group demonstrated a real-time PCR analysis of 1000 bacillus spores [78]. We performed a systematic study on a sensitivity of micro-PCR assay in plastic devices [76]. We demonstrated a feasibility of amplifying template concentrations as low as 10 *E. coli* cells (50 fg of DNA) in presence of blood. Similarly, we demonstrated PCR multiplexing within the same reactor chamber for four different bacteria species.

As described in the previous section of sample preparation, the purified bead–antibody–cell complexes were isolated from the blood sample solution and trapped in the PCR chamber of the integrated microfluidic device as a result of the cell preconcentration and purification steps. Following influx of the PCR reagents to the PCR chamber and closing of all the paraffin-based valves, thermal cell lysis and two-primer asymmetric PCR were performed to amplify an *E. coli* K12-specific gene fragment and achieve single-stranded DNA amplicon in the presence of beads. A pair of *E. coli* K12-specific primers were used to amplify a 221 bp *E. coli* K12-specific gene (MG1655) fragment, with forward primer: 5′ AAC GGC CAT CAA CAT CGA ATA CAT 3′ and reverse primer: 5′ GGC GTT ATC CCC AGT TTT TAG TGA 3′. The PCR reaction mixture (20 μL) consisted of tris-HCl (pH 8.3) 10 mM, KCl 50 mM, $MgCl_2$ 2 mM, gelatin 0.001%, dNTPs 0.4 mM each, bovine serum albumin (BSA) 0.1%, forward primer 0.05 μM, reverse primer 5 μM, and AmpliTaq DNA polymerase (Applied Biosystems, Foster City, CA) 5 units.

Note that during the PCR thermal cycling, the PCR chamber was sandwiched between two Peltier elements, whereas all the paraffin microvalves were located at a distance of 10 mm away from the heating zone, in order to prevent bivalve actuation during PCR thermal cycling. PCR was performed with an initial denature step at 94°C for 4 min followed by 35 cycles of 94°C for 45 s, 55°C for 45 s, and 72°C for 45 s, with a final extension at 72°C for 3 min. After PCR was completed, the normally closed microvalves were opened, allowing the hybridization buffer and the PCR product to be pumped into the microarray detection chamber, where acoustic mixing (cavitation microstreaming) of the target and DNA hybridization reaction occurred.

### *3.4.3 DNA Microarray Detection*

Following the PCR, two paraffin-based microvalves, one regulating the channel between the PCR chamber and the hybridization buffer storage chamber and the other one between the PCR and the microarray chamber, were opened. The PCR products were then transported to the eSensor microarray chamber along with hybridization buffer (20 μL) using Pump 3 as shown in Fig. 3.1a. The hybridization buffer contains 1X hybridization buffer stock, 7% fetal calf serum, 1 mM hexanethiol, and 0.5 μM signaling probe [29,30]. The microarray chamber was incubated at 35°C and the electrochemical signals from the eSensor microarray were measured at 0 min, 15 min, 30 min, and 1 h.

An e-Sensor™ chip that is an electrochemical detection based low-density microarray device [29, 30] was used to detect the electrochemical hybridization signals corresponding to the redox-reaction of the ferrocene-labeled signaling probes that hybridized with the target DNA (PCR product) bound to the immobilized probes DNA. The e-Sensor chip is a printed circuit board (PCB) chip consisting of an array of gold electrodes modified with a multicomponent self-assembled monolayer (SAM) that includes presynthesized oligonucleotide (DNA) capture probes that are covalently attached to the electrode through an alkyl thiol linker.

When the amplicon solution containing target DNA was introduced into the detection chamber, specific capture probes on an electrode surface encountered complementary DNA from the sample and hybridization occurred. Capture probe and signaling probe bound the target in a sandwich configuration. Binding of the target sequence to both the capture probe and the signaling probe connected the electronic labels to the electrode surface [30]. This added a circuit element to the bioelectronic circuit on that electrode and presence of hybridized (double-stranded) DNA could be detected using alternative current voltammetry (ACV).

Because the incoming target itself was not labeled, the washing step (to remove excessive nonbound target) prior to the signal collection was not required and a continuous monitoring of the binding process with a quantitative measurement of the target accumulation was possible. This electrochemical detection (ECD) based microarray platform eliminates the need for an expensive and sensitive laser-based optical system and fluorescent reagents. By replacing the large optical reader with a small ECD reader, the array can be incorporated into a portable or handheld detector instrument for in-field biological detection and clinical diagnostics.

To enhance the rate of DNA hybridization, a cavitation microstreaming technique was implemented in the hybridization chamber [24]. Hybridization typically relies on diffusion of DNA target to surface-bound DNA probes, and is often a lengthy rate-limiting process (6 to 20h). The bulk of the target solution is at a considerable distance, on the molecular length scale, from the reaction site on the chip surface. For example, the diffusion coefficient of a 250bp DNA fragment in water at room temperature is approximately $2 \times 10^{-7}$ $cm^2/s$, and thus its diffusion time along a length of 500μm is approximately 100min. This inefficient diffusion greatly limits the throughput of sample analyses and can be overcome by micromixing techniques.

Hybridization kinetics experiments were performed to study the hybridization kinetic enhancement induced by cavitation microstreaming [24]. Fig. 3.8 summarizes the hybridization kinetics results for an acoustic mixing-enhanced device and a diffusion-based device [24]. Each data point is the mean value obtained from four electrodes with identical DNA capture probes in the same eSensor device. The target DNA was obtained by PCR amplification from human genomic DNA. Because of the homogeneous nature of the assay, results for the probe coverage were obtained at different time points, giving a current-time curve for the hybridization kinetics.

Note that the *y*-axis in the Figure is the measurement of the faradaic current from the electrodes. The faradaic current is directly proportional to the number of ferrocene moieties in proximity to the electrode surface that in turn is proportional to the number of target nucleic acid molecules hybridized with the probes [30]. The results show that for static (diffusion-based) hybridization, the hybridization signal evolved slowly and exhibited an initial linear increase. It took approximately 6h for the static sample to reach the saturated level (not shown here). Moreover, the standard deviation of each data point shows that the static hybridization has relatively large electrode-to-electrode variation.

For the hybridization assay coupled with cavitation microstreaming, the signal increased rapidly, showing additional acceleration and uniform signal distribution (small standard deviations) compared to the pure diffusion-based device. After 40 minutes of hybridization, the sample in the mixing-enhanced device reached a saturated faradaic current value. The relative rate of hybridization in the two devices is given by the ratio of the time it took for the signal to reach one-half of the saturated value. Hybridization in the mixing-enhanced device occurred at approximately 5.3 times the rate of that in the diffusion-based device.

Cavitation microstreaming not only provides rapid lateral mass transport of fluidic elements, but also enhances vertical mass transport of target DNA in the

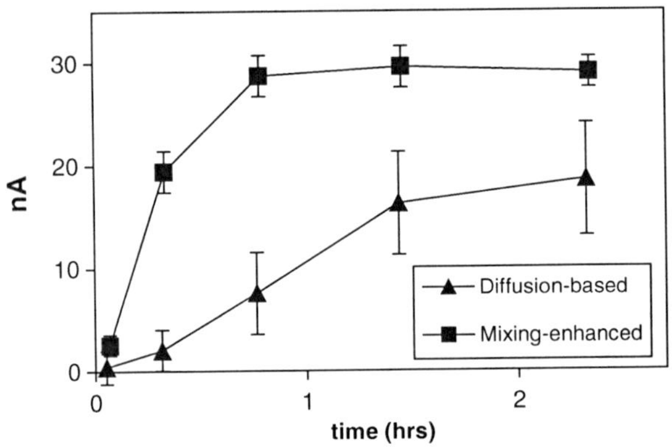


**Fig. 3.8** Hybridization kinetics study on static hybridization versus mixing-enhanced hybridization (5 kHz square wave and 10 $V_{pp}$) performed in eSensor devices. Each data point is the mean value obtained from four electrodes with identical DNA capture probes in the same device (reproduced with permission from [24], copyright 2003, American Chemical Society)

solution towards the DNA probes on the chip surface. The combination of rapid lateral and vertical fluid movements results in rapid transport of targets in solution to the diffusion boundary layer and thus allows for continuous replenishment of fresh DNA targets around the probes that have been depleted of complementary targets. As a result, the hybridization rate is enhanced. Moreover, the rapid fluid movements can enhance the transport of target within the diffusion boundary layer to the probes on the chip surface by reducing the thickness of the diffusion boundary layer.

Cavitation microstreaming in a shallow hybridization chamber reduces the thickness of the diffusion boundary layer by 2.5-fold. Targets are in close proximity to the probes immobilized on the chip surface, resulting in fast hybridization due to short diffusion length. Furthermore, the rapid lateral fluidic movement as observed in the fluidic dye experiments (Fig. 3.3) ensures a homogeneous mixture of targets and sufficient fluid exchange across the large surface area on the chip and thus allows for uniform hybridization signals to be achieved.

Uniformity of hybridization signal is critical especially for high-density microarray and/or detection of low-abundance targets. Lack of lateral flow convection could lead to nonhomogeneous array performance and hybridization differences that are independent of differences in target concentration. Although the hybridization kinetics enhancement using acoustic microstreaming is not as significant as in cases of the flow-through approach [19] and electronic DNA hybridization [14, 79], a distinct advantage that cavitation microstreaming has over the above two methods is its rapid lateral mass transport that significantly enhances uniformity of hybridization. Moreover, cavitation microstreaming requires a very simple mixing apparatus compared to most existing chamber micromixers [22, 36, 80], and thus can be easily implemented in biochip devices. Other advantages of cavitation microstreaming include low power consumption (~2 mW) and low cost.

### 3.4.4 Pathogenic Bacteria Detection

Pathogenic bacteria detection from a whole blood sample was performed using the described integrated microfluidic device (Fig. 3.1). *E. coli* K12 cells inoculated in a whole rabbit blood were used as a model for demonstration. The electrochemical signal corresponding to the hybridized *E. coli* K12-specific gene as shown in Fig. 3.9a was obtained after 1 h hybridization [1]. The whole analysis from loading the blood sample and different reagents into the storage chambers of the biochip to obtaining the hybridization results took 3.5 h. The durations for the different operations were as follows: sample preparation, 50 min; PCR amplification, 90 min; pumping and valving, 10 min; hybridization, 60 min. An on-chip assay from 1 mL of whole blood sample containing $10^3$ *E. coli* K12 cells was also performed demonstrating positive recognition of the *E. coli* K12-specific gene but with low signal-to-noise ratio (data not shown here).

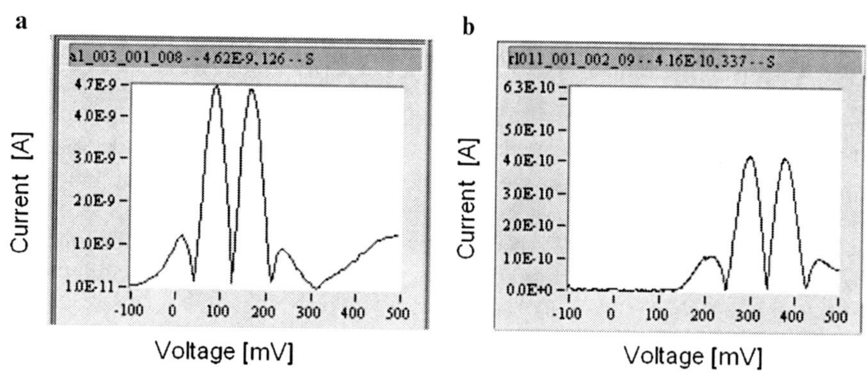


**Fig. 3.9** Electrochemical measurement results obtained from the integrated biochips. (a) Detection of $10^6$ *E. coli* K12 cells from 1 mL of rabbit blood. (b) Genotyping identification of HFE-C gene from 1.4 μL of human blood. During the hybridization process, alternating current voltammetry was used to detect the surface-bound redox-molecules attached to the secondary probes. Fourier transformation was applied to the signal, and the fourth harmonic of the transformation, together with the redox-current value, was displayed on screen (reproduced with permission from [1], copyright 2004, American Chemical Society)

The typical cell capture efficiency using Dynabeads in our experiments was about 40%. In the presence of $1.3 \times 10^7$ (20 μL) Dynabeads, the amplification efficiency was reduced by 50% as compared to the control PCR reaction performed without Dynabeads present. It is believed that the chip assay sensitivity can be improved by assay optimizations, which include: (1) choice of smaller beads (100 nm in diameter as compared to 2.8 μm for the M-280 Dynabeads) that can provide for higher cell capture efficiency due to increased surface-to-volume ratio [60, 61]; and (2) use of paramagnetic beads with lower PCR inhibition rate as compared to the M-280 Dynabeads.

### *3.4.5 Single Nucleotide Polymorphism Assay*

In the hematochromatosis-associated single-nucleotide polymorphism assay performed in our integrated device, the rare target capture and preconcentration steps were omitted because PCR amplification was performed directly from diluted blood samples [81, 82]. Thermal cell lysis followed by an asymmetric PCR amplification was performed to amplify DNA sequences containing the sites for HFE-C polymorphism. After PCR, the hybridization buffer and the amplification product were pumped into the microarray chamber, followed by 1 h hybridization and electrochemical signal scanning. Genotyping result is shown in Fig. 3.9b [1]. The whole analysis from loading the blood sample and different reagents on the chip to obtaining the genotyping results took 2.7 h: cell thermal lysis/PCR, 90 min, pumping and valving, 10 min, and hybridization, 60 min.

## 3.5 Discussion

In this chapter, we have presented a fully integrated microfluidic device that used a DNA microarray as a back-end detection technology. Many other integrated devices demonstrated in the past used capillary electrophoresis (CE; [2, 12, 13] or real-time PCR approaches [83] as detection technologies. The CE technique does not provide information on the fragment sequence that is available through the use of PCR and/or hybridization methods.

Multiplexing using real-time PCR is challenging because a maximum of four different fluorescent markers have been employed to date in a microchip. This allows a limited number of single nucleotide polymorphisms (SNP) that can be identified at a time [84]. Unlike CE and real-time PCR approaches, the eSensor microarray platform provides a solution for highly multiplexed DNA analysis [30, 85] with the capability of processing a complex heterogeneous sample (i.e., PCR product containing denatured blood).

The ability of our system to successfully perform genotyping with as little as 1.4 μL of blood (corresponding to about 10,000 white blood cells) demonstrates its utility for clinical diagnostic applications. Moreover, the electrochemical detection system provides excellent specificity because of the usage of a "sandwich" assay [30]. Our detection system can measure real-time hybridization without the need of a washing step, therefore the assay can be terminated once the genotype is called by the instrument, saving valuable time. Furthermore, the assay performed in our microfluidic device is flexible and has broad applicability due to this system proficiency in detection of both high-abundance and low-abundance DNA from complex biological samples.

The key microfluidic components in the device, including paraffin-based microvalves, cavitation microstreaming mixers, and electrochemical pumps, are simple in design, inexpensive, and easy to fabricate and integrate into a complex microfluidic system, as compared with most of the existing microvalves, micromixers, and micropumps. The use of a cavitation microstreaming technique to achieve rapid and homogeneous on-chip microfluidic mixing not only increases the target cell capture efficiency in the sample preparation process but also allows hybridization assays to be performed in less than one third of the time normally needed without mixing enhancement.

The use of integrated microfluidic components with low power consumption, along with the employment of the electrochemical microarray detection, suggests that handheld operation is feasible for the device. The choice of inexpensive, robust microfluidic technologies, coupled with plastic chip fabrication and standard PCB process, facilitates an easy commercialization path for this technology. Although technical challenges remain, including increasing cell capture efficiency, detection sensitivity, and assay optimization, the integrated platform shown here provides a solution for genetic analysis of complex biological fluidic samples in the fields of point-of-care genetic analysis and disease diagnosis.

## 3.6 Conclusion

We have developed a self-contained disposable microfluidic device for fully integrated genetic assays. The on-chip analysis started with the preparation process of a whole blood sample, which included magnetic bead-based target cell capture, cell preconcentration and purification, and cell lysis, followed by PCR amplification and electrochemical DNA microarray-based detection. Crude biological sample and reagent solutions were loaded into the device, and electrochemical signals corresponding to genetic information were the primary outputs.

The device is capable of handling a large volume (mL) of initial sample to accommodate analysis of rare targets in the sample. The mL volumes were reduced 100-fold when the assay reached the DNA amplification stage. All microfluidic mixers, valves, and pumps are integrated on the chip, but use very simple and inexpensive approaches in order to reduce chip complexity. Both low-abundance and high-abundance DNA detections from blood samples were demonstrated using the devices.

**Acknowledgment** The authors thank Drs. Jiangning Yang and Ralf Lenigk at Motorola Labs and colleagues at Motorola Life Science for all the technical support. This work has been sponsored in part by NIST ATP contract #1999011104A and DARPA contract #MDA972-01-3-0001.

## References

1. Liu, R. H., J. Yang, et al. (2004). Self-contained, fully integrated biochip for sample preparation, polymerase chain reaction amplification, and DNA microarray detection. *Anal. Chem.* 76: 1824–1832.
2. Burns, M.A., B.N. Johnson, et al. (1998). An integrated nanoliter DNA analysis device. *Science* 282(5388): 484–487.
3. Kricka, L. (1998). Miniaturization of analytical systems. *Clin. Chem.* 44(9): 2008–2014.
4. Wilding, P., L.J. Kricka, et al. (1998). Integrated cell isolation and polymerase chain reaction analysis using silicon microfilter chambers. *Anal. Biochem.* 257(2): 95–100.
5. Shi, Y.N., P.C. Simpson, et al. (1999). Radial capillary array electrophoresis microplate and scanner for high-performance nucleic acid analysis. *Anal. Chem.* 71(23): 5354–5361.
6. Emrich, C.A., H.J. Tian, et al. (2002). Microfabricated 384-lane capillary array electrophoresis bioanalyzer for ultrahigh-throughput genetic analysis. *Anal. Chem.* 74(19): 5076–5083.
7. Liu, R.H., K. Dill, et al. (2006). Integrated microfluidic biochips for DNA microarray analysis. *Expert Rev. Molec. Diagnostics* 6: 253–261.
8. Kovacs, G.T.A. (1998). *Micromachined Transducers Sourcebook*. Boston, WCB McGraw-Hill.
9. Manz, A., D.J. Harrison, et al. (1992). Planar chips technology for miniaturization and integration of separation techniques into monitoring systems: Capillary electrophoresis on a chip. *J. Chromatogr.* 593: 253–258.
10. Harrison, D.J., K. Fluri, et al. (1993). Micromachining a miniaturized capillary electrophoresis-based chemical analysis system on a chip. *Science* 261: 895–897.
11. Woolley, A.T. and R.A. Mathies (1994). Ultra-high-speed DNA fragment separations using microfabricated capillary array electrophoresis. *Proc. Natl. Acad. Sci. U.S.A.* 91: 11348.
12. Woolley, A.T., D. Hadley, et al. (1996). Functional integration of PCR amplification and capillary electrophresis in a microfabricated DNA analysis device. *Anal. Chem.* 68(23, Dec. 1996): 4081–4086.
13. Waters, L.C., S.C. Jacobson, et al. (1998). Microchip device for cell lysis, multiplex PCR amplification and electrophoretic sizing. *Anal. Chem.* 70: 158–162.
14. Edman, C., D. Raymond, et al. (1998). Electric field directed nucleic acid hybridization on microchips. *Nucl. Acids Res.* 25: 4907–4914.
15. Radtkey, R., L. Feng, et al. (2000). Rapid, high fidelity analysis of simple sequence repeats on an electronically active DNA microchip. *Nucleic Acids Res.* 28: E17.
16. Erickson, D., X. Liu, et al. (2004). Electrokinetically controlled DNA hybridization microfluidic chip enabling rapid target analysis. *Anal. Chem.* 76(24): 7269–7277.
17. Fan, Z.H., S. Mangru, et al. (1999). Dynamic DNA hybridization on a chip using paramagnetic beads. *Anal. Chem.* 71: 4851–4859.
18. Chee, M., R. Yang, et al. (1996). Accessing genetic information with high-density DNA arrays. *Science* 274: 610–4.
19. Cheek, B.J., A.B. Steel, et al. (2001). Chemiluminescence detection for hybridization assays on the flow-thru chip, a three-dimensional microchannel biochip. *Anal. Chem.* 73: 5777–5783.
20. Lenigk, R., R.H. Liu, et al. (2002). Plastic biochannel hybridization devices: A new concept for microfluidic DNA arrays. *Anal. Biochem.* 311(1): 40–49.
21. Wei, C.W., J.Y. Cheng, et al. (2005). Using a microfluidic device for 1 microl DNA microarray hybridization in 500 s. *Nucleic Acids Res.* 33(8): e78.

22. McQuain, M.K., K. Seale, et al. (2004). Chaotic mixer improves microarray hybridization. *Anal. Biochem*. 325(2): 215–26.
23. Liu, J., B.A. Williams, et al. (2006). Enhanced signals and fast nucleic acid hybridization by microfluidic chaotic mixing. *Angew. Chem. Int. Ed*. 45: 3618–3623.
24. Liu, R.H., R. Lenigk, et al. (2003). Hybridization enhancement using cavitation microstreaming. *Anal. Chem*. 75: 1911–1917.
25. Wilding, P., M.A. Shoffner, et al. (1994). PCR in a silicon microstructure. *Clin. Chem*. 40(9): 1815–8.
26. Anderson, R.C., X. Su, et al. (2000). A miniature integrated device for automated multistep genetic assays. *Nucleic Acids Res*. 28(12): E60.
27. Liu, Y.J., C.B. Rauch, et al. (2002). DNA amplification and hybridization assays in integrated plastic monolithic devices. *Anal. Chem*. 74(13): 3063–3070.
28. Yuen, P., L. Kricka, et al. (2001). Microchip module for blood sample preparation and nucleic acid amplification reactions. *Genome Res*. 11: 405–412.
29. Farkas, D.H. (2001). Bioelectronic DNA chips for the clinical laboratory. *Clinical Chem*. 47(10): 1871–1872.
30 Umek, R.M., S.W. Lin, et al. (2001). Electronic detection of nucleic acids - A versatile platform for molecular diagnostics. *J. Molec. Diagnostics* 3(2): 74–84.
31. Liu, R.H., M. Stremler, et al. (2000). A passive micromixer: 3-D C-shape serpentine microchannel. *J. Microelectromech. Syst*. 9(2): 190–197.
32. Branebjerg, J., P. Gravesen, et al. (1996). Fast mixing by lamination. *MEMS '96*, San Diego, CA.
33. Oddy, M.H., J.G. Santiago, et al. (2001). Electrokinetic instability micromixing. *Anal. Chem*. 73 (24): 5822–5832.
34. Stroock, A.D., S.K.W. Dertinger, et al. (2002). Chaotic mixer for microchannels. *Science* 295(5555): 647–651.
35. Moroney, R.M., R.M. White, et al. (1995). Ultrasonically induced microtransport. *MEMS '95*, The Netherlands.
36. Zhu, X. and E.S. Kim (1998). Microfluidic motion generation with acoustic waves. *Sensors and Actuators: A. Physical* 66 (1–3): 355–360.
37. Liu, R.H., J. Yang, et al. (2002). Bubble-induced acoustic micromixing. *Lab On A Chip* 2(3): 151–157.
38. Nyborg, W.L. (1958). Acoustic streaming near a boundary. *J. Acoust. Soc. Am*. 30: 329–339.
39. Henning, A.K., J. Fitch, et al. (1997). A thermopneumatically actuated microvalve for liquid expansion and proportional control. Technical Digest of *TRANSDUCERS '97: the 1997 International Conference on Solid-State Sensors and Actuators*, Chicago.
40. Selvaganapathy, P., E.T. Carlen, et al. (2003). Electrothermally actuated inline microfluidic valve. *Sensors Actuators A* 104: 275–282.
41. Jerman, H. (1994). Electrically-activated, normally-closed diaphragm valves. *J. Micromech. Microeng*. 4: 210–216.
42. Barth, P.W., C.C. Beatty, et al. (1994). A robust normally closed silicon microvalve. *IEEE Solid-State Sensors and Actuator Workshop*, Hilton Head, SC.
43. Ray, C.A., C.L. Sloan, et al. (1992). A silicon-based shape memory alloy microvalve. *Proc. Materials Research society Symposium* 276: 161–166.
44. Huff, M.A. and M.A. Schmidt (1992). Fabrication, packaging, and testing of a wafer-bonded microvalve. *IEEE Solid-State Sensor and Actuator Workshop*, Hilton Head Island, SC, IEEE Electron Devices Society.
45. Shoji, S., B. V. d. Schoot, et al. (1991). Smallest dead volume microvalves for integrated chemical analyzing systems. Tech. Digest, *Transducers '91: the 1991 International Conference on Solid-State Sensors and Actuators*, San Francisco.
46. Beebe, D.J., J.S. Moore, et al. (2000). Functional structures for autonomous flow control inside microfluidic channels. *Nature* 404: 588–590.
47. Liu, R.H., Q. Yu, et al. (2002). Fabrication and characterization of hydrogel-based microvalves. *J. Microelectromech. Syst*. 11: 45–53.

48. Unger, M.A., H. Chou, et al. (2000). Monolithic microfabricated valves and pumps by multilayer soft lithography. *Science* 288: 113–116.
49. Liu, R.H., J. Bonanno, et al. (2002). Thermally actuated paraffin microvalves. *Proceedings of Micro Total Analysis Systems 2002*, Kluwer Academic, Dordrecht, The Netherlands: 163–165.
50. Yu, C., S. Mutlu, et al. (2003). Flow control valves for analytical microfluidic chips without mechanical parts based on thermally responsive monolithic polymers. *Anal. Chem.* 75: 1958–1961.
51. Zengerle, R., S. Skluge, et al. (1995). A bidirectional silicon micropump. *Sensors Actuators A-Physical* 50: 81–86.
52. Su, Y.C., L. W. Lin, et al. (2002). A water-powered osmotic microactuator. *J. Microelectromech. Syst.* 11(6): 736–742.
53. Bohm, S., W. Olthuis, et al. (1999). An integrated micromachined electrochemical pump and dosing system. *J. Biomed. Microdevices* 1(2): 121–130.
54. Fodor, S.P., J.L. Read, et al. (1991). Light-directed, spatially addressable parallel chemical synthesis. *Science* 251(4995): 767–773.
55. Fortina, P., K. Delgrosso, et al. (2000). Simple two-color array-based approach for mutation detection. *Euro. J.Hum. Genet.* 8: 884–894.
56. Cheng, J., P. Fortina, et al. (1996). Microchip-based devices for molecular diagnosis of genetic diseases. *Mol. Diagn.* 1(3): 183–200.
57. Cheng, J., M.A. Shoffner, et al. (1996). Chip PCR. II. Investigation of different PCR amplification systems in microbabricated silicon-glass chips. *Nucleic Acids Res.* 24(2): 380–385.
58. Gascoyne, P. R. C. and J. Vykoukal (2002). Particle separation by dielectrophoresis. *Electrophoresis* 23: 1973.
59. Cheng, J., E.L. Sheldon, et al. (1998). Isolation of cultured cervical carcinoma cells mixed with peripheral blood cells on a bioelectronic chip. *Anal. Chem.* 70(11): 2321–2326.
60. Ward, M.D., J. Quan, et al. (2002a). High gradient magnetic separation microchannels for integrated microfluidic devices." *Euro. Cell Mater. J.* 3: 123–125.
61. Ward, M.D., J. Quan, et al. (2002b). Metal-polymer hybrid microchannels for microfluidic high gradient separations. *Euro. Cell Mater. J.* 3: 123.
62. Saito, M., F. Kitamura, et al. (2002). Ultrasonic manipulation of locomotive microorganisms and evaluation of their activity. *J. Appl. Phys.* 92(12): 7581–7586.
63. Cheng, J., L.J. Kricka, et al. (1998). Sample preparation in microstructured devices. *Microsyst. Techn. Chem. Life Sci.* 194: 215–231.
64. Taylor, M.T., P. Belgrader, et al. (2001). Fully automated sample preparation for pathogen detection performed in a microfluidic cassette. *Micro Total Anal. Syst.* 670–672.
65. Kopp, M., A. De Mello, et al. (1998). Chemical amplification: Continuous-flow PCR on a chip. *Science* 280: 10461048.
66 Belgrader, P., W. Benett, et al. (1999). PCR detection of bacteria in seven minutes. *Science* 284(5413): 449–450.
67. Boone, T., H. Hooper, et al. (1998). Integrated chemical analysis on plastic microfluidic devices. *Solid-State Sensor and Actuator*, Hilton Head Island, SC.
68. Yu, H., P. Sethu, et al. (2000). A miniaturized and integrated plastic thermal chemical reactor for genetic analysis. *uTAS 2000*, The Netherlands, Kluwer Academic.
69. Kricka, L., P. Fortina, et al. (2002). Fabrication of plastic microchips by hot embossing. *Lab on a Chip* 2: 1–4.
70. Lagally, E.T., P.C. Simpson, et al. (2000). Monolithic integrated microfluidic DNA amplification and capillary electrophoresisanalysis system. *Sensor Actuators B* 63: 138–146.
71. Giordano, B., J. Ferrance, et al. (2001). Polymerase chain reaction in polymeric microchips: DNA amplification in less than 240 seconds. *Anal. Biochem.* 291: 124–132.
72. Lagally, E.T., I. Medintz, et al. (2001). Single-molecule DNA amplification and analysis in an integrated microfluidic device. *Anal. Chem.* 73: 565–570.
73. Khandurina, J., T.E. Meknight, et al. (2000). Integrated system for rapid PCR-based DNA analysis in microfluidic devices. *Anal. Chem.* 72: 2995–3000.

74. Alonso-Amigo, G. (2000). Polymer microfabrication for microarrays, microreactors and microfluidics. *J. Assoc. Lab Aut.* 5: 96–101.
75. Grodzinski, P., R.H. Liu, et al. (2001). Development of plastic microfluidic devices for sample preparation. *Biomed. Microdevices* 3(4): 275.
76. Yang, J., Y. Liu, et al. (2002). High sensitivity PCR assay in plastic micro reactors. *Lab on a Chip* 2: 179–187.
77. Cheng, J., L.C. Waters, et al. (1998). Degenerate oligonucleotide primed-polymerase chain reaction and capillary electrophoretic analysis of human DNA on microchip-based devices. *Anal. Biochem.* 257(2): 101–106.
78. Northrup, M. A., et al. (1998). A miniature analytical instrument for nucleic acids based on micromachined silicon reaction chambers. *Anal. Chem.* 70: 918–922.
79. Sosnowski, R., E. Tu, et al. (1997). Rapid determination of single base mismatch mutations in DNA hybrids by direct electric field control. *Proc. Natl. Acad. Sci.* 94: 1119–1123.
80. Moroney, R.M., R.M. White, et al. (1991). Microtransport induced by ultrasonic lamb waves. *Appl. Phys. Lett.* 59: 774–776.
81. Panaccio, M., M. Georgesz, et al. (1993). Folt PCR - A simple PCR protocol for amplifying DNA directly from whole-blood. *Biotechniques* 14(2): 238 ff.
82. Burckhardt, J. (1994). Amplification of DNA from whole-blood. *PCR-Methods and Applications* 3(4): 239–243.
83. Ibrahim, M.S., R.S. Lofts, et al. (1998). Real-time microchip PCR for detecting single-base differences in viral and human DNA. *Anal. Chem.* 70(9): 2013–2017.
84. Klein, D. (2002). Quantification using real-time PCR technology: applications and limitations. *Trends Molec. Med.* 8(6): 257–260.
85. Umek, R.M., S.S. Lin, et al. (2000). Bioelectronic detection of point mutations using discrimination of the H63D polymorphism of the Hfe gene as a model. *Molec. Diagnosis* 5(4): 321–328.

# Chapter 4
# Integrated Microfluidic Devices for Automated Microarray-Based Gene Expression and Genotyping Analysis

**Robin H. Liu, Mike Lodes, H. Sho Fuji, David Danley, and Andrew McShea**

**Abstract** Microarray assays typically involve multistage sample processing and fluidic handling, which are generally labor-intensive and time-consuming. Automation of these processes would improve robustness, reduce run-to-run and operator-to-operator variation, and reduce costs. In this chapter, a fully integrated and self-contained microfluidic biochip device that has been developed to automate the fluidic handling steps for microarray-based gene expression or genotyping analysis is presented. The device consists of a semiconductor-based CustomArray® chip with 12,000 features and a microfluidic cartridge. The CustomArray was manufactured using a semiconductor-based in situ synthesis technology. The microfluidic cartridge consists of microfluidic pumps, mixers, valves, fluid channels, and reagent storage chambers. Microarray hybridization and subsequent fluidic handling and reactions (including a number of washing and labeling steps) were performed in this fully automated and miniature device before fluorescent image scanning of the microarray chip. Electrochemical micropumps were integrated in the cartridge to provide pumping of liquid solutions. A micromixing technique based on gas bubbling generated by electrochemical micropumps was developed. Low-cost check valves were implemented in the cartridge to prevent cross-talk of the stored reagents. Gene expression study of the human leukemia cell line (K562) and genotyping detection and sequencing of influenza A subtypes have been demonstrated using this integrated biochip platform. For gene expression assays, the microfluidic CustomArray device detected sample RNAs with a concentration as low as 0.375 pM. Detection was quantitative over more than three orders of magnitude. Experiment also showed that chip-to-chip variability was low indicating that the integrated microfluidic devices eliminate manual fluidic handling steps that can be a significant source of variability in genomic analysis. The genotyping results showed that the device identified influenza A hemagglutinin and neuraminidase

R.H. Liu
Osmetech Molecular Diagnostics, Robin.Liu@osmetech.com

M. Lodes, H.S. Fuji, D. Danley, A. McShea
e-mails: CombiMatrix Corp., Sfuji@Combimatrix.com
DDanley@Combimatrix.com
Ameshea@Combimatrix.com

K. Dill et al. (eds.), *Microarrays: Preparation, Microfluidics, Detection Methods, and Biological Applications*,

subtypes and sequenced portions of both genes, demonstrating the potential of integrated microfluidic and microarray technology for multiple virus detection. The device provides a cost-effective solution to eliminate labor-intensive and time-consuming fluidic handling steps and allows microarray-based DNA analysis in a rapid and automated fashion.

## 4.1 Introduction

Microarray assays generally involve multistage sample processing and fluidic handling, which are typically labor-intensive and time-consuming. In most gene expression and genotyping assays, the array needs to be washed thoroughly to remove nonspecific binding of biotinylated cRNA target following hybridization of target in the sample solution to its complementary oligonucleotides probes on the microarray chip surface. Different salt concentrations of washing buffers are used to ensure satisfactory stringency. For indirect labeling, a labeling step is subsequently performed. Another washing is performed to remove excessive labeling reagents before the slide is ready for scanning.

All the above processes involve many manual steps (handling arrays, moving, agitating racks, etc.) with frequent run-to-run and operator-to-operator variation. The combination of these factors can lead to variability in array results. Automation of these processes would improve robustness and reduce costs. Robotic workstations have been developed to automate the whole hybridization and posthybridization process, but these benchtop instruments are generally expensive for most research and clinical diagnostic applications. It is therefore desirable to develop a cost-effective method to integrate and automate the microarray processing in a single and miniature device using microfluidic technology.

Microfluidics lab-on-a-chip technology has proven to be useful for integrated, high-throughput DNA analysis [1]. Microfluidic devices can offer a number of advantages over conventional systems, for example, their compact size, disposable nature, increased utility, and a prerequisite for reduced concentrations of sample reagents. Miniaturized assemblies can be designed to perform a wide range of tasks that range from on-chip liquid pumping and handling to detection of DNA. Integration of several assay functions on a single device leads to assay automation and elimination of operator involvement as a variable. Microfluidic systems provide a real potential for improving the efficiency of techniques applied in drug discovery and diagnostics.

Various materials have been used in the fabrication of microfluidic lab-on-a-chip devices. Lithographic techniques, adapted from semiconductor technology, have been used to build chips in glass [2] and silicon [3]. Unconventional lithography techniques such as soft lithography [4] have been used to fabricate reproducible microstructures of biological materials offering a multitude of possibilities to explore as molecular diagnostic tools [5]. Recently, with increasing emphasis on disposable devices, the use of plastics and plastic fabrication methods has become popular [6–9].

Most of the integrated microfluidic works demonstrated to date have been in the area of on-chip capillary electrophoresis (CE) [10–14], polymerase chain reaction (PCR) [15, 17], and sample preparation [18–19], among many others [1], and there are only a few reports on combining microfluidics with DNA microarrays [19–21]. The microfluidic lab-on-a-chip devices with capabilities of on-chip sample processing and detection provide a cost-effective solution to direct sample-to-answer biological analysis in self-contained and closed systems that minimize the possibility of sample contamination. Such devices will be increasingly important for rapid diagnostic applications in hospitals and in-field biothreat detection.

In this chapter, we report on the development of a self-contained and fully integrated microfluidic array device that automates hybridization and posthybridization processes for microarray gene expression and genotyping assays that involve multistage sample processing and fluidic handling. A brief description of the semiconductor-based CustomArray® platform is included. The integrated device design, fluidics, and developments of the key microfluidic components, such as pumps, valves, and mixers, are described. The demonstrations of gene expression study of the human leukemia cell line (K562) and genotyping detection and sequencing of influenza A subtypes using this integrated biochip platform are presented.

Development of integrated microfluidic array devices for rapid detection and identification of influenza virus is important in the face of concerns over an influenza pandemic. Timely acquisition of information on the influenza A subtypes that are circulating in human and animal populations is crucial for the global surveillance program to effectively monitor disease outbreaks [22]. Knowledge of the exact strain, origin of the strain, and probable characteristics of the virus is essential for surveillance of a disease outbreak and preventing the spread of the disease.

Identification of a virus subtype can be realized by molecular identification of the subtype of viral hemagglutinin (HA) and neuraminidase (NA) genes. Viruses with any combination of the 16 HA and 9 NA subtypes can infect aquatic birds whereas few subtypes have been found to infect humans [23]. However, interspecies transmission can occur after recombination or mixing of subtypes in birds or pigs [24–26]. In addition, new human strains of virus can arise by reassortment or antigenic shifts when two or more subtypes are circulating in the human population [27, 28].

Maintenance of a subtype in the human population can also occur by antigenic drift [27], which occurs when genetic mutations of the HA and NA genes create virons that escape immune surveillance. These mutations arise as a result of viral polymerase infidelity. Therefore, in addition to identification of the circulating subtype, specific knowledge of the genetic makeup of the virus is required in many situations. For example, the avian H5N1 virus ("Bird Flu") has significant potential for further recombination with common human strains (such as H3N2) or other nonhuman strains common in avian populations (H7 and H9 strains). The H5N1 subtype is also difficult to identify because of the lack of sensitivity and specificity of many of the commercial tests, such as viral detection (cell culture) and serological techniques [29–31].

In addition, genotype Z, the dominant H5N1 virus genotype, contains a mutation that is associated with resistance to amantadine and rimantadine. Because of the high susceptibility in humans and resistance to antivirals of this isolate, neuraminidase inhibitors must be given within 48 hours of onset of illness to be effective. Thus, rapid and specific identification of this subtype and accurate sequence information are crucial for proper treatment.

Rapid subtype identification of flu is not always straightforward. Simple serological tests on infected individuals are an ineffective tool for monitoring viruses undergoing a high rate of mutation or rapid recombination. Reverse transcription-polymerase chain reaction (RT-PCR) assays have better sensitivity but are problematic in scenarios where new strains of virus emerge or mixtures of viruses exist. DNA microarrays have recently become an acceptable technology for screening samples for the presence or absence of a large variety of viruses simultaneously and identifying the genotype of an unknown specimen ([32, 35]).

Microarrays are particularly useful for molecular detection and identification of influenza viruses because of their genetic and host diversity and the availability of an extensive sequence database [32, 35, 36]. For example, DNA microarrays have been recently used for identification of influenza A hemagglutinin and neuraminidase subtypes [37]. In this chapter, we report on an integrated biochip platform for rapid genotyping detection and sequencing of influenza A subtypes.

## 4.2 Semiconductor-Based DNA Microarray

The microfluidic array device consists of a CombiMatrix microarray chip (CustomArray) and a microfluidic plastic cartridge, as shown in Fig. 4.1 [38]. The CustomArray is a 1 in. × 3 in. alumina slide with an 11 × 25 mm silicon chip affixed in a cavity in the ceramic package (Fig. 4.2; [39]). A key aspect of the CustomArray platform is a semiconductor-based microarray that allows the manufacture of high-density microelectrode arrays that vary from a density of 1000 to >100,000 electrodes/cm$^2$.

Utilization of active circuit elements in the design permits the selection and parallel activation of individual electrodes in the array to perform in situ oligonucleotide synthesis of customized content on the chip [40, 41]. The oligonucleotides were synthesized on an array of electrodes using phosphoramidite chemistry under electrochemical control. The electrochemical reaction generated at specific electrodes on the chip produced protons, which in turn removed the blocking group on the oligonucleotide strand undergoing synthesis on the electrodes, allowing subsequent DNA synthesis to take place.

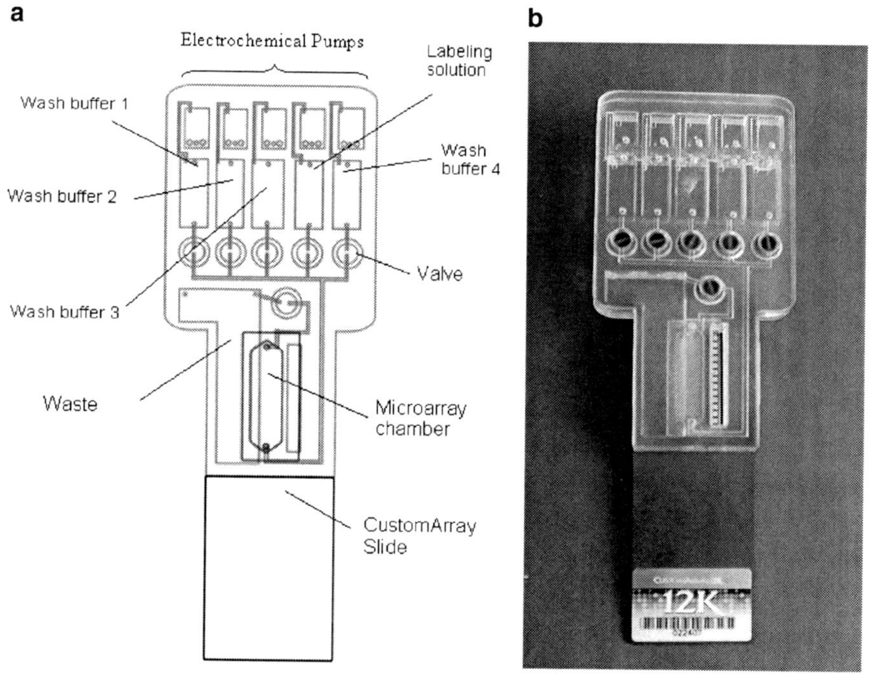


**Fig. 4.1** (a) Schematic of the microfluidic array device. (b) Photograph of the integrated device that consists of a plastic fluidic cartridge and a CombiMatrix CustomArray chip (reproduced with permission from [38], copyright 2006, American Chemical Society)

The silicon chip was manufactured using a commercial mixed signal complementary metal oxide semiconductor (CMOS) process. The microarray chip used in this work has a 56 × 224 array of electrodes located in the center of the chip providing a total of 12,544 electrodes, each with a size of 44 μm in diameter (Fig. 4.2). All the electrodes on the chip are individually addressable, so that unique reactions can be carried out at each individual site. CMOS integrated circuit technology was utilized to create active circuit elements and digital logic on the chip that allowed complex functions to be implemented. These functions include a high-speed digital interface for efficient communication to the chip, data writing and reading from the electrode array, and the setting of appropriate electrical states at the electrode to perform in situ oligonucleotide synthesis.

The oligonucleotide was in situ synthesized on the chip using phosphoramidite chemistry under electrochemical control [40, 41]. The chip surface was coated with a proprietary membrane layer that facilitated the attachment and synthesis of biomolecules in a matrix above the platinum electrode surface. During DNA synthesis, the blocking DMT (dimethoxytrityl) group of the phosphoramidite on the

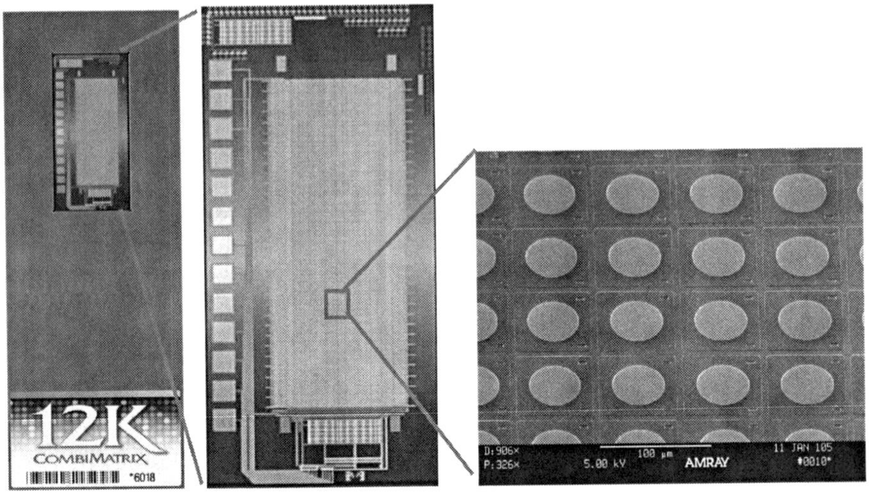


**Fig. 4.2** Photographs and SEM image showing the commercial 12 K CustomArray with 12,544 microelectrodes mounted in a 1 × 3 in. slide. Each microelectrode has a size of 44 μm in diameter. The 13 silver pads on the left of the silicon array chip provide electrical connections required for communicating with the microelectrode array (large rectangle in the middle of the chip; reproduced with permission from [39], copyright 2006, Future Drugs Ltd.)

selective electrode surface was removed by the acid ($H^+$) that was produced by the electrochemical reaction when these electrodes were turned on [40]. An activated nucleotide reagent was then introduced and reacted with the free hydroxyl groups on these electrodes. The chip was subsequently washed, followed by capping, and then an oxidation step to stabilize the central phosphorous atom. The process continued with deprotection of certain electrodes and a coupling step. Using this in situ synthesis method, unique oligomers of 35–40 bases were synthesized at each electrode.

For the gene expression study, the array was designed with a variety of genes expressed by the K-562 leukemia cell line as well as a system of spiked-in control transcripts generated from segments of the *Escherichia coli* (*E. coli*) bacteriophage lambda genome (#NC_001416). The spiked-in control transcripts were used to determine sensitivity, reproducibility, and dynamic range characteristics of microfluidic array devices. Probes were created against various genes involved with immune system pathways, as well as a number of housekeeping genes. The microarray was designed with four replicates of each probe distributed across the array to allow measurement of the variability within the array.

The design of the influenza A subtyping array probes was based on the viral sequence data obtained from the GenBank database and from the Influenza Sequence

Database [42]. For HA serotypes, 1614 animal and 1937 human isolates were selected, and for NA serotypes, 552 human and 831 animal isolates were selected. Both datasets were treated in the same way following a modification of the method of Wang et al. where probe uniqueness was based on subtype differences [43].

For each sequence, nonoverlapping appended primers were made, tiling the entire sequence. These oligonucleotides were designed to have similar annealing stability as judged by a nearest neighbor thermodynamic model [44], and were designed to have a $T_m$ of 50°C. Probes that had significant secondary structure ($T_m$ > 40°C) were taken out of the set. Finally, probes from only the first 500 bp of sequence were used (bp 50 to 500).

After tiling and culling, there were 23,568 HA probes and 15,191 NA probes left. Each sequence and probe was grouped and labeled by its serotype and databases were generated from the compiled sequences of HA and NA isolates. Probes were selected to be exclusive to a given subtype as judged by pairwise BLASTN [45]. The details of probe sequences and the number of probes selected for each subtype are described by Lodes et al. [37]. A poly $T_{10}$ spacer was added to the 3′ end of all probes to avoid surface inhibition. Probe design files for array synthesis were generated with Layout Designer (CombiMatrix Corp., Mukilteo, WA). Oligonucleotide microarrays were synthesized on semiconductor microchips containing over 12,000 independently addressable electrodes. After the in situ synthesis of the microarray, the oligonucleotide probes on the chip were phosphorylated with T4 polynucleotide kinase for 30 min at 37°C.

The sequencing microarray chips were designed using sequences that were representative of each subtype of interest [37]. For our experiments, we chose sequences that represent subtype H9N2 (GenBank accession numbers AF156378 and AF222654, respectively). Probes were tiled by one nucleotide to cover the sequences of interest and probes were designed with a $T_m$ of approximately 55°C. Four probes were designed for each nucleotide to be examined that were identical except for the 5′ nucleotides, which contained either an A, C, T, or G (see Fig. 4.3; [38]). The chips with this design were hybridized with both HA and NA gene targets prepared from Influenza A isolates from infected quail and chicken (A/Quail/Hong Kong/G1/97 (H9N2) and A/Chicken/Hong Kong/NT17/99 (H9N2)).

## 4.3 Microfluidic Cartridge

Coupling with the microarray slide is a microfluidic plastic cartridge that consists of five micropumps, six microvalves, five chambers for the storage of different buffers and reagents, a microarray hybridization chamber, and a waste chamber, as shown in Fig. 4.1. The plastic cartridge measures 40 × 76 × 10 mm and has channels and chambers that range from 500 μm to 3.2 mm in depth and 0.5 to 8.5 mm in width. The prototype of the plastic cartridge consists of multiple

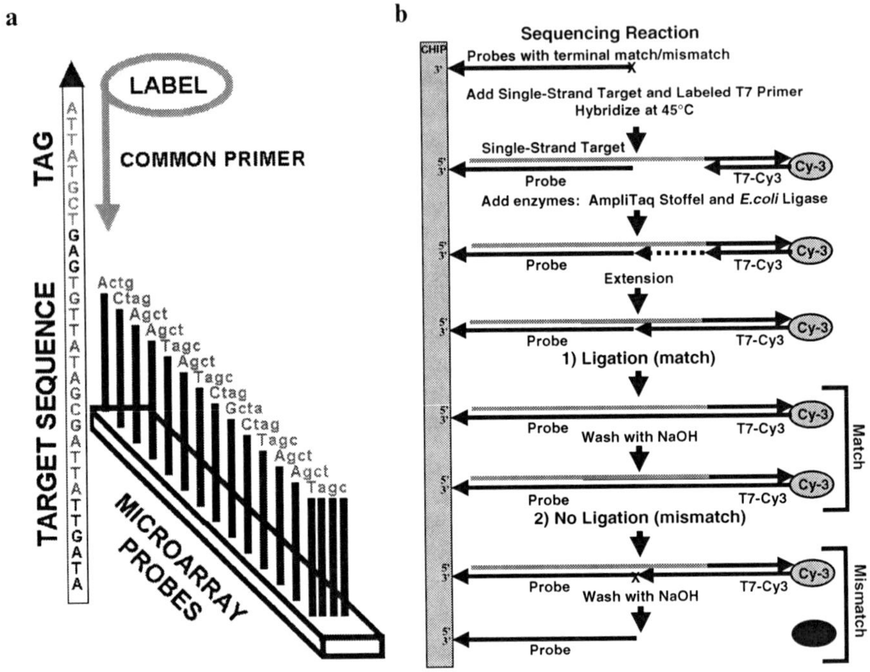


**Fig. 4.3** (a) Diagram of the strategy for microarray DNA sequencing. Four target-specific (antisense) probes, for each base of sequence, were identical except for the 5′ terminal residue (terminate in bases A, C, G, and T). After hybridization of probes and target DNA, annealed Cy3-labeled primer, a mixture of enzymes, buffer, and dNTPs were added to the array and incubated for 30 min at 37°C. The labeled primer (orange) was extended to matching probes and probes that are a perfect match to the target (blue) were ligated whereas mismatches (red) are not. The array was finally washed with 0.1 N NaOH and scanned for fluorescence. (b) Diagram of the strategy for enzyme-based microarray sequencing. Sequencing probes as described above were hybridized with single-strand target that contained a 3′ tag sequence and a labeled oligonucleotide that was complementary to the tag. After washing, an enzyme mix containing Amplitaq polymerase Stoffel fragment and *E. coli* DNA ligase was added to the annealed complex. The labeled primer was extended on the target template to the probe. Ligation to the probe occurred when target and probe sequences matched (1), and no ligation occurred when target and probe did not match/mismatch (2). Stringent washing removed any signal that was not ligated to the probe (reproduced with permission from [38], copyright 2006, American Chemical Society)

layers of acrylic materials that are laminated and assembled using double-sided adhesive tapes.

All the layers, including five layers of acrylic sheets with various thicknesses ranging from 0.5 to 5.7 mm (MacMaster-Carr, Atlanta, GA) and four layers of double-sided adhesive tapes (Adhesive Research Inc., Glen Rock, PA), were machined using a $CO_2$ laser machine (Universal Laser Systems, Scottsdale, AZ).

The 5.7 mm thick acrylic layer had six valve seats in which six duckbill check valves (Vernay Laboratories Inc., Yellow Springs, OH) were glued using an epoxy. These check valves were normally closed and could be opened when the upstream pumping pressure exceeded the cracking pressure of the valves. The valves were used to retain the liquid solutions in their storage chambers and prevent cross-talk of the solutions between two adjacent chambers. The implementation of these check valves did not require a microfabrication process, and their operation required no actuation. As a result, they are less expensive and easier to integrate and operate than most conventional microvalves [46–49].

After assembly, stainless steel wires with 0.5 mm diameter were inserted into the electrochemical pumping chambers followed by sealing with an epoxy. Each electrochemical pumping chamber was then loaded with 50 μL of 1 M $Na_2SO_4$ solution to form an electrochemical pump. The electrolyte loading holes were subsequently sealed using an adhesive tape (Adhesive Research Inc., Glen Rock, PA). The electrochemical micropumps also served as an actuation source for micromixing in the array chamber. The venting hole of the waste chamber was sealed with a hydrophobic membrane vent (Sealing Devices, Lancaster, NY) that allowed gas molecules to pass through and the liquid solution retained in the waste chamber. The plastic cartridge was then bonded with the microarray chip using a double-sided adhesive tape (Adhesive Research Inc., Glen Rock, PA). The tape with a thickness of 0.5 mm was machined with a pattern of the hybridization chamber using the $CO_2$ laser machine.

### 4.3.1 Fluidic Architecture and Operation

The operation of the microfluidic device is as follows. A sample solution was loaded in the array chamber using a pipette. Other solutions required for assays were separately loaded in different storage chambers. For gene expression and subtyping assays, these solutions included four washing buffer solutions and a labeling solution. For sequencing assays, the solutions included a ligase buffer, an enzyme solution containing T4 DNA ligase and Taq polymerase, a NaOH solution, and a wash buffer solution. After sealing the loading ports using a sealing tape (Adhesive Research Inc., Glen Rock, PA), the device was then inserted into an instrument, which provided hybridization heating, temperature sensing, and electrical power for liquid pumping and mixing. The instrument measured 140 × 200 × 200 mm. It consisted of a clamping manifold, a printed circuit board, and a power supply.

The microfluidic device was inserted into the manifold where a thin-film heating element (Minco Corp., Minneapolis, MN) was physically pressed on the microarray slide of the device to provide the heating of the array chamber during the hybridization process. The thin-film heating element consisted of a temperature sensor that provided the temperature feedback to the control circuit board. A flexible cable connector was used to connect the circuit board with the electrical pins for the electrochemical pumps in the cartridge. The board provided electronic control of the hybridization heating, temperature sensing, and electrical power for liquid pumping and mixing.

The on-chip assay process started with a hybridization step in the microarray hybridization chamber, followed by subsequent washing and a posthybridization process. The pumping of liquid solutions was performed using the integrated electrochemical micropumps that operated with a DC current of 8.6 mA. The pump generated a pressure that was used to open the normally closed check valve once the pressure exceeded the cracking pressure of the valve (i.e., 1 psi). During each pumping step, a mixing procedure as described in the following section was implemented. The device was then removed from the instrument. The microarray chip was detached from the microfluidic plastic cartridge before it was scanned using a commercial fluorescent microarray scanner. The fluorescent hybridization signals were detected on the chip and analyzed.

A key aspect of the microfluidic platform is that the device was placed vertically during operation in order to take advantage of the fluid gravity to remove the air bubbles from the system without the use of porous hydrophobic vents [20]. For example, the hybridization chamber was designed with a depth of 600 μm and a width of 6.5 mm, and fluid volume was on the order of tens of μL. The Reynolds number for the fluid flow was less than 10 [50]. Fluid gravity played an important role in fluidics in the hybridization chamber when the chamber was placed vertically. In this chamber where the liquid solutions and gas bubbles entered from the lower portion, buoyant force allowed gas bubbles to travel quickly to the upper portion of the chamber, leaving the chamber bubble-free.

All of the microfluidic components used in this integrated device, including micropumps, microvalves, and micromixers, were designed to be simple, low-cost, and easy to fabricate and integrate into the plastic cartridge, resulting in a cost-effective, manufacturable, and disposable device. These micropumps, microvalves, and micromixers are described in the following sections.

### *4.3.2 Micropumps*

A micropump that can transport liquid solutions with a volume of hundreds of μL is one of the most essential components required in this integrated microfluidic device. Two of the key requirements of such a micropump are single use and low cost. Most traditional pressure-driven membrane-actuated micropumps did not meet the requirements inasmuch as they generally suffer from complicated designs and fabrication, as well as high cost [51–53].

In our device, we utilized integrated electrochemical pumps that relied on electrolysis of water between two electrodes in an electrolyte solution (1 M

$Na_2SO_4$) to generate gases when a DC current was applied. Electrolysis-based pumping techniques that used the generated gases to displace fluids have been previously demonstrated [21, 54–56]. In our device, stainless steel wires instead of platinum wires were used as electrodes, resulting in a reduced cost of the device. Note that the $Na_2SO_4$ solution was not allowed to come into contact with the array chip to prevent contamination of the hybridization with electrode breakdown products. The electrochemical pumps generated gas ($H_2$ and $O_2$) that was used to move liquid solutions from chamber to chamber in the device.

Flow experiments demonstrated that the pumping flowrate, $Q$ (μL/min), ranging from 5.5 to 100 μL/min, was in linear proportion with the DC current, $i$ (mA), ranging from 0.43 to 8.6 mA. Six data points were used to determine the linear regression model. Each data point is the mean value obtained from four pumping rate measurements with identical DC current. The data fit into the linear least squares regression equation $Q = 11.727\ i - 0.0854$, with correlation coefficient ($R^2$) of 0.9993, as shown in Fig. 4.4 [39]. The pumping rate used in this work was 100 μL/min.

It was observed that a yellow product was generated in the electrolyte solution during the electrolysis reaction, indicating that the stainless steel corroded. The corrosion did not pose any problem because the whole cartridge was disposed of after use. This pumping mechanism did not require a membrane and/or check valves in the design. As a result, the fabrication and operation were simpler than most conventional micropumps [51–53].

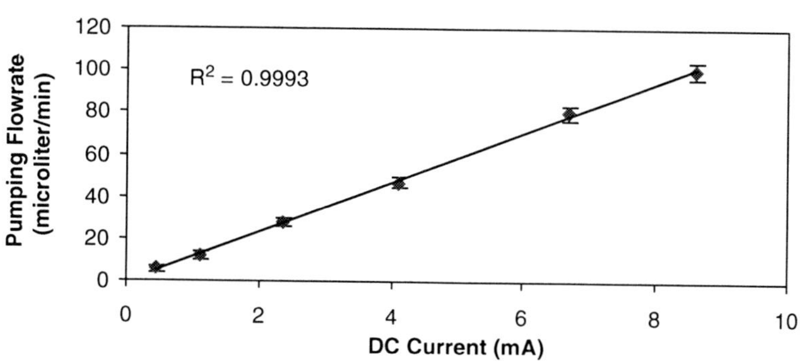


**Fig. 4.4** Measurement of the liquid pumping rate as a function of the applied DC current in an electrochemical pump that relies on the electrolysis of water between two stainless steel electrodes in an electrolyte solution (50 μL of 1 M $Na_2SO_4$). Each data point is the mean value obtained from four pumping rate measurements with identical DC current. The pumping rate was determined by measuring the time required to pump a certain amount of liquid solution (e.g., 200 μL) from the storage chamber to the waste chamber at each DC current (reproduced with permission from [39], copyright 2006, Future Drugs Ltd.)

### 4.3.3 Micromixing

The electrochemical micropumps were also excellent sources to provide gas bubbles to enhance micromixing in the hybridization chamber during the washing and labeling steps. Because the Reynolds number for the fluid flow in the hybridization chamber was less than 10, fluid flows at such a low Reynolds number were predominantly laminar [50]. As a result, mixing of materials between streams in the array chamber was confined to molecular diffusion. The rapid mixing produced by turbulent flows is usually not available because the Reynolds number is below the critical value for transition to turbulence [50]. A pure diffusion-based mixing process can be very inefficient and often takes a long time, particularly when the solution streams contain macromolecules that have diffusion coefficients orders of magnitude lower than that of most liquids.

Inasmuch as the residence time of fluid elements in the flow was smaller than the diffusion time, pure diffusion failed to provide homogeneous mixing of confluent reagent solutions in the array chamber. Therefore, an efficient micromixer was required to enhance micromixing in this device during the washing and labeling steps. We have developed a simple and easy-to-operate mixing technique that is based on the continuous bubbling effect. During the mixing steps, gas bubbles generated from the electrochemical micropump entered into the hybridization chamber from the lower portion. Buoyant force allowed gas bubbles to travel quickly to the upper portion of the chamber. During this traveling process, the two-phase flow resulted in flow recirculation in the liquid solution around the bubbles [20]. As a result, the mixing was enhanced in the chamber.

As shown in Fig. 4.5, fluidic experiments with dye solutions demonstrated the efficiency of the mixing enhancement using this bubbling effect. The fluidic experiment showed that the flow recirculation around the continuously pumped bubbles produced homogeneous solutions in the chamber and thus facilitated a uniform reaction on the array surface during the reaction process (e.g., ligation, extension, or labeling) or washing steps. The assay experiments with DNA hybridization described in the following sections showed high sensitivity, low background signals, and uniform hybridization signals on the microfluidic arrays, which suggested that on-chip microfluidic mixing was uniform and efficient and the bubbling mixing technique had a relatively low shear strain rate and was thus biofriendly to the hybridized DNA on the array. The use of gas bubbles to enhance mixing in the microfluidic chamber proved to be a simple but effective micromixing technique without the use of any external actuation methods such as acoustic agitation [21, 57] or physical rotation.

### 4.3.4 Microvalves

It is necessary to incorporate microvalves in the cartridge to ensure robustness of the design and efficient isolation of liquid solutions in their storage chambers.

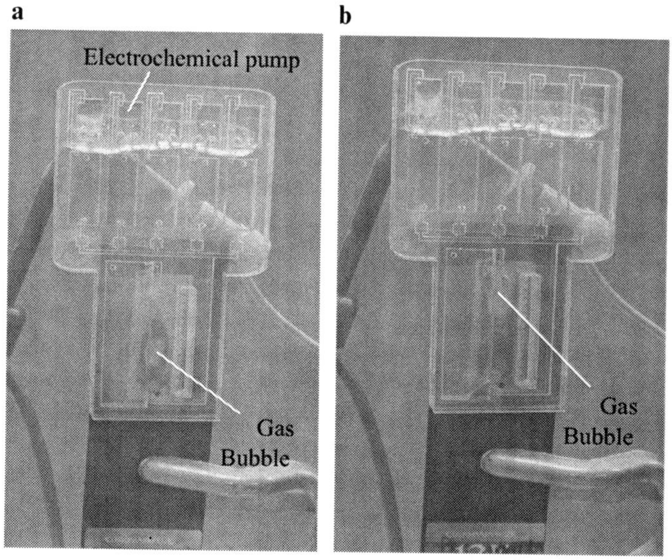


**Fig. 4.5** Photographs showing the bubbling mixing process in the hybridization chamber of a microfluidic array device that was placed vertically. A red color dye was used to visualize the mixing effect. (a) A gas bubble generated using an electrochemical micropump was directed through a microchannel and introduced into the hybridization chamber from the lower portion of the chamber. The hybridization chamber was initially filled with DI water followed by injection of a small amount of red food dye at the lower portion of the chamber before the introduction of the bubble. (b) When the gas bubble traveled to the upper portion of the chamber, flow recirculation was observed in the liquid solution around the bubble, resulting in a dye mixing enhancement in the chamber (reproduced with permission from [38]), copyright 2006, American Chemical Society)

Moreover, a normally closed valve between the array chamber and waste chamber could prevent evaporation of the hybridization solution during the hybridization process at 45°C, which would otherwise result in reduction of the hybridization solution and an increase in the salt concentration of the hybridization solution that would in turn lead to increased nonspecific adhesion of DNA to the array and loss of hybridization stringency.

In the design reported here, six commercially available duckbill check valves (VA 3426, Vernay Laboratory Inc., Yellow Springs, OH) were integrated. The duckbill is a precision, one-piece elastomeric check valve that allows flow in one direction and checks flow in the opposite direction. These normally closed check valves are made of silicone and could be easily glued on the valve seats in the cartridge using epoxy. Once the upstream pumping pressure exceeded the cracking pressure of the valves, the valves were opened. The implementation of these check valves did not require a microfabrication process, and their operation required no actuation. As a result, they were less expensive and easier to integrate and operate than most conventional microvalves [46–49].

## 4.4 Gene Expression Assay

The gene expression sample solution consisted of a complex background sample and spiked-in control transcripts. The complex background sample was prepared from human leukemia, chronic myelogenous (K-562 cell line) poly A+ RNA (Ambion, Austin, TX) utilizing Ambion's MessageAmp aRNA Kit. Biotin was double incorporated using biotin-11-CTP (PerkinElmer, Boston, MA) and biotin-16-UTP (Roche Diagnostics, Mannheim, Germany). Varying concentrations of spiked-in biotin-cRNA control transcripts were combined with a constant amount (150 nM) of K-562 biotin-cRNA complex background such that final concentration of spiked-in control transcripts would range from 1 to 1000 pM in the hybridization. The biotin–cRNA mixtures were fragmented in a 1X fragmentation solution (40 mM trisacetate, pH8.1, 100 mM KOAc, 30 mM MgOAc) at 95°C for 20 minutes.

The fragmented cRNA sample was added to a hybridization solution (6X SSPE, 0.05% Tween-20, 20 mM EDTA, 25% DI Formamide, 0.05% SDS, 100 ng/μL sonicated salmon sperm DNA) and denatured for 3 minutes at 95°C. The sample was placed briefly on ice followed by centrifugation at 13,000 xg for 3 minutes. During the on-chip assay, 95 μL of the hybridization sample solution was loaded into the hybridization chamber. Other solutions including (1) 200 μL of 3X SSPE, 0.05% Tween-20; (2) 200 μL of 0.5X SSPE, 0.05% Tween-20; (3) 200 μL of 2X PBST, 0.1% Tween-20; (4) 200 μL of a labeling solution; and (5) 200 μL of 2X PBST, 0.1% Tween-20, were separately loaded in the storage chambers. The labeling solution contained streptavidin-Cy5 (Molecular Probes, Eugene, OR) that was diluted in a blocking solution (2X PBS, 0.1% Tween-20, 1% Acetylated BSA) to a final concentration of 1 μg/mL. Hybridization was carried out for 18 hours at 45°C.

Following hybridization, the array in the hybridization chamber was washed for 2 minutes with 3X SSPE, 0.05% Tween-20. On-chip washings continued with 0.5X SSPE, 0.05% Tween-20 for 2 minutes and 2X PBST, 0.1% Tween-20 for 2 minutes. The labeling solution was then pumped into the hybridization chamber and incubated for 30 minutes at room temperature. Note that the cartridge was protected from light using an external cover in the instrument to prevent photobleaching of the fluorescent dye.

The final washing step was performed by flowing 2X PBST through the hybridization chamber. Subsequently, the cartridge was separated from the microarray chip with the use of a razor blade. The microarray chip was imaged on an Axon Instruments (Union City, CA) GenePix 4000B –5 μm resolution laser scanner. Imaging was performed while the array was wet with 2X PBST under a LifterSlip™ glass cover slip (Erie Scientific, Portsmouth, NH). Probe fluorescence on the microarray was analyzed and quantified using Microarray Imager software (CombiMatrix Corp., Mukilteo, WA). To study chip-to-chip variability, gene expression assay with the same protocol was performed in three different microfluidic array devices.

Fig. 4.6 shows the fluorescent scanning image of a section of the microfluidic array and the hybridization analysis of the phage lambda spiked-in control transcripts on the microfluidic array [39]. Each data point shown in Fig. 4.6b represents the mean of the normalized probe intensities for the spiked-in control

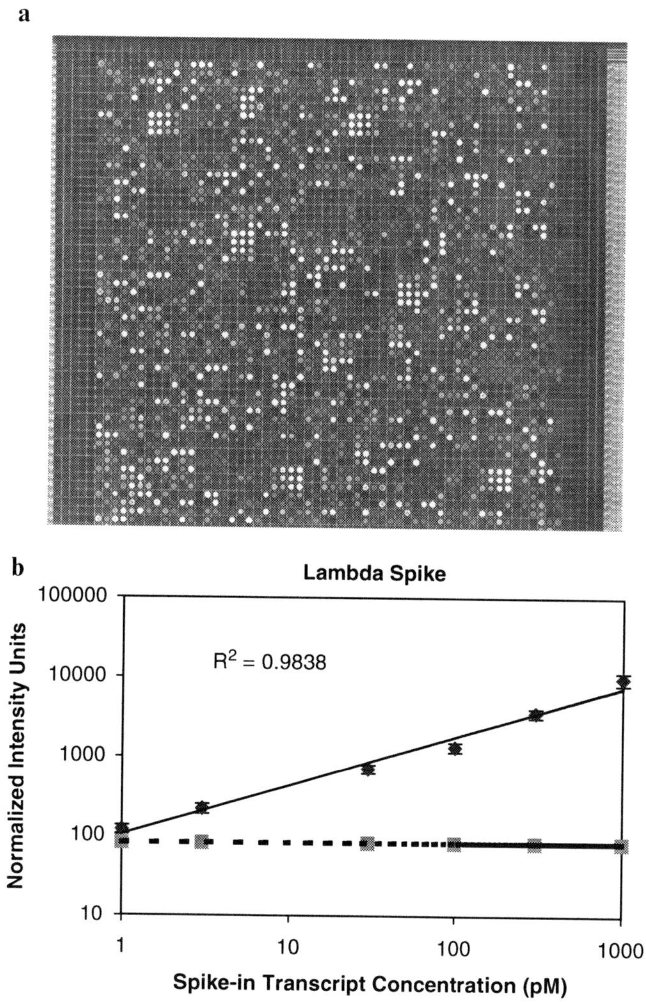


**Fig. 4.6** (a) Fluorescent scanning image showing a section of the microfluidic array after hybridization and posthybridization processes. (b) Hybridization analysis demonstrates the sensitivity and linear dynamic range of the microfluidic array in log scale. Each data point represents the mean of the normalized probe intensities for the spiked-in control transcripts across the array plotted against the corresponding concentrations. Error bars indicate the standard deviation across the array at each data point. Sensitivity was determined to be signal detectable above the average of the negative control signals (the bottom dot line) plus three standard deviations (reproduced with permission from [39], copyright 2006, Future Drugs Ltd.)

transcripts across the array plotted against the corresponding concentrations. Error bars indicate the standard deviation across the array at each data point. In determining the cutoff for sensitivity, the signal was considered significant if greater than three standard deviations above the average of the negative control signals. The result showed that the dynamic range of the microfluidic platform covered three orders of magnitude.

Further studies with lower spiked-in control transcript concentrations showed that the detection limit of the microfluidic array device was 0.375 pM (results not shown here). Because each measurement was made with replicate probes that were spaced across the array to allow measurement of the variability within the array, it was possible to get an accurate representation of reproducibility at each spiked-in concentration. Error bars in Fig. 4.6b represent the standard deviation across the replicate probes, and indicate that hybridization signals are uniform across the whole array. The low background signals and uniform hybridization signals suggest that on-chip microfluidic washing and labeling are uniform and efficient.

The result of the cartridge-to-cartridge reproducibility study is shown in Fig. 4.7. Given that three microfluidic array devices in this study received the same concentration of background target (K-562 cRNA), interarray comparison could be demonstrated by comparing the probes specific to genes expressed by this sample. Scatterplots comparing these probe intensities (raw data) on three different arrays against their

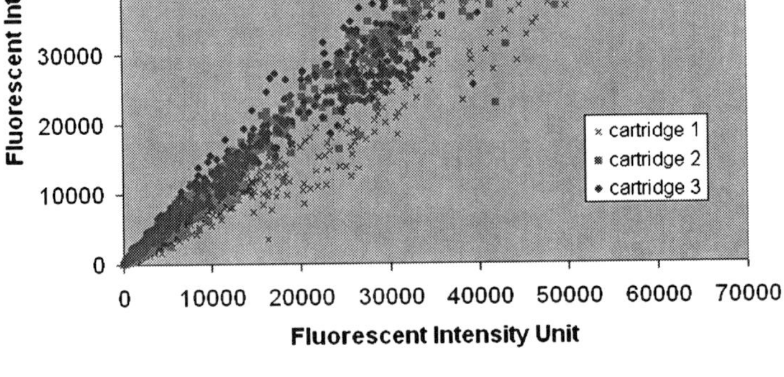


**Fig. 4.7** Scatterplot of the raw fluorescent intensities comparing data from three microfluidic array devices against their average intensities ($x$-axis) and showing cartridge-to-cartridge reproducibility. These three microfluidic array devices received the same concentration of the background target (K-562 cRNA). Both $x$- and $y$-axes are in fluorescent intensity units (reproduced with permission from [39]), copyright 2006, Future Drugs Ltd.)

average intensities are demonstrated in Fig. 4.7. The median covariance (CV) of the unnormalized probe intensities across these three arrays was 14%, indicating that the cartridge-to-cartridge variability is low. This intercartridge CV was calculated by averaging all the CVs of the unnormalized fluorescent intensities of identical probes on the three arrays.

Similarly, the interchip CV of the un-normalized probe intensities across three conventional microarray chips with regular hybridization chambers that have no integrated microfluidic components was approximately 14.5%. For these conventional microarray chips, all the fluidic handling and processes were carried out manually using pipettes by the same user, and the sequence and composition of the buffers for manual processing were identical to those used for the automated microfluidic processing. The comparison results indicate that the cartridge-to-cartridge variability is slightly better than or equivalent to the manual chip-to-chip variability.

## 4.5 Subtyping Assay

The influenza virus subtype reference samples were prepared as follows. The influenza viruses were isolated using Madin–Darby canine kidney (MDCK) cells supplemented with 1 μg/mL L-(tosylamido-2-phenyl) ethyl chloromethyl ketone (TPCK)-treated trypsin. The samples were first added to monolayers of MDCK cells and incubated for 1 h at 37°C to allow viral adsorption to the cells. The inoculum was decanted, Eagle's minimum essential medium supplemented with 0.2% bovine serum albumin was added, and monolayers were incubated for 3–5 days at 37°C.

After cytopathic effects appeared, influenza virus was confirmed by using hemagglutination of chicken erythrocytes and RT-PCR against the HA gene. A panel of reference influenza A virus RNA samples for 15 HA and 9 NA subtypes was developed by conducting hemagglutination inhibition (HI) assays with a panel of reference antisera against HA subtypes 1 through 15 and NA subtypes 1 through 9. Reference virus sequences were confirmed for each HA and NA subtype. Viral RNA was extracted from supernatants of cultures of infected cells by using the RNeasy Mini Kit (Qiagen, Chatsworth, CA) according to the manufacturer's instructions. Reverse transcription and PCR amplification were carried out under standard conditions by using influenza-specific primers. PCR products were purified with QIAquick PCR purification kits (Qiagen, Chatsworth, CA) and sequencing reactions were performed at the Hartwell Center for Bioinformatics and Biotechnology at St. Jude Children's Research Hospital (Memphis, TN).

The target for both subtyping and sequencing studies is a 500~600 bp amplicon that contains the 5′ end of either the HA or NA gene and was amplified from first-strand cDNA. First-strand cDNA was produced from influenza virus RNA (5–20 ng/μL) with SuperScript II reverse transcriptase (Invitrogen, Carlsbad, CA) and a tagged universal primer, <u>TAATACGACTCACTATAGG</u>**AGCAAAAGCAGG** (tag sequence is underlined and universal influenza sequence is in bold).

Amplifications were accomplished with a 10 μM forward tag primer [GCATCCTAATACGACTCACTATAGG] and specific reverse primer, and 2 to 5 μL of first-strand cDNA per 100 μL reaction [37]. The reaction conditions consisted of a 5 min denaturation at 94°C, followed by 40 cycles of a 30 s 94°C denaturation step; a 30 s 55°C annealing step; and a 30 s 72°C extension; and finally a 10 min extension at 72°C. The resulting PCR product was cleaned with a Qiagen QIAquick PCR purification kit and eluted in 100 μL of distilled water.

A second, one-way amplification resulted in single-stranded target. One-way amplifications were accomplished with the respective specific reverse primer only and 2 to 5 μL of cleaned amplification product from the first amplification. The reaction conditions were similar to those described above with 50 cycles of amplification. The resulting product was purified with a Qiagen QIAquick PCR purification kit and eluted in 100 μl of distilled water. This step resulted in tagged, single-stranded target for hybridization in sequencing assays.

For subtyping assays, biotinylated single-strand target for standard hybridizations was produced as described above for one-way PCR, however, biotin-14-dCTP (Invitrogen, Carlsbad, CA) was incorporated into the product during amplification. Prior to hybridization, the single-stranded target was heated to 95°C for 10 min and then placed on ice. Ten × T4 ligase buffer was then added to bring the solution to 1 × concentration and, finally, a 5′ labeled T7 oligonucleotide (Cy-3 or Cy-5, Integrated DNA Technologies, Inc., Coralville, IA) was added to a concentration of 1 μM.

During the on-chip subtyping assay, 95 μL of the hybridization sample solution were loaded into the array chamber in the cartridge. Other solutions including (1) 200 μL of 6x SSPE, 0.05% Tween-20; (2) 200 μL of 3x SSPE, 0.05% Tween-20; (3) 200 μL of 2X PBST, 0.1% Tween-20; (4) 200 μL of a labeling solution; and (5) 200 μL of 2X PBS, 0.1% Tween-20, were separately loaded in the storage chambers 1–5 (Fig. 4.1A). The labeling solution contained streptavidin-Cy5 (Molecular Probes, Eugene, OR) that was diluted in a blocking solution (2X PBS, 0.1% Tween-20, 1% Acetylated BSA) to a final concentration of 1 μg/mL. Hybridization was carried out for one hour at 45 °C in the microarray chamber.

Following hybridization, the array was washed with 6X SSPE, 0.05% Tween-20. On-chip washings continued with 3X SSPE, 0.05% Tween-20 and 2X PBS, 0.1% Tween-20, respectively. The labeling solution was then pumped into the hybridization chamber and incubated for 30 minutes at room temperature. The final washing step was performed by flowing 2X PBST through the hybridization chamber. During each pumping step, a bubbling mixing procedure was implemented. The microarray chip was then detached from the microfluidic plastic cartridge before it was scanned. The total on-chip processing time was approximately 1 hour and 50 minutes. Subtype identification was accomplished by averaging HA and NA subtype intensity values and graphing in Microsoft Excel.

The broad-scan microarray was designed to identify influenza A subtypes based on unique probesets for HA subtypes 1 through 15 and NA subtypes 1 through 9. After hybridization of biotinylated H2 and N7-specific target and labeling with streptavidin-Cy5, high fluorescent signal intensity demonstrated that the subtype specific target DNA annealed to the correct probes, as shown in Fig. 4.8

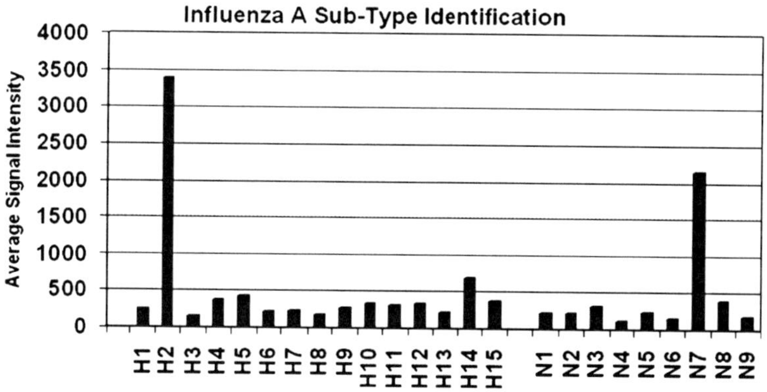


**Fig. 4.8** Influenza subtype analysis demonstrates the robustness of the microfluidic system. The correct Influenza A hemagglutinin (H2) and neuraminidase (N7) subtypes were identified with this automated system with results equal to or better than standard manual techniques (reproduced with permission from [38]), copyright 2006, American Chemical Society)

[38]. By averaging subtype probe signal intensities and graphing the results, the correct influenza A subtypes were identified using the microfluidics device. In general, average positive probe signal was at least four times greater than average negative probe signal.

Variable signal intensities from the negative probes are the result of cross-hybridizations that are due to the close relationship among subtype sequences. Results using the microfluidics device were comparable to the manual control results (where the average signal intensities for H2 and N7 are 3400 and 2500, respectively, and the negative probe and background signals are below 600), suggesting that on-chip microfluidic washing and labeling are efficient and compatible with manual washing and labeling processes. Because there were no replicate probes in the current subtyping array, it was difficult to study the hybridization uniformity. The Influenza A subtyping array used in this study has previously been shown to positively identify all 15 hemagglutinin and all 9 neuraminidase subtypes [38]. Previous study with gene expression analysis in the microfluidic array devices showed high detection sensitivity (375 fM) and uniform hybridization signals [58], further indicating that the on-chip fluidic handling (washing and reaction) is efficient and can be automated with no loss of performance.

## 4.6 Sequencing Assay

In many cases, the identification of an influenza A subtype is not sufficient for making decisions on vaccine development, patient treatment, and general surveillance, and the specific sequence is needed. For example, mutations in sequence coding for the hemagglutinin receptor binding site of subtype H5N1 have been shown to be important for replication of avian viruses in humans [59]. In this study, viral HA and NA subtype sequencing was accomplished with a simple and rapid enzymatic assay in a microfluidics device.

A mixture including DNA polymerase and ligase extended a labeled common primer on single-stranded target DNA to the 5′ end of the hybridized HA and NA sequencing probes and then ligated the extended primer to probes that matched the target sequence (Fig. 4.3). A 0.1 N NaOH wash, a stringent washing procedure, removed any unligated signal and after scanning the array, intensity data were exported to an Excel worksheet. Sequence information was extracted with a routine designed to associate the correct base with the highest signal from sets of four probes that were tiled by one nucleotide to cover the sequence of interest (Fig. 4.3b). The preparation of the sequencing sample target has been described in the above section for the subtyping assay.

During the on-chip sequencing assay, 95 µL of the hybridization sample solution were loaded into the array chamber in the cartridge. Other solutions include (1) 200 µL of 1 × *E. coli* ligase buffer; (2) 200 µL of a mixture containing 155 µl of $dH_2O$, 18 µl of 10 × *E. coli* ligase buffer, 3 µL of 10 mM dNTP, 2 µL (20 units) of AmpliTaq DNA polymerase, Stoffel fragment (Applied Biosystems, Foster City, CA), and 2 µL (20 units) of *E. coli* ligase; (3) 200 µL of 0.1 N NaOH buffer; and (4) 200 µL of 2X PBS, 0.1% Tween-20, were separately loaded in the storage chambers 1–4 (Fig. 4.1a). The 5th storage chamber was left empty in the device. Hybridization was carried out for one hour at 45°C.

Following the hybridization process, the ligase buffer solution was first pumped through the array chamber, removing the sample mixture into the waste chamber and washing the array. The DNA ligase/polymerase mixture was subsequently pumped through the array chamber followed by a 30 minute incubation at 37°C. Once the extension and ligation were completed, the 0.1 N NaOH buffer was pumped through the array chamber at room temperature to denature and wash the array. The 2X PBST (phosphate-buffered saline-Tween 20) buffer was subsequently pumped through the array chamber to ensure a thorough washing. During each pumping step, a bubbling mixing procedure as described in the following section was implemented. The total on-chip processing time was approximately 1 hour and 45 minutes.

The device was then removed from the instrument. The microarray chip was separated from the microfluidic plastic cartridge with the use of a razor blade before it was scanned using a commercial fluorescent scanner (GenePix 4000B, Molecular Devices, Sunnyvale, CA). Imaging was performed while the array was wet with 2X PBST under a LifterSlip™ glass cover slip (Erie Scientific, Portsmouth, NH). Image intensities were analyzed and quantified using Microarray Imager software (CombiMatrix Corp., Mukilteo, WA). HA and NA subtype DNA sequence information was generated from sequencing array intensity data with a Microsoft Excel routine designed to interrogate units of four data points and then associate the most intense signal with the nucleotide represented by that probe. Sequence strings were then used to search the GenBank nonredundant database with BLASTN.

The results of hemagglutinin (H9) and neuraminidase (N2) subtype sequencing in the microfluidic cartridge are shown in Fig. 4.9 [38]. With such an array chip that has over 12,000 features, we were able to sequence over 500 nt, depending upon the length of the target DNA, of both the HA and NA genes with an accuracy of over 90%. In this study, the sequencing accuracies of the HA and NA genes using the microfluidic device are 91% and 94%, respectively, as compared to 95% for the HA gene and 93% for the NA gene using the conventional manually handling array (i.e., the control).

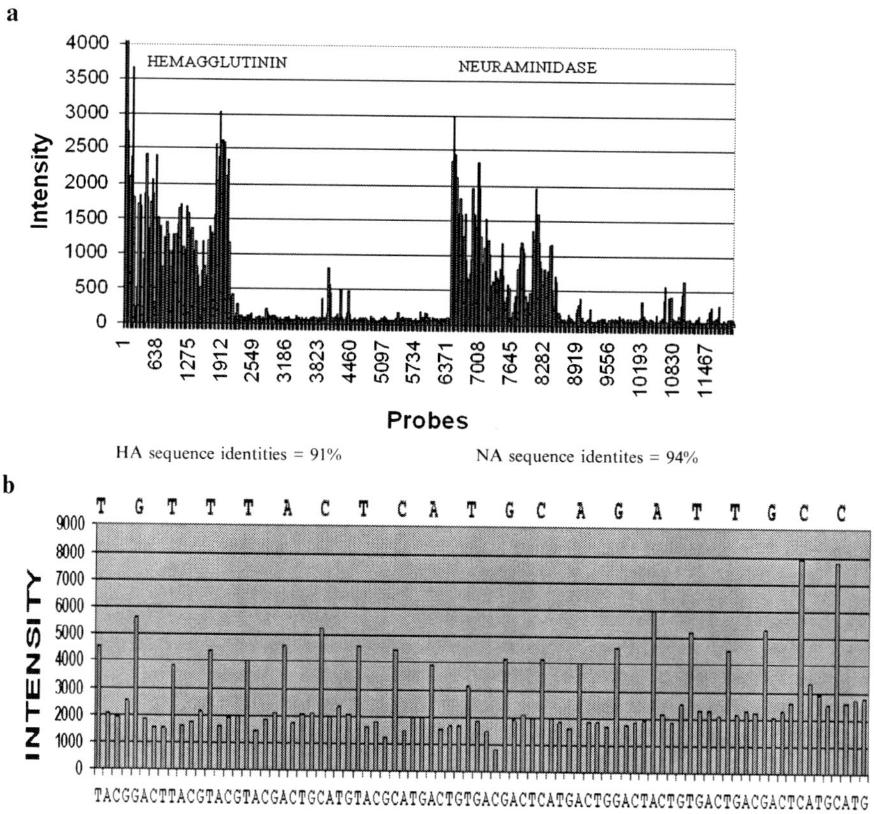


**Fig. 4.9** (a) Probe hybridization intensity values for hemagglutinin H9 and neuraminidase N2 sequencing using the automated cartridge. Although the entire sequences of the HA (approximately 1700 bp) and NA (approximately 1400 bp) genes were encoded on the array, the targets were synthesized from and hybridized to the 5′ 500 nucleotides of the genes as demonstrated by the hybridization patterns (underlined). After extraction of data, resulting sequences for the HA and NA genes were 91% and 94%, respectively, identical to the correct GenBank sequence data. (b) Sequence information is obtained by associating the DNA bases with the greatest signals from groups of four sequencing probes. An expanded section of the neuraminidase gene is shown here. DNA bases above the figure indicate the sequence at the 5′ end of probes and bases below the figure indicate the sequence extracted from the data (reproduced with permission from [38], copyright 2006, American Chemical Society)

The results indicated that the performance of the microfluidic array device was comparable to that of conventional manual handling. Inaccurate sequence calls were generally the result of difficulty to detect mismatches, such as A/G or T/G mismatches or secondary structure within the probe sequences. This artifact of the sequencing array can be resolved by including replicates of the sequencing probes. The average values will reduce the signal from ligated mismatches and other artifacts on the chip. This approach to sequencing is rapid, requiring less than two hours on chip to sequence a target.

## 4.7 Discussion

DNA microarray technology has become one of the most promising analytical tools in molecular biology. It has been widely used for studying mRNA levels and examining gene expression in biological samples [60–64]. Investigators rely on data produced by microarray experiments to assess changes in gene expression levels among various experimental tissues and treatments. The applications of microarrays for gene expression profiling [60] include pathway dissection [65], drug evaluation [62, 63], discovery of gene function [61], classification of clinical samples [66–68], and investigation of splicing events [69], among many others [64].

The highly parallel nature of microarrays has made them invaluable tools for monitoring gene expression patterns of numerous genes simultaneously. The development of new detection methods, simplified methodologies, and broad application to molecular diagnostics are rapidly migrating microarray technologies into the arena of genetic-based diagnostics and personalized medicine [70]. Array-based methods are now being applied in a diagnostic setting and offer the possibility of analyzing multiple analytes and even multiple samples in a highly parallel and high-throughput manner [71]. Because of their sensitivity, specificity, and accuracy, DNA microarrays have become an acceptable technology for screening samples for the presence or absence of a large variety of viruses simultaneously and identifying the genotype of an unknown specimen [32–35]. Microarrays are particularly useful for molecular detection and identification of influenza viruses because of their genetic and host diversity and the availability of an extensive sequence database [32, 35, 36].

The integrated microfluidic array device reported in this chapter automates microarray assays that involve multistage sample processing and fluidic handling that are in general labor-intensive and time-consuming and have significant potential to be errorprone [72]. Automation of the hybridization and posthybridization process allows more stringent processing control over the microarrays and eliminates variations in array data caused by subtle, day-to-day differences in protocol and manual handling [58]. There is more control over a variety of parameters including hybridization temperature and time, washing time and speed, mixing/agitation speed, and labeling time. Although some commercial robotic workstations have also been developed to automate microarray processing, such instruments are generally too expensive and tend to limit the application of microarrays to high-budget applications. These bulky workstations are complicated to operate and often require high cost for maintenance.

In contrast, the self-contained and fully integrated microfluidic array devices reported here are disposable and require simple portable instruments for operation. The plastic microfluidic cartridges can be fabricated using injection molding (instead of laser machining reported here) that would lead to low-cost microfluidic devices. The cost of a microfluidic plastic cartridge component is small compared to the cost of the microarray chip. In order to reduce the overall cost of the device (primarily of the microarray chip) for diagnostic applications, we are currently developing different methods or technologies.

For example, the array is currently being redesigned to have four identical subarray sectors so that four assays can be run on the same array chips. Moreover, the reuse of the array by stripping the hybridized DNA/RNA targets using chemicals before the next use has been developed (www.combimatrix.com). These strategies reduce the per-assay costs so that the arrays are comparable in cost to other diagnostic technologies (e.g., real-time PCR). Furthermore, the fluorescent detection method used for the semiconductor-based microarrays reported in this work can be replaced with the electrochemical detection-based method. Using an electrochemical detector, we could further reduce the costs of equipment by eliminating the expensive optical scanning instrument.

The integration of microfluidics adds new significant functionalities to the conventional microarray platform. Although gene expression, sequencing, and subtyping assays were demonstrated in this study, this integrated microfluidic platform can potentially be applied to many other assays. Although the sample and reagent consumption in the current microfluidic device design is tens to hundreds of μL, it is believed that the design could be further miniaturized so that the sample and reagent consumption could be significantly reduced. It is also possible to integrate front-end sample preparation and DNA amplification into the same platform, which would lead to a fully integrated system with direct sample-to-answer capability [21].

The microfluidic components in the device, including electrochemical pumps, duckbill check valves, and bubbling mixer, are simple in design, inexpensive, and easy to fabricate and integrate into a complex microfluidic system, as compared with most of the existing microfabricated micropumps, microvalves, and micromixers. Because the power consumption of the electrochemical pumps is low (~mW), handheld field use of the integrated microfluidic components is feasible.

The combination of integrated microfluidics with electrochemical detection-based microarray would allow one to perform gene expression study or identify influenza subtypes and sequence samples of interest rapidly and cost-effectively. The readout will no longer be visual and rely on fluorescence that often suffers from photobleaching issues, but rather be electrochemical relying on the intrinsic electronic functionality of the silicon integrated circuit chip. The electrochemical detection will be a key differentiation of the integrated microfluidic array device because the chip will no longer need to be removed from the microfluidic cartridge for fluorescent scanning. The analysis workflow will also be improved as the stitching or templating of the fluorescent image will not be required. The integrated microfluidic platform provides a step towards fulfilling the promise of rapid, automated genetic analysis from complex sample fluids in cost-effective and portable instruments.

## 4.8 Conclusion

The use and applications of microarray technology have grown since its invention. More recent advances indicate that microfluidics can be integrated along with microarrays to fabricate multifunction devices. In this chapter, a self-contained and disposable microfluidic array device for integrated gene expression and genotyping assays is reported. The device automated and integrated hybridization, sample processing, and fluidic handling steps that are considered labor-intensive and time-consuming in the regular manual handling process. All microfluidic components such as micropumps, microvalves, and micromixers are integrated on the microfluidic cartridge, but use simple and inexpensive approaches in order to reduce device complexity. Gene expression study of the human leukemia cell line (K562) and genotyping detection and sequencing of influenza A subtypes have been demonstrated using this integrated biochip platform. It is believed that this integrated microfluidic platform can potentially be applied to many other assays for DNA analysis.

**Acknowledgment** The authors thank Tai Nguyen, Kevin Schwarzkopf, Tony Siuda, Alla Petrova, Kia Peyvan, Michael Bizak, Jeff Kemper, Al Pierce, and Mike Slota for technical support and useful discussions. This work has been sponsored by DoD contract #1999011104A.

## References

1. Kelly, R.T. and A.T. Woolley (2005). Microfluidic systems for integrated, high-throughput DNA analysis. *Anal. Chem.* **77**: 97A–102A.
2. Harrison, D.J., A. Manz, et al. (1992). Capillary electrophoresis and sample injection systems integrated on a planar glass chip. *Anal. Chem.* **64**: 1926–1932.
3. Wilding, P., J. Pfahler, et al. (1994). Manipulation and flow of biological-fluids in straight channels micromachined in silicon. *Clin. Chem.* **40**(1): 43–47.
4. Xia, Y.N. and G.M. Whitesides (1998). Soft lithography. *Ann. Rev. Mater. Sci.* **28**: 153–184.
5. Piner, R.D., J. Zhu, et al. (1999). "Dip-pen" nanolithography. *Science* **283**(5402): 661–663.
6. Becker, H., W. Dietz, et al. (1998). Microfluidic manifolds by polymer hot embossing for micro total analysis system applications. *uTas 98*, Banff, Canada; Dordrecht, Kluwer Academic.
7. Alonso-Amigo, M.G. and H. Becker (2000). Microdevices fabricated by polymer hot embossing. *Abstracts of Papers of the American Chemical Society* **219**: 468–COLL.
8. Grodzinski, P., R.H. Liu, et al. (2001). Development of plastic microfluidic devices for sample preparation. *Biomed. Microdevices* **3(4)**: 275.

9. Boone, T., Z.H. Fan, et al. (2002). Plastic advances microfluidic devices. *Anal. Chem.*.**74**(3): 78A–86A.
10. Harrison, D.J., K. Fluri, et al. (1993). Micromachining a miniaturized capillary electrophoresis-based chemical-analysis system on a chip. *Science* **261**(5123): 895–897.
11. Woolley, A.T., D. Hadley, et al. (1996). Functional integration of PCR amplification and capillary electrophresis in a microfabricated DNA analysis device. *Anal. Chem.* **68**(23, Dec. 1996): 4081–4086.
12. Burns, M.A., B.N. Johnson, et al. (1998). An integrated nanoliter DNA analysis device. *Science* **282**(5388): 484–487.
13. Waters, L.C., S.C. Jacobson, et al. (1998). Microchip device for cell lysis, multiplex pcr amplification and electrophoretic sizing. *Anal. Chem.* **70**: 158–162.
14. Emrich, C.A., H.J. Tian, et al. (2002). Microfabricated 384-lane capillary array electrophoresis bioanalyzer for ultrahigh-throughput genetic analysis. *Anal. Chem.* **74**(19): 5076–5083.
15. Ibrahim, M.S., R.S. Lofts, et al. (1998). Real-time microchip PCR for detecting single-base differences in viral and human DNA. *Anal. Chem.* **70**(9): 2013–2017.
16. Kopp, M., A.D. Mello, et al. (1998). Chemical amplification: Continuous-Flow PCR on a chip. *Science* **280**: 1046–1048.
17. Lagally, E.T., I. Medintz, et al. (2001). Single-molecule DNA amplification and analysis in an integrated microfluidic device. *Anal. Chem.* **73**: 565 – 570.
18. Taylor, M.T., P. Belgrader, et al. (2001). Lysing bacterial spores by sonication through a flexible interface in a microfluidic system. *Anal. Chem.* **73**(3): 492–496.
19. Yuen, P.K., L.J. Kricka, et al. (2001). Microchip module for blood sample preparation and nucleic acid amplification reactions. *Genome Res.* **11**(3): 405–412.
20. Anderson, R.C., X. Su, et al. (2000). A miniature integrated device for automated multistep genetic assays. *Nucleic Acids Res.* **28**(12): e60.
21. Liu, R.H., J. Yang, et al. (2004). Self-contained, fully integrated biochip for sample preparation, polymerase chain reaction amplification, and DNA microarray detection. *Anal. Chem.* **76**: 1824–1832.
22. Hay, A., V. Gregory, et al. (2001). The evolution of human influenza viruses. *Phil. Trans. R. Soc. Lond.* **B356**: 1861–1870.
23. Fouchier, R.A., V. Munster, et al. (2005). Characterization of a novel influenza A virus hemagglutinin subtype (H16) obtained from black-headed gulls. *J. Virol.* **79**(5): 2814–2822.
24. Scholtissek, C., H. Burger, et al. (1985). The nucleoprotein as a possible major factor in determining host specificity of influenza H3N2 viruses. *Virology* 147(2): 287–294.
25. Hoffmann, E., J. Stech, et al. (2001). Universal primer set for the full-length amplification of all influenza A viruses. *Arch. Virol.* **146**: 1–15.
26. Lipatov, A.S., E.A. Govorkova, et al. (2004). Influenza: Emergence and control. *J. Virol.* **78**(17): 8951–8959.
27. Webby, R.J. and R.G. Webster (2001). Emergence of influenza A viruses. *Phil. Trans. R. Soc. Lond.* **B356**: 1815–1826.
28. Mizuta, K., N. Katsushima, et al. (2003). A rare appearance of influenza A(H1N2) as a reassortant in a community such as Yamagata where A(H1N1) and A(H3N2) co-circulate. *Microbiol. Immunol.* **47**(5): 359–361.
29. Ueda, M., A. Maeda, et al. (1998). Application of subtype-specific monoclonal antibodies for rapid detection and identification of influenza A and B viruses. *J. Clin. Microbiol.* **36(2)**: 340–344.
30. Allwinn, R., W. Preiser, et al. (2002). Laboratory diagnosis of influenza-virology or serology? *Med. Microbiol. Immunol. (Berl)* **191(3–4)**: 157–160.
31. Amano, Y. and Q. Cheng (2005). Detection of influenza virus: Traditional approaches and development of biosensors. *Anal. Bioanal. Chem.* **381(1)**: 156–184.
32. Li, J., S. Chen, et al. (2001). Typing and subtyping influenza virus using DNA microarrays and multiplex reverse transcriptase PCR. *J. Clin. Microbiol.* **39(2)**: 696–704.
33. Ellis, J.S. and M.C. Zambon (2002). Molecular diagnosis of influenza. *Rev. Med. Virol.* **12(6)**: 375–389.

34. Ivshina, A.V., G.M. Vodeiko, et al. (2004). Mapping of genomic segments of influenza B virus strains by an oligonucleotide microarray method. *J. Clin. Microbiol.* **42(12)**: 5793–5801.
35. Kessler, N., O. Ferraris, et al. (2004). Use of the DNA flow-thru chip, a three-dimensional biochip, for typing and subtyping of influenza viruses. *J. Clin. Microbiol.* **42(5)**: 2173–2185.
36. Sengupta, S., K. Onodera, et al. (2003). Molecular detection and identification of influenza viruses by oligonucleotide microarray hybridization. *J. Clin. Microbiol.* **41(10)**: 4542–4550.
37. Lodes, M.J., D. Suciu, et al. (2006). Influenza A subtype identification and sequencing with semiconductor-based oligonucleotide microarrays. *J. Clin. Microbiol.* **44**: 1209–1218.
38. Liu, R.H., M.J. Lodes, et al. (2006). Validation of a fully integrated microfluidic array device for influenza A subtype identification and sequencing. *Anal. Chem.* **78**: 4184–4193.
39. Liu, R.H., K. Dill, et al. (2006). Integrated microfluidic biochips for DNA microarray analysis. *Expert Rev. Molec. Diagnostics* **6**: 253–261.
40. Oleinikov, A.V., M.D. Gray, et al. (2003). Self-assembling protein arrays using electronic semiconductor microchips and in vitro translation. *J. Proteome Res.* **2**: 313.
41. Dill, K., D.D. Montgomery, et al. (2004). Immunoassays and sequence-specific DNA detection on a microchip using enzyme amplified electrochemical detection. *J. Biochem. Biophys. Meth.* **59**: 181–187.
42. Macken, C., H. Lu, et al. (2001). The value of a database in surveillance and vaccine selection. *Options for the Control of Influenza IV*. N. C. A. W. H. A.D.M.E. Osterhaus. Amsterdam, Elsevier Science: 103–106.
43. Wang, D., A. Urisman, et al. (2003). Viral discovery and sequence recovery using DNA microarrays. *PLoS Biol.***1**(2): 257–260.
44. Allawi, H.T. and J. Santa Lucia Jr. (1999). Nearest-neighbor thermodynamics and NMR of DNA sequences with internal A.A, C.C, G.G, and T.T mismatches. *Biochemistry* **38**: 3468–3477.
45. Altschul, S.F., T.L. Madden, et al. (1997). Gapped BLAST and PSI-BLAST: A new generation of protein database search programs. *Nucleic Acids Res.* **25**: 3389–3402.
46. Ray, C.A., C.L. Sloan, et al. (1992). A silicon-based shape memory alloy microvalve. *Proc. Mater. Res. Soc. Symposium* **276**: 161–166.
47. Jerman, H. (1994). Electrically-activated, normally-closed diaphragm valves. *J. Micromech. Microeng.* **4**: 210–216.
48. Beebe, D.J., J.S. Moore, et al. (2000). Functional structures for autonomous flow control inside microfluidic channels. *Nature* **404**: 588–590.
49. Liu, R.H., Q. Yu, et al. (2002). Fabrication and characterization of hydrogel-based microvalves. *J.Microelectromech.Syst.* **11**: 45–53.
50. Liu, R.H., M. Stremler, et al. (2000). A passive micromixer: 3-D C-shape serpentine microchannel. *J. Microelectromech. Syst.* **9(2)**: 190–197.
51. Zengerle, R., S. Skluge, et al. (1995). A bidirectional silicon micropump. *Sensors Actuators A-Physical* **50**: 81–86.
52. Unger, M.A., H. Chou, et al. (2000). Monolithic microfabricated valves and pumps by multilayer soft lithography. *Science* **288**: 113–116.
53. Su, Y.C., L.W. Lin, et al. (2002). A water-powered osmotic microactuator. *J. Microelectromech. Syst.* **11**(6): 736–742.
54. Richter, G. (1975). Device for Supplying Medicines. U.S. Patent, 3,894,538.
55. Bohm, S., W. Olthuis, et al. (1999). An integrated micromachined electrochemical pump and dosing system. *J. Biomed. Microdevices* **1(2)**: 121–130.
56. Munyan, J.W., H.V. Fuentes, et al. (2003). Electrically actuated, pressure-driven microfluidic pumps. *Lab Chip* **3**: 217–220.
57. Liu, R.H., R. Lenigk, et al. (2003). Hybridization enhancement using cavitation microstreaming. *Anal. Chem.* **75**: 1911–1917.
58. Liu, R.H., T. Nguyen, et al. (2006). A fully integrated miniature device for automated gene expression DNA microarray processing. *Anal. Chem.* **78**: 1980–1986.

59. Iwatsuki-Horimoto, K., R. Kanazawa, et al. (2004). The index influenza A virus subtype H5N1 isolated from a human in 1997 differs in its receptor-binding properties from a virulent avian influenza virus. *J. Gen. Virol.* **85**: 1001–1005.
60. Schena, M., D. Shalon, et al. (1995). Quantitative monitoring of gene expression patterns with a complementary DNA microarray. *Science* **270**: 467–470.
61. Chu, S., J. DeRisi, et al. (1998). The transcriptional program of sporulation in budding yeast. *Science* **282**: 699–705.
62. Gray, N.S., L. Wodicka, et al. (1998). Exploiting chemical libraries, structure and genomics in the search for kinase inhibitors. *Science* **218**: 533–538.
63. Hughes, T.R., M.J. Marton, et al. (2000). Functional discovery via a compendium of expression profiles. *Cell* **102**: 109–126.
64. Schena, M. (2000). *Microarray BiochipTechnology*. Natick, MA, Eaton.
65. Roberts, C.J., B. Nelson, et al. (2000). Signaling and circuitry of multiple MAPK pathways revealed by a matrix of global gene expression profiles. *Science* **287**: 873–880.
66. Khan, J., R. Simon, et al. (1998). Gene expression profiling of alveolar rhabdomyosarcoma with cDNA microarrays. *Cancer Res.* **58**: 5009–5013.
67. Golub, T.R., D.K. Slonim, et al. (1999). Molecular classification of cancer: class discovery and class prediction by gene expression monitoring. *Science* **286**: 531–537.
68. Perou, C.M., S.S. Jeffrey, et al. (1999). Distinctive gene expression patterns in human mammary epithelial cells and breast cancers. *Proc. Natl Acad. Sci. USA* **96**: 9212–9217.
69. Hu, G.K., S.J. Madore, et al. (2001). Predicting splice variants from DNA chip expression data. *Genome Res.* **11**: 1237–1245.
70. Dill, K. and A. McShea (2005). Recent advances in microarrays. *Drug Discovery Today: Technol.* **2**(3): 261–266.
71. Gershon, D. (2005). DNA microarrays. *Nature* **437**: 1195–1200.
72. Dobbin, K.K., D.G. Beer, et al. (2005). Interlaboratory comparability study of cancer gene expression analysis using oligonucleotide microarrays. *Clin. Cancer Res.* **11**: 565–572.

# Part III
# Statistical Data Evaluation on Microarrays

# Chapter 5
# Intensity Concentration Relationships for Electrochemical Detection

## Latin Square and Mixture Study Analysis

**Mervyn Thomas**

## 5.1 Introduction

The currently accepted technology for polynucleotide quantitation on gene chips involves labelling the assayed polynucleotides with a fluorescent marker, hybridising the sample to the gene chip, and then imaging the gene chip under fluorescence. The background-corrected intensity of fluorescence is assumed to be proportional to the concentration of the target polynucleotide.

CombiMatrix have developed a novel technology for quantification of binding to their custom gene chip platform. The sensor is based on the horseradish peroxidase coupled electrochemical properties of each gene chip cell. The technology is inherently cheaper, more reliable, more sensitive, and more robust than fluorescence-based optical sensors of gene expression binding.

CombiMatrix have noted that the response between polynucleotide concentration and reading intensity is nonlinear. That is, there is no longer a simple proportionality between background-corrected fluorescence and polynucleotide concentration. Whilst this is actually irrelevant to the majority of potential gene expression uses of the technology, it is a barrier to market acceptance.

Any gene chip may contain technical replicates, cells on the same chip which have identical probes. The purposes of such technical replicates are:

1. To reduce variability and increase sensitivity
2. To increase robustness by identifying failed probes

Whilst these benefits are real, increasing the number of technical replicates reduces the real estate available for different probes, and increases the cost of whole genome analysis, perhaps requiring multiple chips per sample. CombiMatrix must therefore provide a convincing rationale for the number of technical replicates to be used on their chips. This issue has direct consequences for marketing the technology.

---

M. Thomas
Empheron Inc.

K. Dill et al. (eds.), *Microarrays: Preparation, Microfluidics, Detection Methods, and Biological Applications*,

### 5.1.1 Objectives

The objectives of this report are:

1. To provide a simple empirical model of the relationship between intensity and polynucleotide concentration
2. To provide a simple means of estimating differences in polynucleotide concentration using differences in measured intensity
3. To provide an initial estimate of the sensitivity of this process
4. To provide a sound basis for deciding on the number of technical replicates

## 5.2 Background Correction and Normalisation Methods

Background correction is applied one chip at a time. We assume a vector **S** of observed chip intensities, and assume that **S** may be represented as the sum of a noise component and a signal component: **S** = $X + Y$ where $X$ is the unobserved signal component and $Y$ is the unobserved noise component. We consider two approaches to background correction: deconvolution and an ad hoc quantile-based approach.

### 5.2.1 Deconvolution

The deconvolution approach follows Bolstad et al. [1], except that the Bolstad et al. algorithm was developed for strictly positive observed probe intensities **S**. The approach assumes that the unobserved signal component $X$ follows an exponential distribution with parameter $\alpha$, and that the unobserved noise component $Y$ follows a normal distribution with mean $\mu$ and standard deviation $\sigma$. $X$ and $Y$ are assumed to be independent.

The adjusted probe expression $\hat{X}_i$ values are given by the formula:

$$\hat{X}_i = z_i + \frac{\sqrt{2}\sigma\, e^{-\frac{z_i^2}{2\sigma^2}}}{\sqrt{\pi}\left(\mathrm{erf}\left(\frac{z_i\sqrt{2}}{2\sigma}\right)+1\right)},$$

where $zi = (Si - \mu) - \alpha\sigma^2$. Here erf represents the error function:

$$\mathrm{erf}(\mathrm{x}) = \frac{2}{\sqrt{\pi}}\int_{-\infty}^{x} e^{-t^2} dt.$$

The derivation of the deconvolution expression is shown in the appendix.

Estimates for μ, σ, and α are obtained in the same way as Bolstad, with the exception that the kernel density estimate is trimmed to take account of erratic values, and the estimate for α is based on a Huber estimate of the mean. This ensures that the estimates are not seriously influenced by the small number of erratic values (vide infra).

### 5.2.2 *Quantile Adjustment*

The quantile adjustment method is an ad hoc algorithm, designed to ensure that all expression measures are strictly positive, and that the expression measure for the lowest intensity probes is close to zero. The formula for the background correction index $\hat{X}_i$ is given by:

$$\hat{X}_i = \begin{cases} c\,|q\beta\,| & \text{if } S_i < q\beta, \\ S_i - q\beta + c\,|q\beta\,| & \text{if } S_i \geq q\beta, \end{cases}$$

where $q\beta$ is the $\beta^{th}$ quantile of $S$ and $c$ is a small positive constant. Results presented in this report were obtained with β set arbitrarily to 0.001 and cset to 0.005. These values were found to give acceptable results for the datasets studied here.

The value selected for β should be adjusted if the number of features per gene chip changes. With 12,000 features per gene chip, setting β = 0.001 trims the smallest 12 expression levels. Some chips do have more than one erratic value, but very few chips have more than two erratic values. Choosing β to ensure that at least four features are trimmed should prevent erratic values influencing the normalisation.

## 5.3 Normalisation

Different chips are usually hybridised with different amounts of RNA, and different instrument settings may change the overall level of response. These effects are likely to produce widespread changes in intensity, essentially across all the features on the gene chip. Normalisation is an attempt to reduce the variability of expression measures, by correcting for these whole-scale differences in overall intensity.

We consider three approaches to normalisation: median normalisation, quantile–quantile normalisation, and loess normalisation.

### 5.3.1 *Median Normalisation*

The simplest normalisation approach is to divide all feature intensities by the median intensity for that chip. This is appropriate if variation in total RNA or instrument settings simply scales all feature intensities linearly. In other gene expression platforms this is usually a reasonable assumption for some sources of variation, but quite wrong for others. For example, with the Affymetrix platform median normalisation copes quite well with variation caused by changes in the intensity of illumination, but performs badly for variation caused by changes in the total RNA concentration.

Median normalisation is applied one chip at a time. That is, the normalisation of one chip does not depend on the values for other chips present in the data set.

### 5.3.2 Quantile–Quantile Normalisation

Quantile–quantile normalisation was described by Bolstad et al. [1]. It is a generalisation of median normalisation (which matches only the 50th quantile of each chip's intensity distribution) in that it matches every quantile of each chip's distribution. Essentially, each chip is forced to have the same empirical cumulative distribution function. Quantile–quantile normalisation normalises all of the gene chips simultaneously. That is, the normalisation performed for any chip depends on the other chips which are being normalised in the same analysis.

Quantile–quantile normalisation has been shown to perform very well with Affymetrix data. It is the most widely used method of normalisation with this technology.

### 5.3.3 Loess Normalisation

Loess normalisation matches the distribution of probe values across chips more completely than median normalisation, but less completely than quantile–quantile normalisation. A target chip is produced by forming the median of each feature, across all chips. That is, $Ti$ = median ($Sij$, $j$ = 1, …, $N$), where $Sij$ represents the observed intensity for feature $I$ on chip $j$. There are assumed to be $N$ chips.

A loess regression [2] is conducted for each chip, using the target chip as a dependent variable and the observed chip values as the predictor variable. The normalised values for each chip are then obtained as the fitted values from the loess regression.

Loess normalisation has been shown to perform well with Affymetrix and other gene chip technologies, but it is computationally expensive. Like quantile–quantile normalisation, loess normalisation addresses all of the chips simultaneously. The normalisation performed for any one chip depends on the other chips which are being normalised.

## 5.4 Preprocessing Options

We have described two methods of background correction (deconvolution and ad hoc quantile adjustment), and three methods of normalisation (median, quantile–quantile, and loess). These choices may be combined to generate six preprocessing options:

1. Quantile background correction, median normalisation
2. Quantile background correction, quantile–quantile normalisation
3. Quantile background correction, loess normalisation
4. Deconvolution background correction, median normalisation
5. Deconvolution background correction, quantile–quantile normalisation
6. Deconvolution background correction, loess normalisation

## 5.5 Latin Square Experiment

### *5.5.1 Experimental Design*

This study was designed to test linearity of response across varying concentrations of transcript spike-ins, both within the same chip and between chips. Ten concentrations of each spike-in were adopted, but the design was constrained by the following factors.

1. Each spike-in can be added at only one concentration on any one chip.
2. Only six distinct lambda phage spike-ins were available.

The maximum number of different spike-in concentrations on any one chip was therefore six. At least ten different concentrations were required to provide an adequate analysis of linearity.

An incomplete block design [3] was adopted, in which each chip represented a block, and spike-in transcript and concentration represented the design factors. The design was constructed as an augmented Latin square [3]. The spike-in transcripts were added to a fixed concentration of reference complex human RNA.

The design ensured that each spike-in is used at all ten of the differing concentrations. It involved ten groups of chips, each of which contained two replicate chips. The details of each group are specified in Table 5.1.

### *5.5.2 Gene Chip Design*

Each of the gene chips had 12,544 features, representing 589 distinct probes. There were three different probes for each of the six lambda phage spike-ins. Each of

**Table 5.1** Experimental Design for Spike-In Experiment

| | Spike-In Concentrations (pM) | | | | | |
|---|---|---|---|---|---|---|
| Group | Lambda 12 | Lambda 9 | Lambda 8 | Lambda 6 | Lambda 5 | Lambda 3 |
| 1 | 0.5 | 2.0 | 8.0 | 32.0 | 64.0 | 128.0 |
| 2 | 128.0 | 0.5 | 2.0 | 8.0 | 32.0 | 64.0 |
| 3 | 64.0 | 128.0 | 0.5 | 2.0 | 8.0 | 32.0 |
| 4 | 32.0 | 64.0 | 128.0 | 0.5 | 2.0 | 8.0 |
| 5 | 8.0 | 32.0 | 64.0 | 128.0 | 0.5 | 2.0 |
| 6 | 2.0 | 8.0 | 32.0 | 64.0 | 128.0 | 0.5 |
| 7 | 6.0 | 12.0 | 24.0 | 48.0 | 96.0 | 198.0 |
| 8 | 198.0 | 6.0 | 12.0 | 24.0 | 48.0 | 96.0 |
| 9 | 96.0 | 198.0 | 6.0 | 12.0 | 24.0 | 48.0 |
| 10 | 48.0 | 96.0 | 198.0 | 6.0 | 12.0 | 24.0 |

these three different probes was present in 21 technical replicates, giving a total of 63 features per spike-in transcript. In addition there were 25 negative control probes, with either 15, 21, or 22 technical replicates. The total number of negative control features was 478.

## 5.6 Methods

### *5.6.1 Linearity of Response*

Data were preprocessed using each of the six options (each combination of the two background correction and the three normalisation algorithms). For each chip, data on the same spike-in were combined using a robust Huber *M* estimate of the mean [7]. Data for the six spike-in transcripts were extracted from the datasets.

The relationship between voltage and spike-in concentration was markedly nonlinear, but a linearising transformation was adopted. The Huber average log2 voltage was plotted against the log10 concentration of the spike-in.[1] Simple Pearson correlation coefficients were calculated between Huber average log voltage and log concentration. Correlation coefficients were calculated for each spike-in, and for all spike-ins.

This analysis was then repeated, but based on subsamples 1, 2, 3, and 5 of the 63 features for each spike-in transcript.

[1] A log2 transformation was used for voltages so as to represent a twofold change by a difference of one on the log2 scale. A log10 transformation was used for concentrations to simplify interpretation of the graphs. The linearity of the response is, of course, unaffected by the choice of base for the logarithm.

### 5.6.2 *Sensitivity*

Whilst we may define sensitivity measures for individual gene chips, realistic and scientifically valid gene expression studies always involve biological replication. In such studies, biological variation is usually the key determinant of experimental sensitivity (in the sense of statistical power), not technical variability of the gene chip. What we require of a gene chip is that its technical variability is small relative to the biological variability. Biological variability, however, will be very different in different scientific contexts.

There is therefore no unique, universally relevant definition of gene chip sensitivity which can be applied in all biological contexts. The appropriate measure of sensitivity depends on the nature of the application.

We need a simple objective measure of gene chip sensitivity, which may be communicated to potential customers without undue difficulty. An obvious candidate is provided by the probability that a spike-in probe mean exceeds the negative control mean. This sensitivity measure is increased as the variance of technical replicates decreases, or the number of technical replicates increases. We calculate sensitivity for a range of technical replicates, based on the observed technical replicate variation. If there is no difference between the background mean and the mean for a given probe, we would expect this sensitivity to be approximately 50%.

It must be stressed that in an experiment with proper biological replication, the probability of detecting differential gene expression should be much greater than the sensitivity calculated in this way.

## 5.7 Results

### 5.7.1 *Linearity of Response*

Table 5.2 shows the correlation between Huber average log voltage and log concentration for each spike-in, after processing with each of the six composite preprocessing algorithms. All of these correlations are very high, indicating a usable linear relationship between Huber average log voltage and log concentration.

Results for loess normalisation and quantile–quantile normalisation are very similar, but results for median normalisation are markedly inferior. For this dataset, both deconvolution and the ad hoc quantile-based background correction give acceptable results.

Fig. 5.1 shows plots of log2 voltage versus log10 concentration for the deconvolution method of background correction. Results are presented for quantile– quantile normalisation and for median normalisation only, because quantile–quantile and loess normalisation give such similar results. The relationship does seem to be approximately linear, although there is some evidence of nonlinearity at the higher concentrations.

**Table 5.2** Correlation of Huber Average Log Voltage with Log Spike-In Concentration

| | | Deconvolution | | | Quantile | | |
|---|---|---|---|---|---|---|---|
| No. Reps | Spike-In | QQ | Lo | Med | QQ | Lo | Med |
| 1 | All | 0.937 | 0.936 | 0.939 | 0.937 | 0.940 | 0.921 |
| 2 | All | 0.965 | 0.953 | 0.945 | 0.961 | 0.963 | 0.955 |
| 3 | All | 0.968 | 0.965 | 0.958 | 0.951 | 0.965 | 0.960 |
| 5 | All | 0.976 | 0.968 | 0.967 | 0.969 | 0.968 | 0.969 |
| All | All | 0.983 | 0.983 | 0.977 | 0.987 | 0.987 | 0.980 |
| 1 | L12 | 0.933 | 0.939 | 0.925 | 0.855 | 0.935 | 0.888 |
| 2 | L12 | 0.969 | 0.964 | 0.941 | 0.936 | 0.952 | 0.956 |
| 3 | L12 | 0.951 | 0.949 | 0.954 | 0.916 | 0.949 | 0.961 |
| 5 | L12 | 0.982 | 0.951 | 0.950 | 0.974 | 0.955 | 0.955 |
| All | L12 | 0.992 | 0.993 | 0.986 | 0.994 | 0.994 | 0.991 |
| 1 | L3 | 0.969 | 0.967 | 0.965 | 0.979 | 0.967 | 0.965 |
| 2 | L3 | 0.980 | 0.981 | 0.959 | 0.983 | 0.979 | 0.977 |
| 3 | L3 | 0.973 | 0.988 | 0.975 | 0.988 | 0.987 | 0.990 |
| 5 | L3 | 0.989 | 0.985 | 0.977 | 0.986 | 0.987 | 0.989 |
| All | L3 | 0.987 | 0.986 | 0.979 | 0.995 | 0.994 | 0.992 |
| 1 | L5 | 0.971 | 0.983 | 0.958 | 0.977 | 0.969 | 0.963 |
| 2 | L5 | 0.982 | 0.974 | 0.977 | 0.990 | 0.985 | 0.980 |
| 3 | L5 | 0.982 | 0.983 | 0.972 | 0.991 | 0.988 | 0.978 |
| 5 | L5 | 0.985 | 0.988 | 0.977 | 0.991 | 0.988 | 0.984 |
| All | L5 | 0.989 | 0.988 | 0.982 | 0.995 | 0.995 | 0.993 |
| 1 | L6 | 0.950 | 0.965 | 0.942 | 0.980 | 0.970 | 0.950 |
| 2 | L6 | 0.987 | 0.981 | 0.978 | 0.980 | 0.988 | 0.988 |
| 3 | L6 | 0.986 | 0.984 | 0.981 | 0.987 | 0.988 | 0.979 |
| 5 | L6 | 0.992 | 0.989 | 0.985 | 0.996 | 0.994 | 0.982 |
| All | L6 | 0.995 | 0.993 | 0.991 | 0.998 | 0.998 | 0.992 |
| 1 | L8 | 0.980 | 0.980 | 0.979 | 0.969 | 0.984 | 0.981 |
| 2 | L8 | 0.986 | 0.979 | 0.993 | 0.988 | 0.984 | 0.988 |
| 3 | L8 | 0.985 | 0.989 | 0.990 | 0.991 | 0.991 | 0.982 |
| 5 | L8 | 0.989 | 0.992 | 0.992 | 0.991 | 0.993 | 0.991 |
| All | L8 | 0.992 | 0.993 | 0.995 | 0.995 | 0.996 | 0.990 |
| 1 | L9 | 0.872 | 0.907 | 0.950 | 0.945 | 0.906 | 0.870 |
| 2 | L9 | 0.945 | 0.917 | 0.924 | 0.969 | 0.944 | 0.926 |
| 3 | L9 | 0.973 | 0.943 | 0.948 | 0.926 | 0.960 | 0.957 |
| 5 | L9 | 0.973 | 0.966 | 0.971 | 0.972 | 0.963 | 0.986 |
| All | L9 | 0.991 | 0.991 | 0.991 | 0.991 | 0.992 | 0.988 |

Note: Results are presented for each combination of background correction and normalisation algorithms. Columns labelled QQ are for quantile–quantile normalisation. Columns labelled Lo are for loess normalisation. Quantile labelled Med are for median normalisation. There are few differences between quantile–quantile and loess normalisation, but median normalisation is consistently inferior.

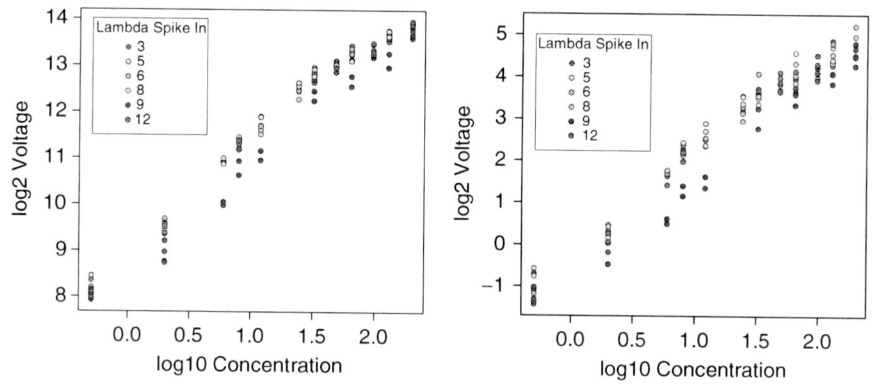


**Fig. 5.1** Plot of Huber average log voltage versus log spike-in concentration following deconvolution. Quantile–quantile normalisation is shown on the left, median normalisation on the right

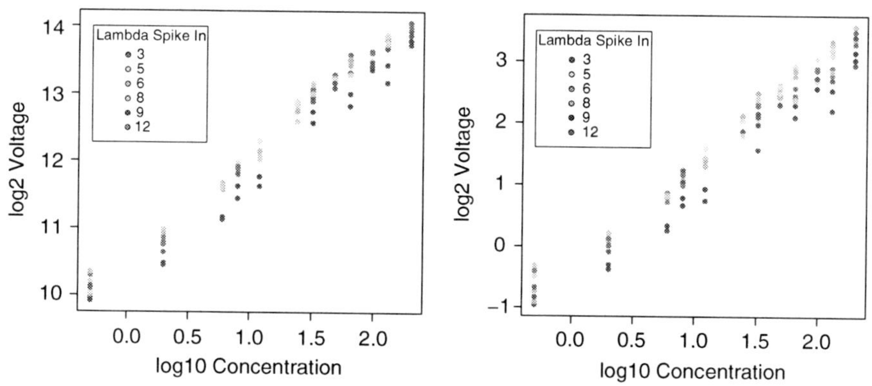


**Fig. 5.2** Plot of Huber average log voltage versus log spike-in concentration following ad hoc background correction. Quantile–quantile normalisation is shown on the left, median normalisation on the right

Fig. 5.2 shows equivalent results for the ad hoc quantile-based background correction. Again the relationship is reasonably linear, although this time any departure from linearity is associated with low concentrations rather than high concentrations.

Figs. 5.3 and 5.4 show the equivalent results when only one technical replicate is selected for each transcript. The relationships are still strong, but any tendency to nonlinearity is lost in the increased variance.

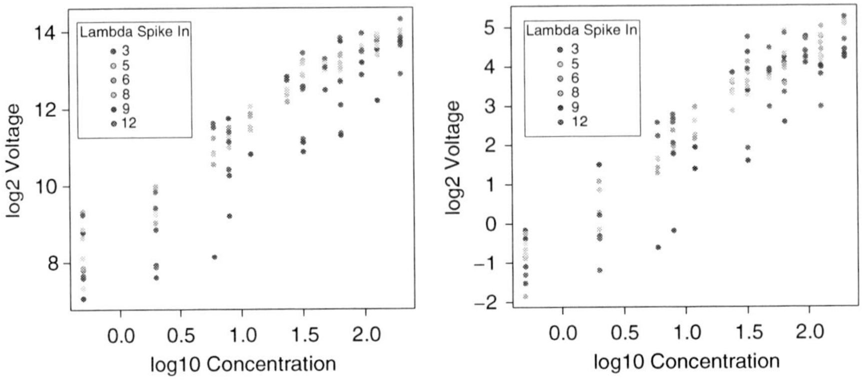


**Fig. 5.3** Plot of Huber average log voltage versus log spike-in concentration following deconvolution. Quantile–quantile normalisation is shown on the left, median normalisation on the right. These results are based on only one probe per transcript

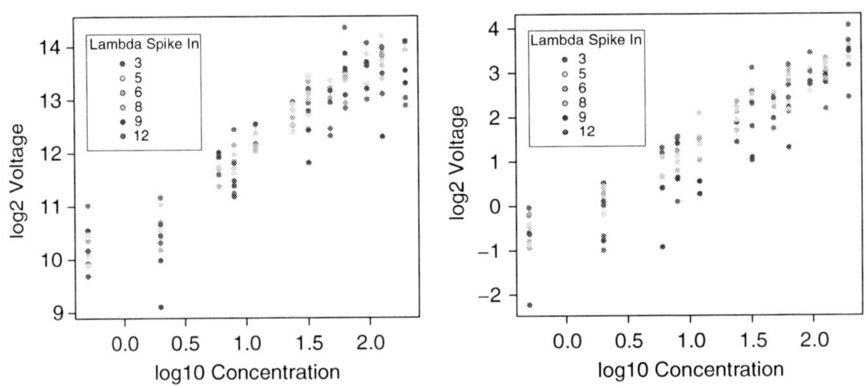


**Fig. 5.4** Plot of Huber average log voltage versus log spike-in concentration following ad hoc background correction. Quantile–quantile normalisation is shown on the left, median normalisation on the right. These results are based on only one probe per transcript

## *5.7.2 Sensitivity*

Fig. 5.5 shows the probability of detecting an increase in mean probe expression over negative control expression as a function of the number of technical replicates, and the probe concentration. The analysis is based on quantile–quantile normalisation. Different plotting symbols are used for each of the lambda phage spike-ins. The smallest sensitivity was 75%, well in excess of the 50% sensitivity to be expected if there is no true detection. Note that there is considerable variability in detection rate between the different spike-ins. The probability of detection increases markedly with the number of technical replicates.

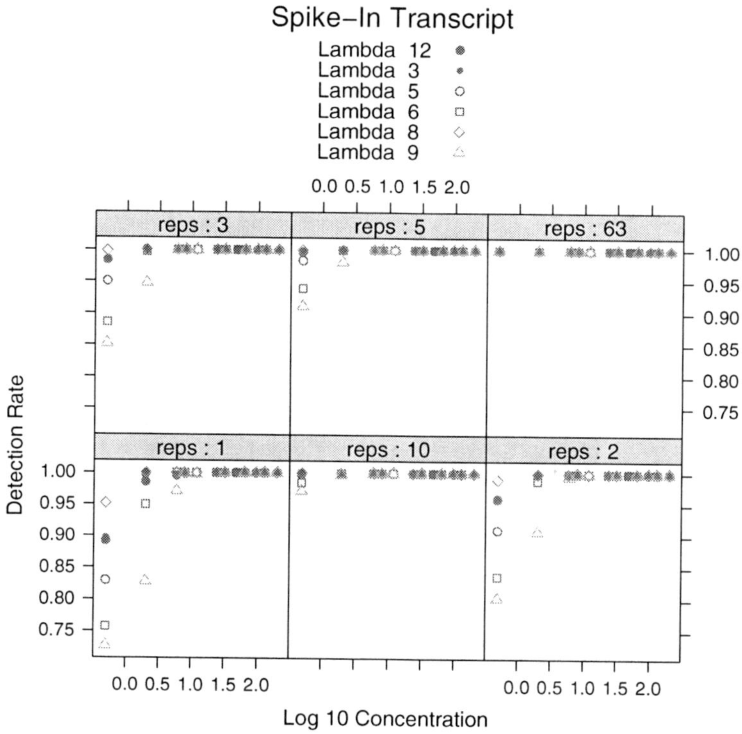


**Fig. 5.5** Sensitivity analysis for deconvolution background correction. The different plotting symbols represent different lambda phage spike-ins. The Lambda 6 and Lambda 9 spike-ins have markedly lower sensitivity than the others

Fig. 5.6 shows similar results for the ad hoc background correction algorithm. This algorithm has a marginally lower sensitivity than the deconvolution algorithm.

### 5.7.3 *Erratic Values*

Approximately one third of the gene chips showed erratic values in one or more features. This is illustrated by Fig. 5.7 which shows the results for gene chips 4000299.G3 and 4000300.G3. It is clear that chip 4000299.G3 has two erratic values.

The total number of erratic values over twenty gene chips each with 12,544 features was 8. This gives a fault rate of approximately $3 \times 10^{-5}$, and approximately 0.35 faults per gene chip.

Although these erratic values will not seriously affect the background correction applied to other features on the chip, they will result in incorrect estimates for the

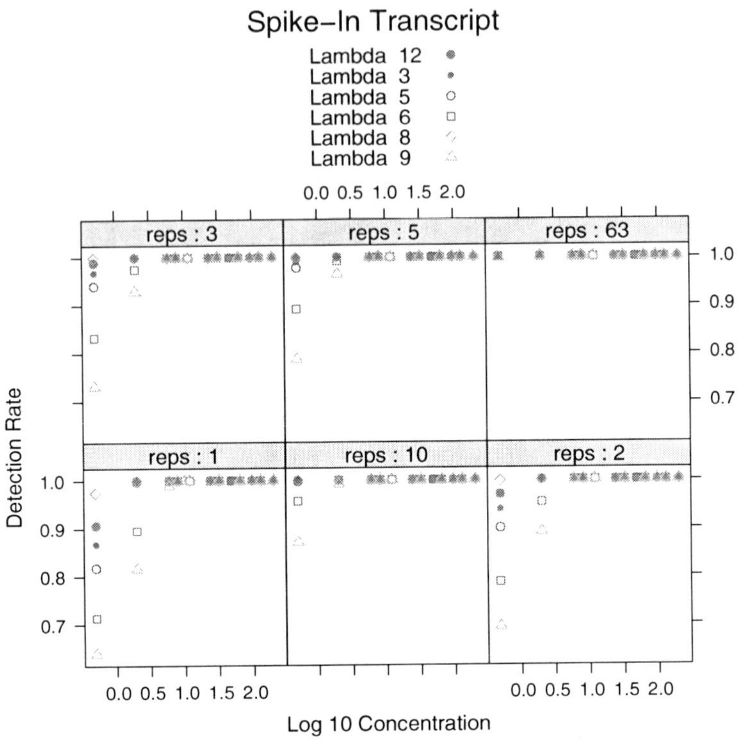


**Fig. 5.6** Sensitivity analysis for ad hoc background correction

feature concerned. This may or may not be a problem, depending on the number of technical replicates.

## 5.7.4 *Discussion*

There is a strong linear relationship between Huber average log voltage and log concentration of the spike-in. This relationship is evident both within chips and between chips. Both deconvolution and the ad hoc algorithm provide an acceptable means of background correction, and a reasonably linear response on the log scale. The deconvolution algorithm is more nearly linear at low concentrations, whilst the ad hoc approach is more nearly linear at high concentrations.

There is some suggestion that sensitivity is better with the deconvolution approach than with the ad hoc background correction. This may be associated with better performance of deconvolution at low signal strength. Chip sensitivity is acceptable, even with no technical replicates for each probe. However, the incidence of erratic values does pose a problem. The following strategies might be adopted for dealing with this.

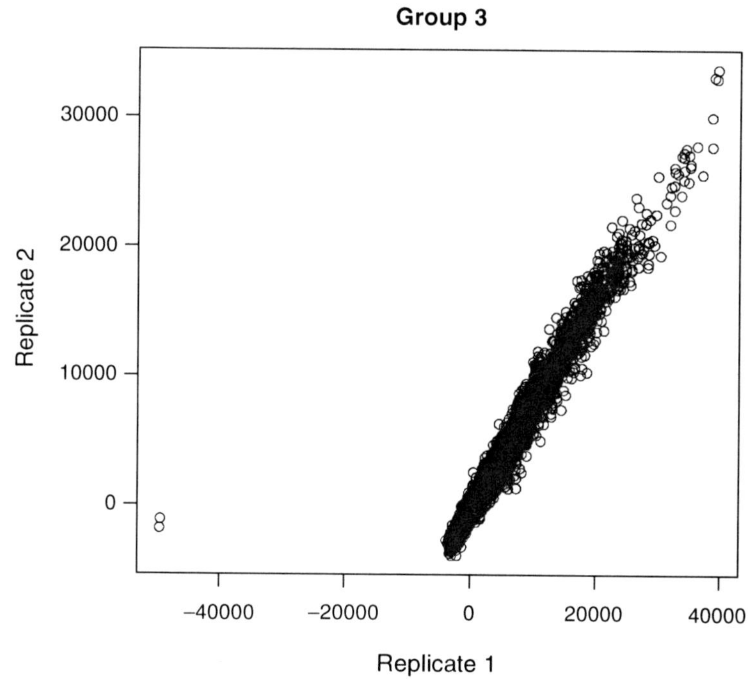


**Fig. 5.7** Plot of replicate gene chips for group three of the Latin square experiment. Two erratic values are clearly visible for replicate one (chip 4000299.G3)

1. Eliminate the problem with better quality control. This is, of course, the best approach if it is economically and technically feasible.
2. Use two replicates per probe. The probability that any probe will have erratic values in two replicates is vanishingly small (approximately $9 \times 10^{-10}$). If one feature for a probe were affected, there would be a huge between technical replicate difference for that probe. Two replicates per probe would then provide a very effective means of detecting the problem. Individual genes could then be dropped from the dataset as a whole.
3. Use three technical replicates per probe. With three technical replicates per probe the problem would be self-correcting. The probability of any two of the three features having erratic values would be very small. A robust average of feature values for the probe would be largely uninfluenced by the one erratic value.
4. Use only one technical replicate, but adopt robust analysis procedures in downstream analysis. This is perfectly feasible, but would not be possible with the generally available gene expression analysis tools. It would call for much greater sophistication in data analysis.

## 5.8 Mixture Experiment

### 5.8.1 Experimental Design

This experiment involves mixtures of two different human reference RNAs. One was a standard complex human reference RNA, the other was RNA extracted from brain tissue, which is expected to show strong differences from the standard complex reference. The design used a fixed total quantity of RNA, split between the two sources as defined in Table 5.3. Each experimental group had two replicate chips.

### 5.8.2 Methods

Three methods of analysis were used to investigate the relationship between gene expression and mixing proportion: principal components analysis, end member analysis, and partial least squares. Of these methods, principal components analysis is the least likely to reveal a strong relationship between gene expression and mixing proportion, and partial least squares is the most likely. The objective of using three methods is not to compare the approaches with each other, but to investigate the relative performance of our background correction and normalisation approaches using postprocessing of varying sophistication.

It is possible that the relative sophistication of the partial least squares approach will accommodate inadequacies in preprocessing, and therefore mask differences in performance between the background correction and normalisation approaches. It should be noted that none of these methods uses any a priori knowledge of which genes are likely to differ between the human reference mRNA mixture and the brain tissue. Any differences in gene expression which are present are diluted by those genes that do not show differential expression.

**Table 5.3** Experimental Design for Mixture Experiment

| Group | Standard Human Reference RNA | Other Reference RNA |
|---|---|---|
| 1 | 0 | 100 |
| 2 | 1 | 99 |
| 3 | 5 | 95 |
| 4 | 10 | 90 |
| 5 | 25 | 75 |
| 6 | 75 | 25 |
| 7 | 90 | 10 |
| 8 | 95 | 5 |
| 9 | 99 | 1 |
| 10 | 100 | 0 |

#### 5.8.2.1 Principal Components

Principal component scores [8] were calculated for each gene chip. The principal components represent the highest variance linear combinations of the genes (subject to orthonormality constraints). They may be interpreted as the most informative possible low-dimensional summary of the gene expression data. The first principal component score was then plotted against the mixing proportion.

#### 5.8.2.2 Linear End Member Analysis

Linear end member analysis [4] was originally developed to model hyperspectral remote sensing images. Each pixel was assumed to be derived from a compositional model, as a linear mixture of different end members. End members were assumed to represent particular 'pure' terrain types (e.g., forest canopy, grassland, etc.). When the end members are known, the mixing proportions in each pixel can be estimated by simple linear regression. The technique is easily adapted to the analysis of gene expression in mixture experiments. The end members represent the gene expression profile of the pure mixture components, and the mixing proportions are estimated.

Let l1 represent the mean normalised gene expression profile for component one of the mixture (i.e., the first end member), and let l2 represent the mean normalised gene expression for component two (i.e., the second end member). Then for a mixture of the two components, with mixing proportion of component two given by p we may write:

$$\mathrm{x} = \mathrm{l}_1 + p(\mathrm{l}_2 - \mathrm{l}_1)$$

$$\hat{p} = \frac{(\mathrm{x} - \mathrm{l}_1)'(\mathrm{x} - \mathrm{l}_1)}{(\mathrm{l}_2 - \mathrm{l}_1)'(\mathrm{l}_2 - \mathrm{l}_1)},$$

where $\hat{p}$ is the linear least squares estimator of $p$.

The end member gene expression profiles l1 and l2 were estimated by the mean of the normalised gene expression profiles for human reference RNA and for brain tissue RNA. These end members were then used to compute the predicted mixing proportion for each of the mixture points, and an $R^2$ statistic was calculated. This analysis was undertaken for every combination of the background correction and normalisation algorithms.

#### 5.8.2.3 Partial Least Squares

Partial least squares [6] is a biased regression technique [5], suitable for use with high-dimensional data. Partial least squares components are low-dimensional summaries of the gene expression data, calculated to have the highest predictive power for some dependent variable. The dependent variable was the known proportion of brain mRNA sample in each mixture.

The success of the predicted relationship was evaluated by cross-validation [9]. Each gene chip was dropped from the dataset in turn, and the partial least squares regression was computed. This was used to generate a predicted value for the dropped gene chip. This process was repeated until every gene chip had been dropped and predicted. At the end of this process, each gene chip had been used to develop the predictor – and each gene chip had been used to test the predictor – but no gene chip was used to fit the predictor and to test it simultaneously. Cross-validation produces approximately unbiased estimates of prediction success (under mild regularity assumptions). The $R^2$ values were calculated using the cross-validated predicted mixing proportions.

### 5.8.3 *Results*

#### 5.8.3.1 Principal Components

Fig. 5.8 shows the first principal component obtained with deconvolution background correction and quantile–quantile normalisation plotted against mixing proportion. The relationship is convincingly linear, with the component scores shifted from the endpoints with mixing proportions as low as 10%.

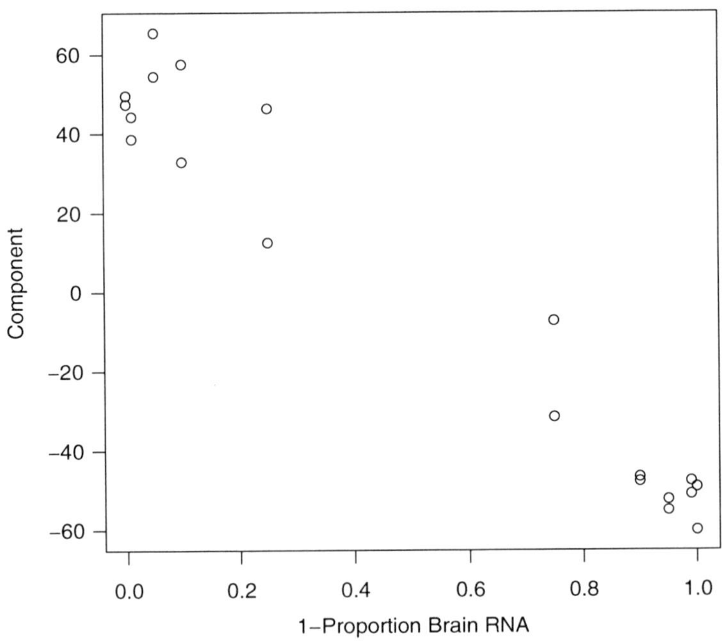


**Fig. 5.8** Plot of first principal component versus mixing proportion for deconvolution background correction and quantile–quantile normalisation

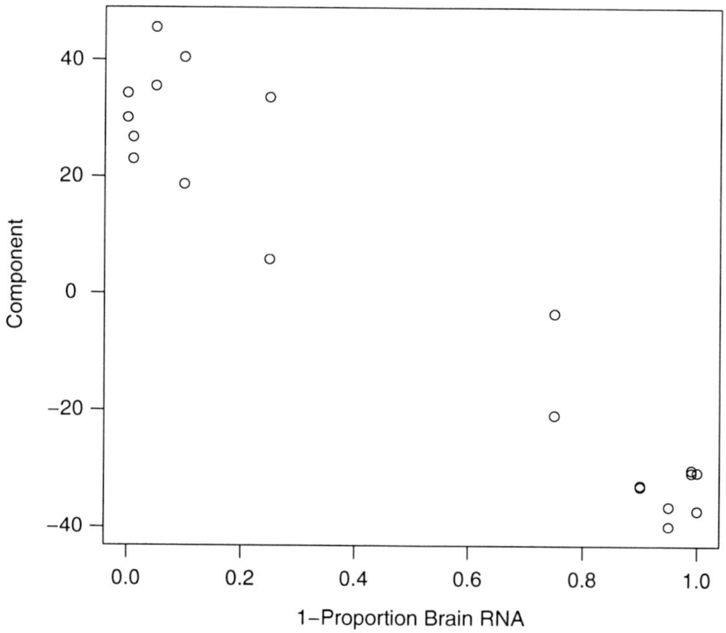


**Fig. 5.9** Plot of first principal component versus mixing proportion for ad hoc background correction and quantile–quantile normalisation

Fig. 5.9 shows similar results for the ad hoc background correction and quantile–quantile normalisation. Again the relationship is strong and linear. The relationship has very slightly lower variance for the deconvolution approach than for the ad hoc approach.

#### 5.8.3.2 Partial Least Squares and Linear End Member Analysis

Table 5.4 shows the $R^2$ values for the partial least squares and end member analysis estimates of mixing proportions. Results are shown for each normalisation algorithm, and for both background correction algorithms. There are strong correlations between the observed and the predicted mixing proportions.

The correlations are better for partial least squares than for the end member analysis. Deconvolution is marginally superior to the ad hoc background correction technique for partial least squares, and quantile–quantile normalisation is superior to median normalisation. Deconvolution is also superior to the ad hoc background correction for the end member analysis, but the differences are more marked than for the partial least squares analysis.

The deconvolution algorithm seems to be very slightly better than the ad hoc background correction, and the quantile–quantile normalisation seems superior to the median normalisation.

**Table 5.4** $R^2$ for Predicted and Observed Mixing Proportions by Background Correction and Normalisation

| BC | Normalisation | | |
|---|---|---|---|
| | Quantile–Quantile | Loess | Median |
| Method | PLS EM | PLS EM | PLS EM |
| Deconvolution | 0.977 0.960 | 0.980 0.953 | 0.961 0.913 |
| Ad hoc | 0.963 0.958 | 0.977 0.950 | 0.903 0.884 |

Note: PLS represents the $R^2$ calculated from a partial least squares analysis; EM represents the $R^2$ calculated from the end member algorithm. Deconvolution represents background correction using deconvolution; ad hoc represents background correction using the ad hoc algorithm. Quantile–quantile, loess, and median are the three normalisation algorithms.

#### 5.8.3.3 Discussion

There are clear relationships between gene expression and mixing proportion, irrespective of the nature of the preprocessing (background correction and normalisation) and the form of the postprocessing (principal components, end member analysis, or partial least squares).

The differences between the preprocessing options are relatively minor, although there is some suggestion that deconvolution performs better than the ad hoc background correction, especially for the weaker methods of post processing. Even the addition of small proportions ($\approx$10%) of brain mRNA produces noticeable changes in gene expression.

#### 5.8.3.4 Conclusions

1. Although the relationship between voltage and concentration is nonlinear, there is a robust linear relationship between Huber average log voltage and log spike-in concentration. The Electrasense machine is clearly capable of generating data which may be used for gene expression analysis.
2. Background correction should be performed either using deconvolution, or using a simple ad hoc algorithm based on quantiles. The deconvolution approach seems marginally better than the ad hoc approach.
3. For these studies, normalisation based on the quantile–quantile algorithm was perfectly acceptable. This algorithm is faster than the loess approach, but superior to the median normalisation approach. There may be datasets for which quantile–quantile normalisation is inappropriate (e.g., chips with many blank features, or with many negative control features).
4. There is reasonable sensitivity based on only one technical replicate per gene chip.
5. The Latin square experiment revealed occasional extreme values. These are thought to be associated with electrical shorts caused by manufacturing problems. Approximately one third of the chips showed at least one erratic value. This

must be addressed, either by enhanced quality control, or the use of two or more technical replicates per probe.

## 5.9 A Derivation of Deconvolution Formula

We have an observed probe reading $S$ defined on the interval $(-\infty, +\infty)$. Following Bolstad [1] we assume that the observed reading is the sum of a normally distributed noise term $Y$ plus an exponentially distributed signal term $X$. We further assume that the unobserved signal component $X$ follows an exponential distribution with parameter $\alpha$, and that the unobserved noise component $Y$ follows a normal distribution with mean $\mu$ and standard deviation $\sigma$. $X$ and $Y$ are assumed to be independent.

Then:

$$S = X + Y,$$

$$fy(y) = \frac{1}{\sqrt{2\pi}\sigma} e^{-\frac{(y-\mu)^2}{2\sigma}},$$

$$fx(x) = \alpha e^{-\alpha x},$$

where $X \in (0, \infty)$ and $(-\infty, +\infty)$. Note that this differs from Bolstad in that the domain of $Y$ is $(-\infty, \infty)$ rather than $(0, \infty)$. This changes the limits of integration, and results in a slightly different deconvolution formula for the conditional expectation of $X$ given $S$.

Assuming that $X$ and $Y$ are independent, the joint distribution of $X$ and $Y$ is:

$$fxy(x, y) = \alpha e^{-\alpha x} \frac{1}{\sqrt{2\pi}\sigma} e^{-\frac{(y-\mu)^2}{2\sigma}}.$$

The joint distribution of $S$ and $X$ is therefore:

$$fxs(x, s) = \frac{\alpha}{\sqrt{2\pi}\sigma} e^{-\alpha x} e^{-\frac{(x-s+\mu)^2}{2\sigma^2}}.$$

and we obtain the marginal distribution of $S$ as

$$fs(s) = \frac{\alpha}{\sqrt{2\pi}\sigma} \int_0^{+\infty} e^{-\alpha x} e^{-\frac{(x-s+\mu)^2}{2\sigma^2}} dx$$

$$= \frac{\alpha}{2} e^{\frac{\sigma^2\alpha^2 - \alpha s + \alpha\mu}{2}} \left[1 - \operatorname{erf}\left(\frac{\alpha\sigma^2 - s + \mu}{\sqrt{2}\sigma}\right)\right],$$

where

$$\operatorname{erf}(x)=\frac{2}{\sqrt{\pi}}\int_{-\infty}^{x} e^{-t^2}\,dt.$$

From this we obtain the conditional distribution of $X$ given $S$:

$$f_{X|S}(x|s)=\frac{f_{XS}(x,s)}{f_S(s)}$$

$$=\frac{\sqrt{2}}{\sigma\sqrt{\pi}}\frac{e^{-\frac{(\alpha\sigma^2+x-s+\mu)^2}{2\sigma^2}}}{1-\operatorname{erf}\left(\frac{(\alpha\sigma^2-s+\mu)\sqrt{2}}{2\sigma}\right)}.$$

Making the substitution $(s-\mu)-\alpha\sigma^2=z$ we obtain the conditional expectation of $X$ given $S$:

$$E_{X|S}\{x|s\}=z+\frac{\sqrt{2}\sigma e^{-\frac{z^2}{2\sigma^2}}}{\sqrt{\pi}\left(\operatorname{erf}\left(\frac{z\sqrt{2}}{2\sigma}\right)+1\right)}.$$

This is different from the conditional expectation derived by Bolstad (because of the different limits of integration). Bolstad, however, recommends an approximation to his adjustment formula, and his approximation is identical to the expression derived here.

## References

1. B.M. Bolstad, R.A. Irizarry, M. Astrand, and T.P. Speed. A comparison of normalization methods for high density oligonucleotide array data based on bias and variance. *Bioinformatics*, 19(2):185–193, 2003.
2. W.S. Cleveland. Robust locally weighted regression and smoothing scatterplots. *Journal of the American Statistical Association*, 74(368):829–836, 1979.
3. W.G. Cochran and G.M. Cox. *Experimental Designs*. John Wiley & Sons, New York, second edition, 1957.
4. A. Cross, J. Settle, N. Drake, and R. Paivinen. Subpixel measurement of tropical forest cover using avhrr data. *International Journal of Remote Sensing*, 12:1119–1129, 1991.
5. T. Hastie, R. Tibshirani, and J. Friedman. *The Elements of Statistical Learning: Data Mining, Inference and Prediction*. Springer, New York, 1st edition, 2001.

6. I. Helland. PLS regression and statistical models. *Scandinavian Journal of Statistics*, 17:97–114, 1990.
7. P. Huber. *Robust Statistics*. John Wiley and Sons, New York, 1981.
8. I. Jolliffe. *Principal Components Analysis*. Springer Verlag, New York, 1986.
9. M. Stone. Crossvalidatory choice and assessment of statistical predictions. *Journal of the Royal Statistical Society. Series B (Methodological)*, 36:111–147, 1974.

# Part IV
# Applications

# Chapter 6
# Genotyping Arrays

**Michael J. Lodes, Dominic Suciu, David Danley, and Andrew McShea**

**Abstract** Although the most common use of DNA microarrays is gene expression profiling, microarrays are also used for many other applications, including genotyping, resequencing, SNP analysis, and DNA methylation assays. Here we describe genotyping arrays for Influenza A subtype identification and for upper respiratory pathogen diagnostics using standard hybridization techniques and we also describe resequencing, SNP, and methylation assays using an enzyme-based strategy [25, 26].

## 6.1 Pathogen Identification

The need for laboratory assays that rapidly identify infectious diseases is substantiated by a number of government initiatives for their development. The Epidemic Outbreak Surveillance (EOS) program was initiated by the U.S. Air Force as an Advanced Concept Technology Demonstration (ACTD) to create a diagnostic assay that would identify 10 to 20 viruses and bacteria that are associated with upper respiratory infections (URI). The Centers for Disease Control has also established its Laboratory Influenza Test program to develop an improved diagnostic assay(s) that detects seasonal flu and novel influenza A viruses.

As with the EOS program, the CDC recognizes the importance of being able to distinguish between flu and other URI infections. A device capable of identifying this combination of bacteria and viruses must be able to overcome complexities associated with potentially highly variable genomes, which is especially relevant

M.J. Lodes, D. Suciu, D. Danley, and A. McShea
CombiMatrix Corporation, 6500 Harbour Heights Parkway, Mukilteo, WA 98275
e-mails: CombiMatrix Corp., mlodes@combimatrix.com
CombiMatrix Corp., dsuciu@combimatrix.com
CombiMatrix Corp., ddanley@combimatrix.com
CombiMatrix Corp., amcshea@combimatrix.com

K. Dill et al. (eds.), *Microarrays: Preparation, Microfluidics, Detection Methods, and Biological Applications*,

to influenza A subtyping. To date, there are 16 identified hemagglutinin [1] and 9 neuraminidase subtypes for Influenza A. This RNA virus has a negative strand, segmented genome, and can infect a broad range of animals including humans. Identification of a virus subtype is typically by serological or molecular identification of the subtype of viral hemagglutinin (HA) and neuraminidase (NA) genes. Viruses with any combination of the hemagglutinin and neuraminidase subtypes can infect aquatic birds whereas fewer subtypes have been found to infect humans. However, interspecies transmission can occur after recombination or mixing of subtypes in birds or pigs [2, 16, 65]. In addition, new human strains of virus can arise by reassortment to accomplish antigenic shift when two or more subtypes infect the same host [3, 4].

Identification of influenza subtypes is routinely accomplished with viral detection (cell culture) and serological techniques such as complement fixation, hemagglutination, hemagglutination inhibition assays, and immunofluorescence methods [51-8]. Traditional methods are generally effective, but involve labor-intensive protocols and highly trained personnel. Because of their speed, specificity and sensitivity, genomic assays are ideal for complementing serological assays for identifying the genotype of an unknown specimen, especially in cases where antigenic tests are not specific enough to differentiate closely related groups [9–14].

Reverse transcription-polymerase chain reaction (RT-PCR) is widely used for virus identification [15–17]. However, a positive amplification can be verified only by subsequent assays to elaborate sequence information. By overcoming this limitation, microarrays and biosensors have become valuable tools for viral discovery, detection, genotyping, and sequencing [6, 9, 10, 13, 17–26].

Although traditional assays for pathogen detection and typing represent the gold standard, they alone cannot meet the future needs of rapid, sensitive, specific, and simple methods. For example, although immunological methods are excellent for determining influenza subtypes, they do not give detailed genetic information or information when antigenic shifts occur. RT-PCR techniques depend on specific primers, which may fail when corresponding viral sequences mutate.

Microarrays offer an excellent solution as a downstream assay to PCR amplification. Semiconductor-based oligonucleotide array technology can be used with fluorescent labels and traditional optical scanning devices or used as a biodetector using electrochemical techniques for analysis [27]. This platform is extremely flexible, allowing array designs to be rapidly and easily modified and synthesized, and thus permitting oligonucleotides of interest to be tested empirically. In addition, the ability to use electrochemical detection with semiconductor microchips eliminates the need for expensive optical scanning equipment [27–29]. The use of an endpoint measurement on the PCR reaction, such as with a real-time PCR system, when preparing the target sample that will be hybridized to the array can be used as a decision point to "gate" the choice of samples to be run (i.e., assays are only run if the PCR reaction is positive).

Two genotyping arrays have been developed: an influenza A subtyping array that contains specific probes for each of 16 hemagglutinin (HA) subtypes and 9 neuraminidase (NA) subtypes and also an upper respiratory pathogen diagnostic array that can be used to detect both bacteria and viruses that produce upper respiratory clinical symptoms. These arrays can be hybridized with target generated from all 16 HA and 9 NA subtypes or from upper respiratory tract pathogens by a one-tube RT-PCR amplification of virus RNA or bacterial DNA.

The arrays were developed with nonoverlapping probes with similar annealing stabilities that were generated from sequence databases. For the influenza subtyping chip, subtype-specific probes were selected from a pool of over 23,000 HA and 15,000 NA sequences and then compared to the database to ensure that each probe was unique to the respective subtype and would hybridize to the maximum number of variant sequences. Subtype-specific probes were also subgrouped to give a finer detail for follow-up analysis, such as cluster analysis and sequence reconstruction. The probes for the upper respiratory arrays were developed in a similar manner and then cross-checked against databases to minimize the possibility of cross-hybridization of target and background from host genomic DNA.

The arrays were validated by hybridizing target (see Fig. 6.1) that was generated from RNA or DNA from all HA and NA subtypes on the influenza chip and all viruses and bacteria on the upper respiratory chip, which includes influenza A and B,

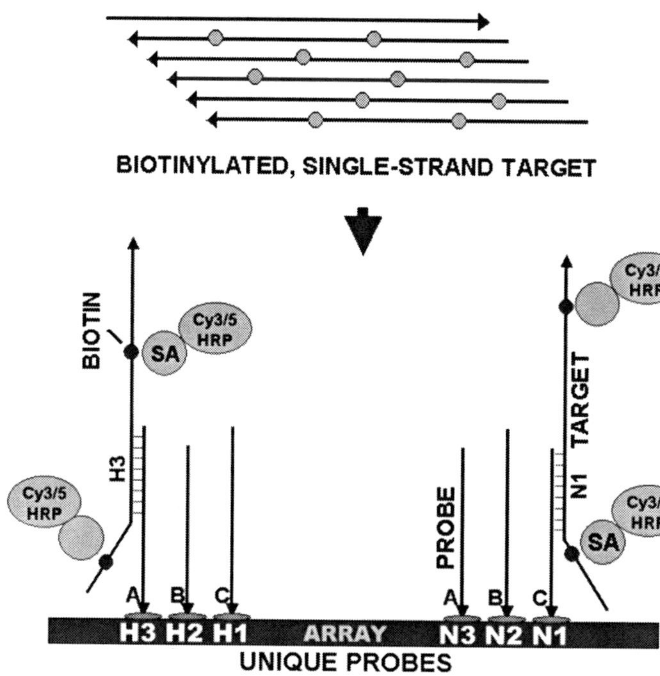


**Fig. 6.1** Diagram of target amplification and hybridization strategy for the identification of influenza A subtype H3N1. Briefly, a one-tube, three-stage RT-PCR reaction results in single-strand cDNA that is first amplified and then copied to produce biotin-labeled, single-strand target. Target is then hybridized to the array for 1 h and then labeled with either Streptavidin-Cy-5, fluorescent scanning, or with streptavidin-HRP, for electrochemical detection

parainfluenza, adenovirus 4, respiratory syncitial virus, coronavirus, *Streptococcus pyogenes*, *Bordetella pertussis*, *Chlamydia pneumoniae*, *and Mycoplasma pneumonia* (see Fig. 6.2a,b). The target sample preparation system used with these arrays is similar to standard RT-PCR-based methods, except that it uses a very redundant/consensus priming system that maximizes the chance that novel strains of influenza will amplify, and thus, minimize false negative results.

Primers and probes for bacteria were developed from conserved genes and then compared to genomes of related organisms. The arrays contain multiple probes that correspond to key distinguishing elements of each organism or subtype. The combination of assay speed (approximately 1 hour hybridization), array sectoring which would allow multiple assays on one array, the potential to strip and reuse the chip up to five times (for conventional hybridizations), and the adaptability to inexpensive electrochemical scanning devices make these arrays a superb adjunct to real-time PCR by supporting multiplex assays and analyses in a single PCR tube.

Rapid identification of upper respiratory pathogens followed by HA and NA subtyping of influenza A viruses, will significantly decrease the time and cost for the identification of potential lethal virus and bacterial strains and lead to better treatment and management of infections. Microarray and biosensor technologies show great promise for virus detection and genotyping and are needed for rapid vaccine development, environmental screening, and the detection of bioterrorism agents [14, 15, 20, 30, 31].

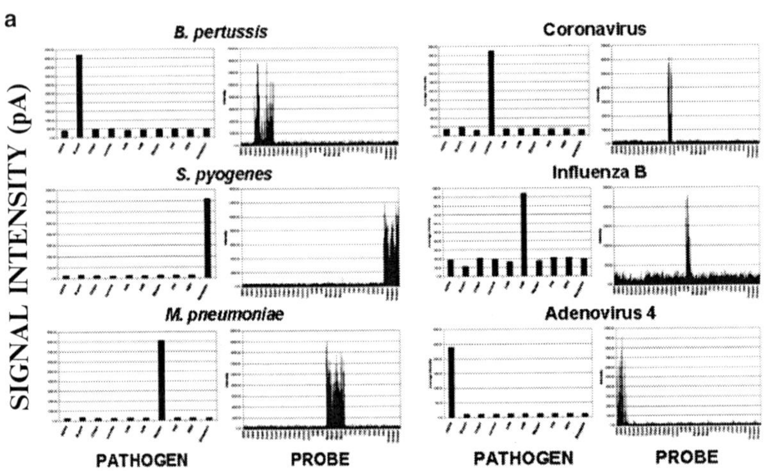


**Fig. 6.2a** Results of upper respiratory pathogen identification using electrochemical detection. High signal intensities indicate the bacterial pathogen (left panels) and viral pathogen (right panels) identity. Probe signal intensities are shown to the right of the typing panels. Identified bacteria include: *Bordetella pertussis*, *Streptococcus pyogenes*, *and Mycoplasma pneumoniae*; and identified viruses include: Coronavirus, Influenza B, and Adenovirus 4 (see [26], PLoS ONE for details)

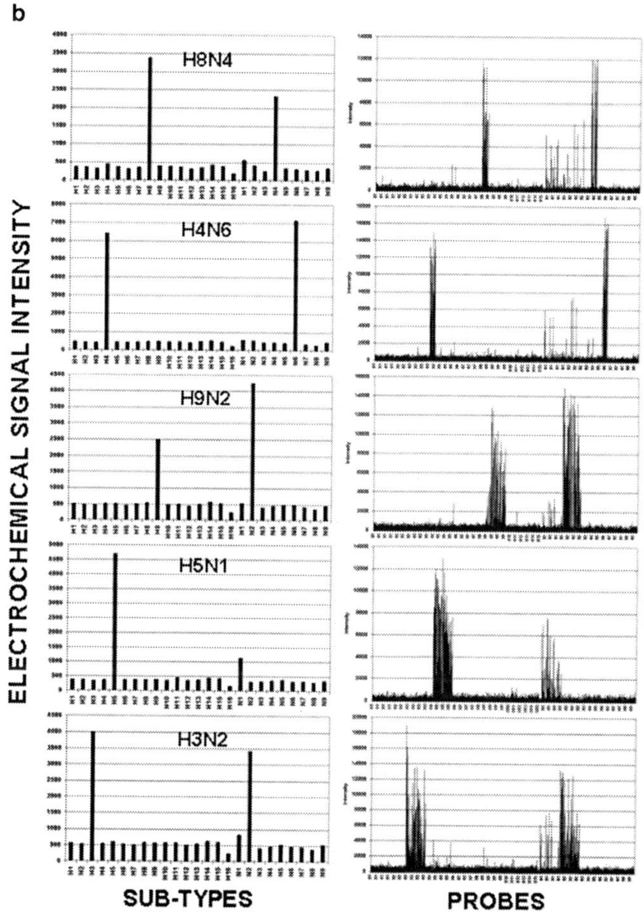


**Fig. 6.2b** Results of influenza A subtyping using electrochemical detection. High signal intensities indicate the hemagglutin (H) and neuraminidase (N) subtypes. Subtype identities are shown in the left panels and probe signal intensities are shown in right panels

## 6.2 Genotyping Assay

Targets were labeled by biotin incorporation during RT-PCR. Briefly, 25 μl reaction mixes included 12.5 μl of reaction buffer (Invitrogen; SuperScript III One-Step RT-PCR kit with Platinum Taq), 2 μl of 5 mM $MgSO_4$, 0.7 μl of 0.4 mM biotin-14-dCTP (Invitrogen), 2 μl primer pool (IDT, Coralville, IA), 0.5 μl enzyme mix, 2 μl RNA or DNA sample (diluted in a solution containing 40.0 μg yeast tRNA (Invitrogen), and 9.6 μg BSA (NEB) per 1.0 ml of $dH_2O$), and 4.3 μl $dH_2O$. Thermal cycling parameters were: (50°C – 30 min) × 1 cycle; (94°C – 4 min) × 1 cycle;

(94°C – 30 s, 56°C – 45 s, 72°C – 45 s) × 40 cycles; (94°C – 30 s, 68°C – 60 s) × 30 cycles and (72°C – 5 min) × 1 cycle. Primers were designed so that the forward primer Tm was approximately 50°C and the reverse primer Tm was approximately 65°C. Influenza subtyping was accomplished with a universal forward primer with a 5′ tag and thus the RT stage of amplification was set to 42°C instead of 50°C (5′CTATAGGAGCAAAAGCAGG). Amplification of target was confirmed by visualizing 3.0 µl of each reaction product on 6% polyacrylamide gels (Invitrogen) and staining with SYBR Green I (Molecular Probes, Invitrogen).

Initially, microarrays were prehybridized for 30 min at 45°C in 50 µl of a solution consisting of 5 ml of 2 × hybridization solution (see below), 1 ml of 50 × Denhardt's solution (Sigma), and 0.5 ml of 1% SDS (Sigma). For hybridization (see Fig. 6.1), PCR reactions from primer pools were combined and mixed 1:1 with 2 × hybridization buffer, which consisted of 6 ml of 20 × SSPE (Ambion, Austin, TX), 0.1 ml of 10% Tween 20 (Sigma), 0.56 ml of 0.5 M EDTA (Ambion), 0.5 ml of 1% SDS (Sigma), and 3.84 ml of $dH_2O$ (Ambion). Microarray hybridization chambers were filled (50 µl volume) and sealed with tape. The arrays were incubated for 1 h at 45°C with rotation in a hybridization oven (Fisher Scientific, Pittsburgh, PA) and washed for 5 min at 45°C with 3X SSPE with 0.05% Tween 20; twice with 2 × PBS with 0.1% Tween 20 (PBST); and then blocked for 5 min with 5 × PBS/Casein (BioFX Laboratories, Owing Mills, MD).

For labeling, microarrays were incubated for 30 min with ExtrAvidin Peroxidase (Sigma) diluted 1:1000 in BSA Peroxidase Stabilizer (BioFX). Arrays were washed twice with 2 × PBST, and twice with pH4 Conductivity Buffer Substrate (BioFX). TMB Conductivity 1 Component HRP Microwell Substrate (BioFX) was added to the array, and it was scanned immediately with an ElectraSense® microarray reader (CombiMatrix Corp.). This instrument measures µA at each of 12,544 electrodes on the array in 25 s and outputs data in picoamps to a simple text file that can be used to create a pseudoimage or can be transferred to and graphed with an Excel macro.

After hybridization to microarrays and data extraction by ECD, graphs of mean subtype intensity values can be used to predict the correct pathogen genotype or subtype. This information can next be broken down into subtype-subgroups. This grouped data can then be used to cluster samples into like groups with alignment software. This software treats each hybridization as an ordered list of intensities, where the value at each position corresponds to the intensity of a unique probe. A distance metric, such as correlation, allows for the determination of the difference between any two hybridization vectors. Next a similarity matrix is created containing the distances between all the hybridizations that are being compared. This similarity matrix can then be used with several clustering programs, such as the BioEdit Sequence Alignment Editor, as diagrammed in Fig. 6.3.

Sequence can also be reconstructed with array probe signal intensity data and then aligned with the most similar sequence in GenBank (Fig. 6.4). Briefly, sequence reconstruction software can take all the existing sequences in the database, align them against the chip probe sequences, and then create a probe profile for each sequence. For each hybridization that is analyzed, data are passed through these probe profiles, and the probe profiles with the highest scores are chosen. Next,

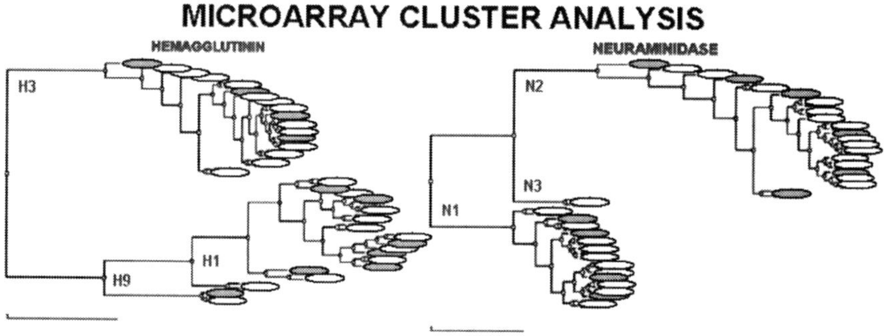


**Fig. 6.3** Cluster analysis: signal intensity data for each sample are compared and probe-to-probe correlations are determined. This data similarity matrix is then used develop an output that shows the relationships of all samples to one another. Clusters have been grouped into the hemagglutinin subtypes H1, H3, and H9 and the neuraminidase subtypes N1, N2, and N3. Green ovals represent known reference samples to which unknown samples are compared

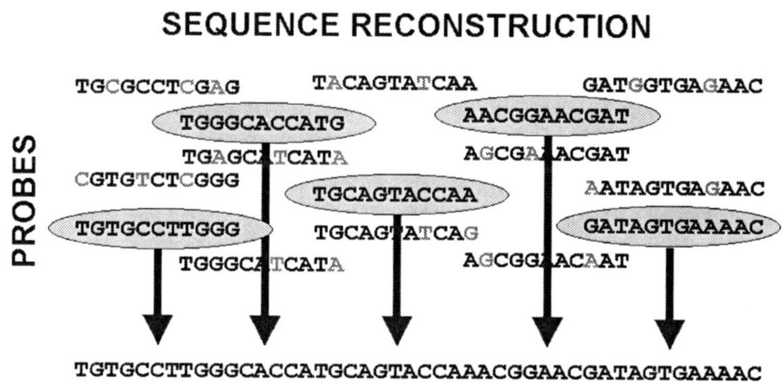


**Fig. 6.4** Sequence reconstruction from microarray data: multiple probes per pathogen, that represent the genetic diversity in genomic databases, are synthesized on an array and probed with an unknown target. Our proprietary software extracts the relevant probe signal and translates the associated probe sequence into a reconstructed sequence

sequence reconstruction proceeds within the small database containing the winning sequences. By limiting the dataset, resulting data have a lower background.

## 6.3 Single Nucleotide Polymorphisms/Resequencing

Oligonucleotide microarrays have also been developed for single nucleotide polymorphism (SNP) analysis and for resequencing of target DNAs. Over 10 million single nucleotide polymorphisms (SNP) have been estimated to occur in the human genome (International HapMap Project: www.hapmap.org/). Many of these polymorphisms have been associated directly or indirectly with genetic diseases including Crohn's disease, ataxia telangiectasia, and Alzheimer's disease. Certain Crohn's disease patients, for example, have been shown to have mutations in one or more of several genes that are associated with increased susceptibility and disease behavior. These genes include the CARD15/NOD2 gene [32, 33], and the MDR1 gene [34]. Also, mutations in the ataxia telangiectasia mutated (ATM) gene have been associated with lymphoma [35] and ataxia telangiectasia, which is characterized by cerebellar and neuromotor degeneration and immune deficiency [36].

The ability to detect mutations in patient genomes allows for a specific diagnosis and therapy and also allows for the prediction of potential disease in patients with a family history of genetic disease. Several technologies have been utilized to detect known SNPs including hybridization on microarrays or in real-time PCR; enzymatic nucleotide extension, cleavage, or ligation; and mass spectroscopy (Reviewed in: [37–40]).

Oligonucleotide microarray-based hybridization analyses of SNPs have been used to screen for both previously characterized sequence variants and for the discovery of new sequence variants (reviewed in [41]). This approach uses either a gain of signal approach, where probes are complementary to sequence changes of interest, and measures gain of hybridization signal to these probes relative to reference samples; or loss of signal, which analyzes loss of hybridization signal to perfect match probes that are complementary to wild-type sequence [41].

Potential difficulties with this approach include the ability to detect heterozygous base changes versus homozygous mutations; intra- and intermolecular structures, such as hairpin and G-rich sequence; and G/C-rich versus A/T-rich sequence. In addition, either suboptimal hybridization conditions for G/C-rich sequence must be used to detect an A/T-rich sequence, or A/T-rich probes must be increased in length to equalize hybridization conditions, thus reducing one's ability to detect changes in the A/T-rich sequence [42]. Hybridization analyses suffer limitations in detecting known mutations and have demonstrated a high accuracy on only 65% of the DNA surveyed due to a high hybridization stringency [43].

Enzymatic procedures have also been used for mismatch discrimination. These approaches include primer extension or minisequencing, and ligation of probes to sequence specific primers, using the genomic sequence as a hybridization template [44–47]. Primer extension, or minisequencing, involves the extension, on single-stranded, amplified genomic DNA, of a specific primer in the presence of polymerase and either fluorescent ddNTPs or 1 ddNTP and 3 dNTPs, and detection with gel or capillary sequencing or MALDI/TOF, respectively [37]. Ligation reactions usually require two adjacent primers to anneal to a genome-derived target. The upstream primer usually contains a label on the 5′ end and the 3′ nucleotide is designed to be opposite the SNP of interest. When the 3′ nucleotide forms a perfect match with the target, the primer, with label, is covalently attached by ligase to the downstream primer.

Detection is by fluorescent display on a microarray or by MALDI/TOF [37, 48–51]. One disadvantage in this procedure is the expense of using labeled specific primers for SNPs being screened.

A microchip-based multiplex SNP assay that combines the sensitivity and specificity of ligation with the cost-effective strategy of using a labeled commonoligonucleotide that is extended to the site of the match/mismatch is described next. This system can also be used for resequencing DNA by interrogating each nucleotide of interest. Briefly, a target sequence is amplified from the patient genomes of interest, with two specific primers, one containing a common tag for reamplification and detection (See Fig. 6.5). The common tag, which is added during amplification, provides the template for primer extension and an antisense sequence for labeled primer hybridization. The procedure consists of an on-chip hybridization of the combined single-stranded PCR product and labeled common primer, followed by a single-step extension/ligation reaction that can be accomplished with a DNA ligase such as *E. coli* DNA ligase and a polymerase such as Taq Stoffel fragment or reverse transcriptase (Fig. 6.6). The selection of appropriate enzymes for this combined reaction is critical. The polymerase must not have strand displacement or exonuclease activity and the ligase must be able to discriminate a mismatch. A final stringent wash step with NaOH removes unligated label and allows viewing of the SNP (Fig. 6.7) or sequence of interest (Fig. 6.8).

In many situations, identification of a pathogen is not sufficient and a specific gene sequence is required. For example, genotype Z, the dominant avian H5N1

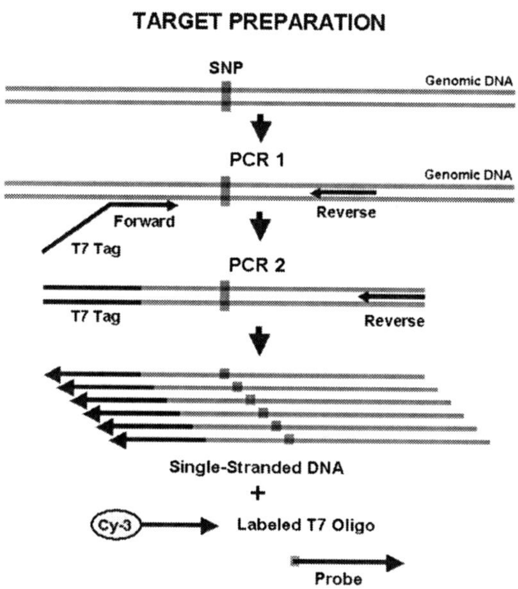


**Fig. 6.5** Diagram of target amplification for SNP analysis and resequencing chips. Briefly, the genomic DNA of interest is amplified with a pair of specific primers. The forward primer contains a tag sequence (T7) for labeling and primer extension with a Cy-5 or biotin labeled oligonucleotide. A second amplification with only the reverse primer results in single-stranded target for hybridization to the array probes

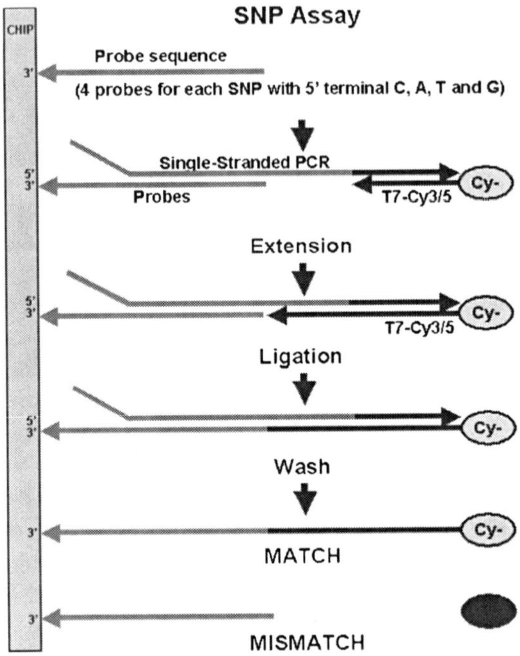


**Fig. 6.6** Diagram of a SNP assay (or resequencing assay). Single-stranded target (see Fig. 6.5) and labeled primer (T7-Cy3/5) are coincubated on the array. After annealing of target and labeled primer, a mixture of polymerase and ligase are added to extend the labeled primers and ligate to probes. The array is then washed with NaOH to remove all unligated label and then scanned for fluorescence (or current for electrochemical detection). Identification of SNPs or generation of sequence is accomplished by detection of signal associated with the correctly ligated primer as only one of four (or two of four in the case of a heterozygous SNP) probes will be ligated to the labeled extended primer

virus genotype currently circulating in Vietnam and Thailand, contains a mutation that is associated with resistance to amantadine and rimantadine [31, 52].

Antiviral therapies generally should be given within 48 hours of onset of illness to be effective against human influenza [31]. Thus rapid and specific identification of this subtype and accurate sequence information is crucial for proper treatment. Also, highly pathogenic strains of H5 and H7 influenza viruses can, in some cases, be distinguished from low pathogenic strains by sequencing the hemagglutinin gene, especially the area encoding the cleavage site [53].

With the enzyme-based SNP/resequencing assay described here, one is able to sequence approximately 500 or more nucleotides of genes of interest. This resequencing array and assay resulted in approximately 95% or more accurately called bases (998 of 1043 bases; [25]). Miscalls were predominately due to strong secondary structures, which can be predicted and avoided before the assay is carried out, and the ligated mismatches A/G, A/A, T/G, G/G, and T/T. These mismatches are generally more difficult to detect because of their low delta G values. For

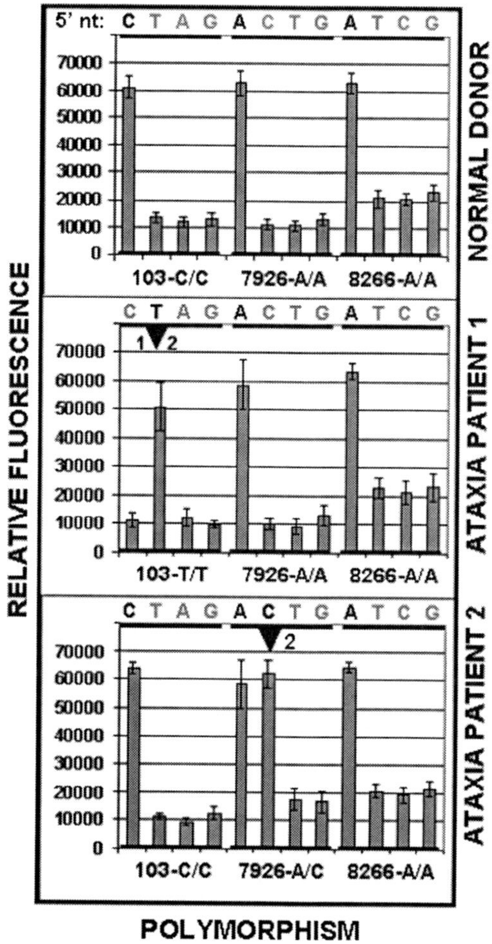


**Fig. 6.7** Results of a SNP assay on Ataxia patient genomic DNA. Three potential SNPs in the human Ataxia-telangiectasia-mutated (ATM) gene were interrogated from two patient genomic DNA samples and one control normal donor sample. Results show the wild-type sequence (first bars) in the normal donor DNA and either homozygous (patient 1, arrowhead) or heterozygous (patient 2, arrowhead) mutations in the ATM patient DNA. Signal intensity, indicated by relative fluorescence, is shown at the left and the locations of polymorphisms within the ATM gene, are shown at the bottom

example, sequence errors resulting from A/G mismatches represented 1.4% of the total errors or 6% of the potential A/G mismatches (15 of 251). Strong secondary structures (hairpins and palindromes) interfere with probe-target hybridization and result in a reduced signal. These sequencing arrays should contain either a consensus subtype sequence or a known subtype sequence that lacks a high degree of secondary structure. Replicate probes for each base of sequence should reduce artifacts due to difficult mismatches by averaging out mismatch signal.

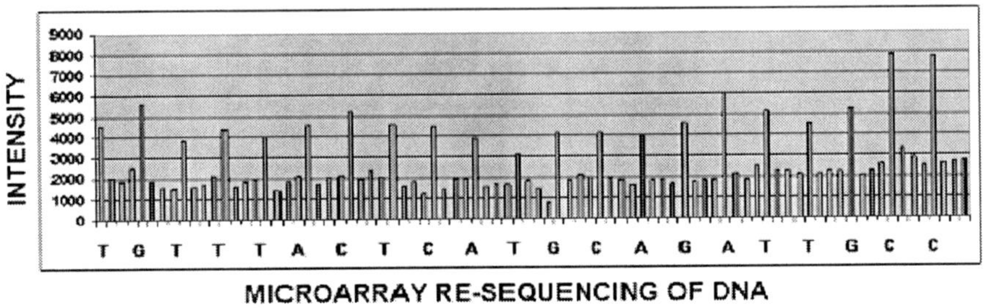


**Fig. 6.8** Example of a segment of DNA sequence from a resequencing array: this assay is essentially a SNP assay where each nucleotide in the sequence of interest is interrogated. Signal intensity is shown at the left and resulting sequence is shown below

Because microarray-based sequencing is based on probe-target hybridization, the target sequence cannot diverge significantly from the arrayed sequence. However, under nonstringent hybridization conditions, internal mismatches between probe and target sequence do not have as great an impact on hybridization and sequencing. This technique is best suited for resequencing, sequencing similar viruses such as seasonal quasispecies complexes, or for surveying mutations in an isolate over time.

## *6.3.1 Resequencing and SNP Assay*

The oligonucleotide probes synthesized on the resequencing microarray chip were 5′ phosphorylated with T4 polynucleotide kinase (PNK, New England Biolabs, Beverly, MA) for 30 min at 37°C. The array was then preblocked for 15 min at 45°C with 6 × SSPE containing 0.05% Tween-20 (SSPET), 2.0 mM EDTA, 5 × Denhardt's solution, and 0.05% SDS. Single-stranded target, with an antisense T7 tag added with the forward primer (5′ TAATACGACTCACTATAGGAG CAAAAGCAGG) during PCR as shown in Fig. 6.5 (see Section 6.1 for details), was heated to 95°C for 10 min, placed on ice and T4 ligase buffer added to 1 × concentration. A 5′ biotin-labeled T7 oligonucleotide (Integrated DNA Technologies, Inc., Coralville, IA) was added to a concentration of 1 μM to provide signal for detection and primer for extension. This solution was added to the chip array hybridization chamber and incubated at 45°C for 1 h.

After washing the array with 2 × PBS-0.5% Tween 20, 2 × PBS and 1 × *E. coli* ligase buffer, a mixture consisting of 1 × *E. coli* ligase buffer, 0.2 mM dNTP, 20 units each of AmpliTaq DNA polymerase, Stoffel fragment (Applied Biosystems, Foster City, CA), and *E. coli* ligase were added to the array and incubated at 37°C for 30 min. The array was washed twice for 2 min each with 0.1 N NaOH at room temperature, blocked and labeled as described in Section 6.1, and then scanned with ElectraSense (CombiMatrix Corp., Mukilteo, WA).

## 6.4 DNA Methylation

The methylation of DNA is a general mechanism for control of transcription in vertebrates. Cytosine residues in vertebrate DNA can be modified by the addition of methyl groups at the 5-carbon position. DNA is methylated specifically at the Cs that precede Gs in the DNA chain (CpG dinucleotides). This methylation is correlated with reduced transcriptional activity of genes that contain high frequencies of CpG dinucleotides in the vicinity of their promoters. Methylation inhibits transcription of these genes via the action of a protein, MeCP2, that specifically binds to methylated DNA and represses transcription [54]. The analysis of methylation patterns is also fundamental for the understanding of cell differentiation, X chromosome inactivation, regulation of developmental programming, aging processes, diseases, and cancer development [55–58].

Methylated DNA can be detected on oligonucleotide microarrays by using several techniques including methylation-specific restriction enzyme (MSRE) analysis [59], differential methylation hybridization (DMH) technique [60], and integrated analysis of methylation by isoschizomers [61]. One can also use standard hybridization to detect mismatches or the enzyme-based technique described in the SNP/resequencing section. For these methods, detection of methylated cytosines in genomic DNA involves PCR amplification of chemically modified DNA in which unmethylated cytosine residues have been converted to uracil by hydrolytic deamination, but methylated cytosine residues remain unconverted. Urea may improve efficiency of bisulphite-mediated sequencing of 5[prime]-methylcytosine in genomic DNA [62].

Briefly, genomic DNA is digested with endonucleases or sonicated to produce short fragments. Fragmented DNA is then denatured with sodium hydroxide and treated with sodium metabisulphite and hydroquinone and cycled at 55°C and 95°C in a thermocycler [63]. The bisulphite treated DNA is further processed by desalting, with a 3 M column, for example, and by elution in methanol, desiccation, and resuspension in water [64]. Finally the DNA is treated with sodium hydroxide to complete the conversion of unmethylated C residues to Ts. The conversion of unmethylated Cs and the nonconverted methylated Cs can be detected with SNP or resequencing arrays. Results of an enzyme-based assay used to detect methylated DNA are shown in Fig. 6.9.

### 6.4.1 *Methylation Assay*

CpG methylation is accomplished by sonicating 2 ug of genomic DNA to product fragments of approximately 500 to 1000 bp. Genomic DNA is diluted to 0.2 ug/ul and sonicated at a power setting of 3 for 10 s. Sonicated DNA is then purified with a Qiagen nucleotide removal column. The DNA is then denatured by the addition of 1/9 vol of freshly prepared 3 M sodium hydroxide and incubation for 15 min

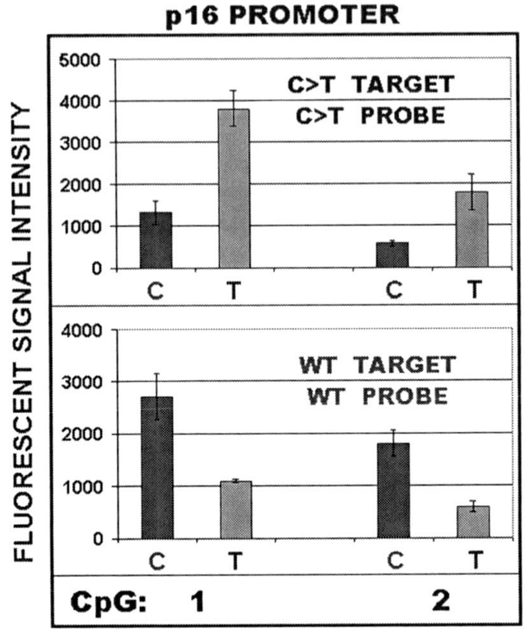


**Fig. 6.9** Illustration of a methylation assay to show the position of CpG dinucleotides. Random donor genomic DNA was sonicated and divided into two groups: wild-type (WT) and converted (C > T). For conversion, unmethylated cytosine residues are converted to uracil by hydrolytic deamination and methylated cytosine residues remain unconverted. The region of genomic DNA containing the CpG residues of interest was amplified from the wild-type and converted DNAs using the strategy illustrated in Fig. 6.5 and then hybridized to arrays designed to identify the wild-type Cytosine (C) residue or the converted Thymidine (T) residue. Fluorescence intensity is shown to the left and the CpG dinucleotides being assayed (1 or 2) are shown below. The top panel illustrates the results of hybridization of the converted target and the bottom panel illustrates results of hybridization of the wild-type target using the protocol described in Fig. 6.6

at 37°C. A 6.24 M urea/2 M sodium metabisulphite (4 M bisulphite) solution is made by dissolving 7.5 g of urea in 10 ml of sterile distilled water, adding 7.6 g of sodium metabisulphite (8.5 g sodium bisulphate), adjusting the pH to 5 with 10 M sodium hydroxide and adding sterile water to a final volume of 20 ml. The urea/bisulphite solution and 10 mM hydroquinone are then added to the denatured DNA to final concentrations of 5.36 M, 3.44 M, and 0.5 mM, respectively.

The reaction is performed in a 0.5 ml PCR tube overlaid with 100 μl of mineral oil and cycled 20 times at 55°C for 15 min followed by denaturation at 95°C for 30 s [63]. The bisulphite-treated DNA is desalted with a 3 M column: Add 9 parts NEN A (0.1 M *Tris* 7.7, 1 mM EDTA, and 10 mM TEA), load onto a 3 M EMPORE column and centrifuge into a 15 ml tube and then wash by centrifugation with 10 parts NEN A. Repeat centrifugation until dry. Elute in 500 ul NEN B

(50% MeOH/dH2O) and dry in a lyophilizer. Resuspend pellet in 90 ul $dH_2O$ and add fresh 3 M NaOH to a concentration of 0.3 M (1:10) and incubate at 37°C for 15 min. Precipitate by adding $NH_4OAc$, pH 7, to the 3 M column-purified material (3 M final) and precipitate with EtOH. Dry with a lyophilizer, resuspend in 100 ul TE buffer, and store at −20°C until analyzed. PCR conditions to produce single-stranded target from methylated DNA and wild-type DNA are described in Section 6.3 as for resequencing arrays.

**Acknowledgments** We would like to thank Jodi Dalrymple and Marty Ross for their contributions to this study.

## References

1. Fouchier R.A., V. Munster, A. Wallensten, T.M. Bestebroer, S. Herfst, D. Smith, G.F. Rimmelzwaan, B. Olsen, and A.D. Osterhaus. (2005) Characterization of a novel influenza A virus hemagglutinin subtype (H16) obtained from black-headed gulls. *J. Virol.* 79(5), 2814–2822.
2. Scholtissek, C., H. Burger, O. Kistner, and K.F. Shortridge. (1985) The nucleoprotein as a possible major factor in determining host specificity of influenza H3N2 viruses. *Virology*. 147(2), 287–294.
3. Mizuta, K., N. Katsushima, S. Ito, K. Sanjoh, T. Murata, C. Abiko, and S. Murayama. (2003) A rare appearance of influenza A(H1N2) as a reassortant in a community such as Yamagata where A(H1N1) and A(H3N2) co-circulate. *Microbiol. Immunol.* 47(5), 359–361.
4. Webby, R.J, and R.G. Webster. (2001) Emergence of influenza A viruses. *Phil. Trans. R. Soc. Lond. B*356, 1815–1826.
5. Allwinn, R., W. Preiser, H. Rabenau, S. Buxboum, M. Sturmer, and H.W. Doerr. (2002) Laboratory diagnosis of influenza—Virology or serology? *Med. Microbiol. Immunol. (Berl)* 191(3–4), 157–160.
6. Amano, Y., and Q. Cheng. (2005) Detection of influenza virus: Traditional approaches and development of biosensors. *Anal. Bioanal. Chem.* 381(1), 156–184.
7. Palmer D.F., M.T. Coleman, W.R. Dowdle, and G.C. Schild. (1975) Advanced laboratory techniques for influenza diagnosis. Immunology Series No. 6. U.S. Department of Health, Education, and Welfare. p. 51–52.
8. Ueda, M., A. Maeda, N. Nakagawa, T. Kase, R. Kubota, H. Takakura, A. Ohshima, and Y. Okuno. (1998) Application of subtype-specific monoclonal antibodies for rapid detection and identification of influenza A and B viruses. *J. Clin. Microbiol.* 36(2), 340–344.
9. Ellis, J.S., and M.C. Zambon. (2002) Molecular diagnosis of influenza. *Rev. Med. Virol.* 12(6), 375–389.
10. Korimbocus J., N. Scaramozzino, B. Lacroix, J.M. Crance, D. Garin, and G. Vernet. (2005) DNA probe array for the simultaneous identification of herpes viruses, enteroviruses, and flaviviruses. *J. Clin. Microbiol.* 43(8), 3779–3787.
11. Schweiger, B., I. Zadow, and R. Heckler. (2002) Antigenic drift and variability of influenza viruses. *Med. Microbiol. Immunol. (Berl)*. 191(3–4), 133–138.
12. Taubenberger, J.K., and S.P. Layne. (2001) Diagnosis of influenza virus: Coming to grips with the molecular era. *Mol. Diagn.* 6(4), 291–305.
13. Wang, D., L. Coscoy, M. Zylberberg, P.C. Avila, H.A. Boushey, D. Ganem, and J.L. DeRisi. (2002) Microarray-based detection and genotyping of viral pathogens. *PNAS.* 99(4), 15687–15692.
14. Zou, S. (1997) A practical approach to genetic screening for influenza virus variants. *J. Clin. Microbiol.* 35(10), 2623–2627.

15. Adeyefa, C.A., K. Quayle, and J.W. McCauley. (1994) A rapid method for the analysis of influenza virus genes: Application to the reassortment of equine influenza virus genes. *Virus. Res.* 32(3), 391–399.
16. Hoffmann, E., J. Stech, Y. Guan, R.G. Webster, and D.R. Perez. (2001) Universal primer set for the full-length amplification of all influenza A viruses. *Arch. Virol.* 146, 1–15.
17. Bodrossy, L., and A. Sessitsch. (2004) Oligonucleotide microarrays in microbial diagnosis. *Curr. Opin. Microbiol.* 7(3), 245–254.
17. Templeton, K.E., S.A. Scheltinga, M.F.C. Beersma, A.C.M. Kroes, and E.C.J. Claas. (2004) Rapid and sensitive method using multiplex real-time PCR for diagnosis of infections by influenza A and influenza B viruses, respiratory syncytial virus, and parainfluenza viruses 1, 2, 3, and 4. *J. Clin. Microbiol.* 42(4), 1564–1569.
18. Baxi, M.K., S. Baxi, A. Clavijo, K.M. Burton, and D. Deregt. (2006) Microarray-based detection and typing of foot-and-mouth disease virus. *Vet. J.* 172(3):473–481.
19. Ivshina, A.V., G.M. Vodeiko, V.A. Kuznetsov, D. Volokhov, R. Taffs, V.I. Chizhikov, R.A. Levandowski, and K.M. Chumakov. (2004) Mapping of genomic segments of influenza B virus strains by an oligonucleotide microarray method. *J. Clin. Microbiol.* 42(12), 5793–5801.
20. Kessler N., O. Ferraris, K. Palmer, W. Marsh, and A. Steel. (2004) Use of the DNA flow-thru chip, a three-dimensional biochip, for typing and subtyping of influenza viruses. *J. Clin. Microbiol.* 42(5), 2173–2185.
21. Li, J., S. Chen, and D.H. Evans. (2001) Typing and subtyping influenza virus using DNA microarrays and multiplex reverse transcriptase PCR. *J. Clin. Microbiol.* 39(2), 696–704.
22. Sengupta, S., K. Onodera, A. Lai, and U. Melcher. (2003) Molecular detection and identification of influenza viruses by oligonucleotide microarray hybridization. *J. Clin. Microbiol.* 41(10), 4542–4550.
23. Wang, D., A. Urisman, Y.-T. Liu, M. Springer, T.G. Ksiazek, D.D. Erdman, E.R. Mardis, M. Hickenbotham, V. Magrini, J. Eldred, J.P. Latreille, R.K. Wilson, D. Ganem, and J.L. DeRisi. (2003) Viral discovery and sequence recovery using DNA microarrays. *PLoS Biol.* 1(2), 257–260.
24. Roth, S.B., J. Jalava, O. Ruuskanen, A. Ruohola, and S. Nikkari. (2004) Use of an oligonucleotide array for laboratory diagnosis of bacteria responsible for acute upper respiratory infections. *J. Clin. Microbiol.* 42 (9), 4268–4274.
25. Lodes, M.J., D. Suciu, M. Elliott, A.G. Stover, M. Ross, M. Caraballo, K. Dix, J. Crye, R.J. Webby, W.J. Lyon, D.L. Danley, and A. McShea. (2006) Use of semiconductor-based oligonucleotide microarrays for influenza A virus subtype identification and sequencing. *J. Clin. Microbiol.* 44(4), 1209–1218.
26. Lodes, M.J., D. Suciu, J.L. Wilmoth, M. Ross, K. Dix, K. Bernards, A.G. Stöver, M. Quintana, N. Iihoshi, W.J. Lyon, D.L. Danley, and A. McShea. (2007) Identification of upper respiratory tract pathogens using electrochemical detection on an oligonucleotide microarray. *PLoS ONE.* 2(9), e924.
27. Dill, K., D.D. Montgomery, A.L. Ghindilis, K.R. Schwarzkopf, S.R. Ragsdase, and A.V. Oleinikov. (2004) Immunoassays based on electrochemical detection using microelectrode arrays. *Biosens. Bioelectron.* 20, 736–742.
28. Ghindilis, A.L., Smith, M.W., Schwarzkopf, K.R., Roth, K., Peyvan, K., Munro, S., Lodes, M.J., Stover, A., Bernards, K., Dill, K., and McShea, A. (2007) CombiMatrix oligonucleotide arrays: Genotyping and gene expression assays employing electrochemical detection. *Biosens. Bioelectron.* 22(9–10):1853–1860.
29. Liu, R.H., M.J. Lodes, T. Nguyen, T. Siuda, M. Slota, H.S. Fuji, and A. McShea. (2006) Validation ao a fully integrated microfluidic array device for influenza A subtype identification and sequencing. *Anal. Chem.* 2006 Jun 15; 78(12), 4184–4193.
30. Macken, C., H. Lu, J. Goodman, and L. Boykin. (2001) The value of a database in surveillance and vaccine selection. in *Options for the Control of Influenza IV*. A.D.M.E. Osterhaus, N. Cox, and A.W. Hampson (Eds.). Amsterdam: Elsevier Science, 2001, 103–106.

31. Yuen, K.Y. and S.S.Y. Wong. (2005) Human infection by avian influenza A H5N1. *Hong Kong Med. J.* 11(3), 189–199.
32. Helio, T., L. Halme, M. Lappalainen, H. Fodstad, P. Paavola-Sakki, U. Turunen, M. Farkkila, T. Krusius, and K. Kontula. (2003) CARD15/NOD2 gene variants are associated with familially occurring and complicated forms of Crohn's disease. *Gut*. 52(4), 558–562.
33. Newman, B., M.S. Silverberg, X. Gu, Q. Zhang, A. Lazaro, A.H. Steinhart, G.R. Greenberg, A.M. Griffiths, R.S. McLeod, Z. Cohen, M. Fernandez-Vina, C.I. Amos, and K. Siminovitch. (2004) CARD15 and HLA DRB1 alleles influence susceptibility and disease localization in Crohn's disease. *Am. J. Gastroenterol.* 99(2), 306–315.
34. Brant, S.R., C.I. Panhuysen, D. Nicolae, D.M. Reddy, D.K. Bonen, R. Karaliukas, L. Zhang, E. Swanson, L.W. Datta, T. Moran, G. Ravenhill, R.H. Duerr, J.P. Achkar, A.S. Karban, and J.H. Cho. (2003) MDR1 Ala893 polymorphism is associated with inflammatory bowel disease. *Am. J. Hum. Genet.* 73(6), 1282–1292.
35. Fang, N.Y., T.C. Greiner, D.D. Weisenburger, W.C. Chan, J.M. Vose, L.M. Smith, J.O. Armitage, R.A. Mayer, B.L. Pike, F.S. Collins, and J.G. Hacia. (2003) Oligonucleotide microarrays demonstrate the highest frequency of ATM mutations in the mantle cell subtype of lymphoma. *PNAS*. 100(9), 5372–5377.
36. Hacia, J.G., B. Sun, N. Hunt, K. Edgemon, D. Mosbrook, C. Robbins, S.P.A. Fodor, D.A. Tagle, and F.S. Collins (1998) Strategies for mutational analysis of the large multiexon ATM gene using high-density oligonucleotide arrays. *Genome Res.* 8, 1245–1258.
37. Kirk, B.W., M. Feinsod, R. Favis, R.M. Kliman, and F. Barany, (2002) Single nucleotide polymorphism seeking long-term association with complex diseases. *Nucleic Acids Res.* 30(15), 3295–3311.
38. Kwok, P.-Y. (2001) Methods for genotyping single nucleotide polymorphisms. *Annu. Rev. Genomics Hum. Genet.* 2, 235–258.
39. Jenkins, S., and N. Gibson. (2002) High-throughput SNP genotyping. *Comp. Funct. Genom.* 3, 57–66.
40. Shi, M.M. (2001) Enabling large-scale pharmacogenetic studies by high-throughput mutation detection and genotyping technologies. *Clin. Chem.* 47(2), 164–172.
41. Hacia, J.G. (1999) Resequencing and mutational analysis using oligonucleotide microarrays. *Nature Genetics*. Supplement 21, 42–47.
42. Urakawa, H., S. El Fantroussi, H. Smidt, J.C. Smoot, E.H. Tribou, J.J. Kelly, P.A. Noble, and D.A. Stahl. (2003) Optimization of single-base-pair mismatch discrimination in oligonucleotide microarrays. *Appl. Environ. Microbiol.* 69(5), 2848–2856.
43. Patil, N., A.J. Berno, D.A. Hinds, W.A. Barrett, J.M. Doshi, C.R. Hacker, C.R. Kautzer, D.H. Lee, C. Marjoribanks, D.P. McDonough, B.T. Nguyen, M.C. Norris, J.B. Sheehan, N. Shen, D. Stern, R.P. Stokowski, D.J. Thomas, M.O. Trulson, K.R. Vyas, K.A. Frazer, S.P. Fodor, and D.R. Cox. (2001) Blocks of limited haplotype diversity revealed by high-resolution scanning of human chromosome 21. *Science*. 294, 1719–1723.
44. Broude, N.E., T. Sano, C.L. Smith, and C.R. Cantor. (1994) Enhanced DNA sequencing by hybridization. *PNAS*. 91, 3072–3076.
45. Dubiley S, E. Kirillov, and A. Mirzabekov. (1999) Polymorphism analysis and gene detection by minisequencing on an array of gel-immobilized primers. *PNAS*. 27(18), e19.
46. O'Meara, D., A. Ahmadian, J. Odeberg, and J. Lundeberg. (2002) SNP typing by apyrase-mediated allele-specific primer extension on DNA microarrays. *NAR*. 30(15), e75.
47. Rickert, A.M., T.A. Borodina, E.J. Kuhn, H. Lehrach, and S. Sperling. (2004) Refinement of single-nucleotide polymorphism genotyping methods on human genomic DNA: Amplifluor allele-specific polymerase chain reaction versus ligation detection reaction-TaqMan. *Anal. Biochem.* 330(2), 288–297.
48. Zhong, X-b., R. Reynolds, J.R. Kidd, K.K. Kidd, R. Jenison, R.A. Marlar, and D.C. Ward. (2003) Single-nucleotide polymorphism genotyping on optical thin-film biosensor chips. *PNAS*. 100(20), 11559–11564.

49. Iannone, MA., J.D. Taylor, J. Chen, M.S. Li, P. Rivers, K.A. Slentz-Kesler, and M.P. Weiner. (2000) Multiplexed single nucleotide polymorphism genotyping by oligonucleotide ligation and flow cytometry. *Cytometry*. 39(2), 131–140.
50. Chen, X., K.J. Livak, and P.-Y. Kwok. (1998) A homogeneous, ligase-mediated DNA diagnostic test. *Genome Res*. 8, 549–556.
51. Consolandi, C., A. Frosini, C. Pera, G.B. Ferrara, R. Bordoni, B., Castiglioni, E. Rizzi, A. Mezzelani, L.R. Bernardi, G. DeBellis, and C. Battaglia. 2004. Polymorphism analysis within the HLA-A locus by universal oligonucleotide array. *Hum. Mutat*. 24(5), 428–434.
52. Chen, H., G. Deng, Z. Li, G. Tian, Y. Li, P. Jiao, L. Zhang, Z. Liu, R.G. Webster, and K.Yu. (2004) The evolution of H5N1 influenza viruses in ducks in southern China. *PNAS*. 101(28), 10452–10457.
53. Lee, C-W., D.E. Swayne, J.A. Linares, D.A. Senne, and E.L. Suarez, (2005) H5N2 Avian influenza outbreak in Texas. *J. Virol*. 79(17), 11412–11421.
54. Cooper, G.M. (2000) *The Cell: A Molecular Approach*. Second edition, Boston University. Sinauer Associates, Sunderland, MA.
55. Schumacher A., P. Kapranov, Z. Kaminsky, J. Flanagan, A. Assadzadeh, P. Yau, C. Virtanen, N. Winegarden, J. Cheng, T. Gingeras, and A. Petronis. (2006) Microarray-based DNA methylation profiling: Technology and applications. *Nucleic Acids Res*. 34(2), 528–542.
56. Gebhard C., L. Schwarzfischer, T.H. Pham, E. Schilling, M. Klug, R. Andreesen, and M. Rehli. (2006) Genome-wide profiling of CpG methylation identifies novel targets of aberrant hypermethylation in myeloid leukemia. *Cancer Res*. 66(12), 6118–6128.
57. Lewin, J., A.O. Schmitt, P. Adorjan, T. Hildmann, and C. Piepenbrock. (2004) Quantitative DNA methylation analysis based on four-dye trace data from direct sequencing of PCR amplificates. *Bioinformatics*. 20(17), 3005–3012.
58. Galm O., J.G. Herman, and S.B. Baylin. (2006) The fundamental role of epigenetics in hematopoietic malignancies. *Blood Rev.*. 20(1), 1–13.
59. Adrien L.R., N.F. Schlecht, N. Kawachi, R.V. Smith, M. Brandwein-Gensler, A. Massimi, S. Chen, M.B. Prystowsky, G. Childs, and T.J. Belbin. (2006) Classification of DNA methylation patterns in tumor cell genomes using a CpG island microarray. *Cytogenet Genome Res*. 114(1), 16–23.
60. Yan P.S., C.M. Chen, H. Shi, F. Rahmatpanah, S.H. Wei, and T.H. Huang. (2002) Applications of CpG island microarrays for high-throughput analysis of DNA methylation. *J. Nutr*. 132(8 Suppl), 2430S–2434S.
61. Hatada, I., M. Fukasawa, M. Kimura, S. Morita, K. Yamada, T. Yoshikawa, S. Yamanaka, C. Endo, A. Sakurada, M. Sato, T. Kondo, A. Horii, T. Ushijima, and H. Sasaki. (2006) Genome-wide profiling of promoter methylation in human. *Oncogene*. 25(21), 3059–3064.
62. Paulin R., G.W. Grigg, M.W. Davey, and A.A. Piper. (1998) Urea improves efficiency of bisulphite-mediated sequencing of 5'-methylcytosine in genomic DNA. *Nucleic Acids Res*. 26(21), 5009–5010.
63. Rein, T., H. Zorbas, and M. DePamphilis. (1997) Active mammalian replication origins are associated with a high-density cluster of mCpG dinucleotides. *Mol. Cell Biol*. 17, 416–426.
64. Clark, S.J., J. Harrison, C.L. Paul, and M. Frommer, (1994) High sensitivity mapping of methylated cytosines. *Nucleic Acids Res*., 22, 2990–2997.
65. Lipatov, A.S., E.A. Govorkova, R.J. Webby, H. Ozaki, M. Peiris, Y. Guan, L. Poon, and R.G. Webster. (2004) Influenza: Emergence and control. *J. Virol*. 78(17), 8951–8959.

# Chapter 7
# Peptide-Based Microarray

**Resmi C. Panicker, Hongyan Sun, Grace Y. J. Chen, and Shao Q. Yao**

**Abstract** The peptide array has come into focus as an emerging screening platform for large-scale protein detection and activity studies. The materials presented in this chapter examine the recent developments in the field of peptide microarrays with special emphasis on the generation and applications of high-density arrays of peptides on glass slides.

## 7.1 Introduction

Undoubtedly, the human genome project has accelerated the pace of discovery and made it possible for the identification of thousands of new genes that are critical to diseases [1]. By studying the dynamic description of gene regulation, proteomics offers a powerful tool to unravel gene functions, holding promises to have a significant impact on our understanding of the molecular composition and function of cells. Furthermore, by studying the whole array of proteins in a cell, tissue, or an organism at any given time, it provides a global integrated view of disease states and cellular processes. A variety of proteomic techniques has been deployed, especially those based on the traditional separation technique, 2D-GE. The ability of this technique to separate thousands of proteins in a specific cell or tissue, including their posttranslationally modified forms, has enabled it to become a major separation technique in protein analysis. And thus it is well suited for the global analysis of protein expression in an organism. However, 2D-GE suffers from a number of long-standing problems, including low throughput, a limited dynamic detection range, poor reproducibility, low sensitivity, as well as difficulties in analyzing hydrophobic, small, and very basic or acidic proteins.

---

S.Q. Yao
Departments of Chemistry and Biological Sciences, Medicinal Chemistry Program of the Office of Life Sciences, National University of Singapore, 3 Science Drive 3, Singapore 117543, Republic of Singapore

K. Dill et al. (eds.), *Microarrays: Preparation, Microfluidics, Detection Methods, and Biological Applications*, 

Incremental improvements in 2D-GE technology, including the use of sensitive staining methods and higher-resolving gels, and sample fractionation prior to 2D-GE, have alleviated, but not eliminated, some of these problems [2, 3]. Difference gel electrophoresis, namely DIGE, [4], multiplexed proteomics approach [5], multi-dimensional protein identification technology [6], isotope-coded affinity tagging [7], and so on, are some other related developments. Most of these technologies require downstream instrumentations such as mass spectrometry in order to identify the proteins of interest individually. They are therefore time-consuming and not easily automatable. In addition to these difficulties, these techniques mostly examine the relative abundance of proteins in a proteome. However, this information is not sufficient to describe the various processes taking place in the cell.

One of the main challenges present in proteomics is to fathom the function and activity of proteins, especially that of enzymes which account for >20% of all drug targets [8]. It is well known that many human diseases are caused by the aberrant regulation of enzymes inside cells [9]. As a matter of fact, proteomics and drug discovery rely more on technologies that will enable the complete characterization and activity studies of enzymes rather than techniques evaluating the relative abundance of proteins in cell. At the Eleventh Asian Chemical Congress held in Seoul, we introduced the term "catalomics" [10] to draw attention to another emerging field in "-omics" in which the high-throughput functional analysis of enzymes is carried out using chemical proteomic approaches (Fig. 7.1). Emerging technologies, especially array-based ones hold the potential to rapidly profile the entire proteome revealing novel protein functions and comprehensive protein interaction networks of an organism.

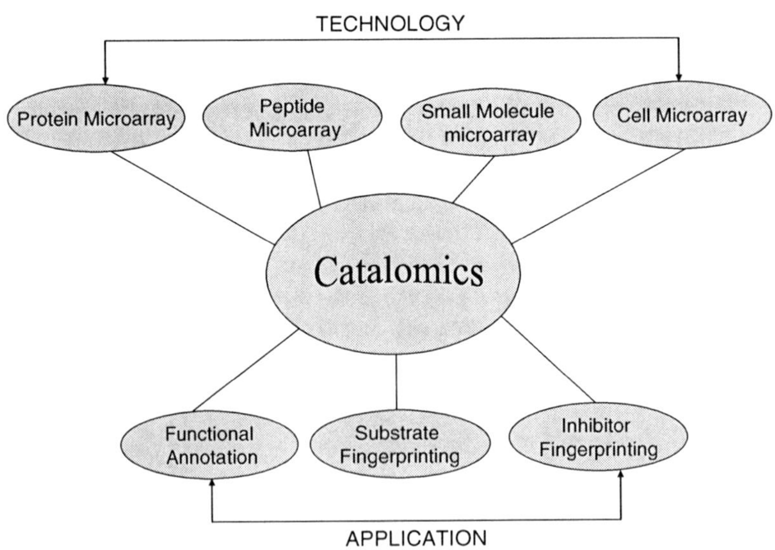


**Fig. 7.1** Overview of catalomics

Microarrays have rapidly attracted the interest of scientists, due to their ability to reproducibly analyze enzyme activities with high fidelity in high-throughput screenings. This development is significant, especially in the postgenomic era, where the research focus is shifting from protein identification to protein functional analysis and from the study of single proteins to that of multiple proteins. The salient feature of these microarray-based platforms lies in their ability to perform simultaneous screening of a variety of molecules in a single step, thus allowing rapid data acquisition and parallel sample comparisons. Another advantage is that the method requires only minute quantities of expensive reagents for most biological assays.

Protein-profiling arrays can be fabricated mainly based on different types of platforms such as protein microarrays, peptide microarrays, small molecule microarrays, and cell arrays [11]. Although the DNA microarray has enabled the high-throughput analysis of genetic materials in a miniaturized format, the development of microarrays of other biomolecules such as proteins, peptides, and small molecules have been much more challenging, especially when these arrays are to be used for enzyme profiling experiments. This is primarily due to the fact that successful enzymatic catalysis requires a number of criteria to be satisfied, including the proper folding/orientation of the immobilized enzyme/substrate, buffer composition of the reaction, the presence of cofactors, pH, temperature, and so on. Unlike DNA, which is highly stable and robust, enzymes are known to lose their functional activity upon immobilization onto a solid surface. Furthermore, there is presently no known technique that can effectively amplify proteins.

Another aspect is that the methods for protein expression do have several limitations. The inevitable chemical, physical, and structural variation among different proteins results in their nonspecific absorption to solid surfaces, thus creating further problems for their immobilization in a microarray. Alternatively, enzyme profiling could be done with a peptide microarray in which potential peptide substrates of enzymes are immobilized. This "sister platform" is currently being exploited for the high-throughput evaluation of complex protein functions. The value of synthetic peptides for the analysis of biological phenomena in general and especially the immune response of higher organisms have been proven for decades.

Peptides are per se privileged as biologically active molecules and they have become accessible in an almost unlimited structural diversity through combinatorial chemistry and molecular biology [12, 13]. Peptides are chemically quite resistant compounds. Unlike proteins they require no problematic folding into stable active conformations. Thus, peptides are perfect probe molecules for a robust screening assay. Their chemistry is well established and routine. Chemical synthesis has virtually no limitation concerning the incorporation of unnatural building blocks such as D-amino acids, organic residues, branched, and cyclized structures. The last decade witnessed a tremendous upswing in the development of peptides displayed in microarray formats as a technological means for various high-throughput enzyme activity studies, or "catalomics." In addition, peptide arrays have also been employed in the rapid discovery of drug candidates as well as in the design and selection of effective substrates, ligands, or inhibitors for a variety of enzymes and antibodies.

This contribution reviews the progressive development of peptide arrays through the past few years, novel chemistries that have facilitated the display of peptides on glass in a high-density array format and finally the applications of peptide arrays in the detection and activity studies of various classes of enzymes.

## 7.2 Origin and Development of Peptide Arrays

Existing microarray-based technologies are based on the idea of the DNA microarray, which found its roots in the form of Southern blots, a technique that has been widely used in traditional biology labs for decades. Invented in 1975 by Ed Southern and coworkers, the technique relies on the transfer of electrophoretically separated DNA onto a membrane before hybridization with specific oligonucleotide probes in order to identify potential gene targets [14]. The year 1984 witnessed the beginning of combinatorial synthesis of peptides when Geysen et al. first reported the peptide library synthesis by multipin technology [15]. Thus the chemical array was first described as a form of combinatorial library method. Peptide libraries were synthesized in 96-well reaction chambers and hence each peptide was pure and well defined. These peptides were used for B-cell epitope-mapping. However, this method was able to generate only a small set of peptides and hence was useful only for low-throughput applications. Several years later, Frank et al. modified this approach by synthesizing and analyzing multiple peptides as spots on cellulose paper, the so-called SPOT libraries method [16]. The array was screened with an enzyme-linked calorimetric assay and individual color beads were then physically isolated for sequencing.

In 1991, Fodor et al. reported the use of the light-directed, spatially addressable parallel chemical synthesis method to generate peptide microarrays [17]. The utility of this novel peptide array was demonstrated by the successful identification of preferred peptide sequences that bind strongly to a fluorescently labeled antibody. Ironically, it was the development of the DNA microarray, not the peptide array, that benefited greatly from Fodor's idea. Further development of this innovative technology in the field of peptide arrays was held back by the poor quality of peptides synthesized on the glass surface, primarily due to the relatively low efficiency in peptide deprotection/coupling steps. The potential extension of this approach to the fabrication of oligonucleotide-based arrays was further illustrated by the light-directed synthesis of a dinucleotide, which later on became the core technology of Affymetrix in the commercialization of the GeneChip™ [18, 19].

Alternatively, Lam et al. described the "one-bead–one-compound" (OBOC) combinatorial peptide library method [20, 21]. An OBOC combinatorial bead library can be considered as a huge spatially separable but nonaddressable chemical microarray because only a single peptide entity is displayed on each bead. This array was screened by an enzyme-linked calorimetric method. Subsequently, beads containing the desired peptides were picked and identified by peptide microsequencing. About the same time, Houghten et al. reported the use of a tea bag technique to synthesize peptide mixtures, and the use of an iterative approach or positional scanning approach to analyze these libraries [22]. Since then, the field of combinatorial chemistry has taken off and has now become a major discipline.

By taking advantage of Brown's spotting method, Schreiber et al. first described the successful generation of high-density microarrays made of nonnucleic acid biomolecules, including peptides [23, 24]. This approach made it possible for potential immobilization and simultaneous screenings of tens of thousands of peptide substrates on a small glass surface. Later, Falsey et al. reported a peptide array which made use of the regiospecific chemical reactions to immobilize *N*-terminal residues on glass slides [25]. Houseman et al. developed a novel class of peptide chips using modified peptides immobilized on a glass surface coated with self-assembled monolayers of alkane dithiols and used antibodies for the detection of peptide phosphorylation in the peptide array [26, 27].

Simultaneously, our group offered two new approaches for modifications of the microarray technique based on the chemistries involved in the native chemical ligation and biotin–avidin interaction [28, 29]. We exploited the resulting peptide array and subsequently developed a fluorescence-based assay for the rapid profiling of kinase activity in a microarray format [28, 29]. Further extension of the strategy by combining the benefits of combinatorial concepts with microarray-based screenings indicated that different forms of combinatorial peptide libraries (e.g., positional scanning, alanine scanning, deletion, etc.) may be introduced on a peptide microarray, thus providing further throughputs of the strategy [30].

Ellman et al. used an elegant approach in which an array of fluorogenic protease substrates were immobilized on aldehyde derivatized glass slides via chemoselective oxime bond formation reaction [31]. A 361-member spatially addressable peptide library was synthesized and immobilized on an aldehyde-derivatized glass slide. Using a similar approach, we recently extended this strategy to generate a series of substrates for different classes of hydrolytic enzymes, and the resulting conjugates have been immobilized on a glass slide to generate a small molecule-based microarray capable of sensitive detection of different hydrolytic enzymes [32]. Another elegant work in site-specific immobilization of peptides onto a microarray is the strategy based on a traceless version of the Staudinger ligation [33].

A novel method for the site-specific immobilization of peptides and oligonucleotides to semicarbazide glass slides was reported by Oliver et al. [34]. Keating et al. first reported the use of an array of coiled-coil peptides to study the dimerization of leucine zipper (bZIP) transcription factors [35]. In their case, the peptides used to generate the corresponding array were made by recombinant approaches, rather than chemical synthesis. Recently, supramolecular hydrogels were reported as an array platform to provide a semiwet environment for the immobilized peptides/proteins, making them more compatible with enzyme assays [36].

Other recent reports in the related fields are the construction of spatially defined arrays of peptoids using photolithographic synthesis by Li et al. [37] and the synthesis of metal chelating hexapeptide on a chip by Cheng et al. [38]. Usui et al. recently presented a novel dry peptide array system as a promising tool for detecting and analyzing target proteins. This system is the first example of an array preparation and assay procedure under dry conditions that uses designed peptides as nonimmobilized capture agents for the detection of proteins [39].

A recent addition to the affinity-based protein immobilization strategy is based on the combined use of the carbohydrate binding module (CBM) and its natural three-dimensional substrate such as cellulose [40]. CBM-fused antibodies or peptides were produced, immobilized on cellulose surfaces, and used for serodiagnosis of human immunodeficiency virus patients. The disadvantage of this approach relies on the fact that proteins must be biotinylated or tagged. Recently, by comparing the catalytic activities of enzymes immobilized on silicon surfaces with and without proper orientation Cha et al. [41], established the crucial role of the probe orientation for a reliable use of protein microarrays as quantitative tools in biomedical research.

Last year also witnessed a search for novel planar surfaces for peptide microarray applications by various groups [42]. Recently, time-resolved SPR measurements in an imaging format have been used to obtain multiplexed kinetic information about protein-peptide adsorption and surface enzyme reactions. In this report, Wegner et al. [43] used peptide chips fabricated on glass slides coated with a thin film of gold. More recently, an elegant work from the Mrksich group has demonstrated microfluidic analysis of kinases using a peptide array [44]. This work presents a brilliant example where integration of microfluidic devices with microarray technologies may potentially provide a means for further miniaturization in high-throughput assays leading to nanoarrays.

## 7.3 Generation of Peptide Arrays

There are two kinds of peptide microarray synthesis methodologies: parallel on-chip synthesis by photolithographic or SPOT methods and spotting array methods. Parallel on-chip synthesis relies on in situ generation of peptides on the chip surface (Fig. 7.2). The spotting array involves presynthesis of an array of compounds and their direct transfer plus immobilization onto a glass or membrane substrate. The latter approach is more versatile and can be applied to generate an array of almost any molecule.

### *7.3.1 Parallel On-Chip Synthesis*

#### 7.3.1.1 Light-Directed Parallel Chemical Synthesis

The very first report about the peptide array was from Fodor's group [17]. They devised a new strategy for the simultaneous generation of thousands of peptides on a glass slide. By adopting various attributes of solid-phase chemistry and photolithography they demonstrated a light-directed spatially addressable parallel

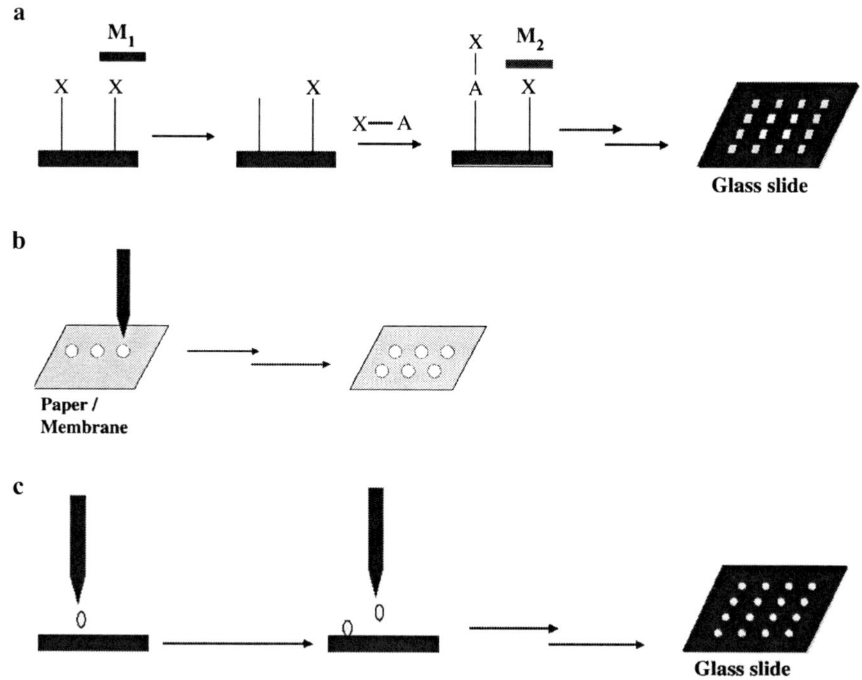


**Fig. 7.2** Schematic representation of (a) in situ synthesis of peptides using photolithographic technique; (b) in situ generation of peptides via SPOT™ synthesis; (c) spotting microarray approach

chemical synthesis to yield a diverse set of chemical products. By using NVOC, a photolabile group, as the *N*-terminal amino acid protecting group, and the glass surface as the solid support, the authors were able to use solid-phase peptide synthesis to chemically generate thousands of peptides on the glass slide simultaneously. Each coupling cycle of the peptide synthesis was precisely controlled by a set of photomasks with predefined configurations that allow for selective deprotection of the *N*-terminal amino group of the growing peptide chain, leading to selective coupling of different amino acids onto different peptides.

Immunobiochemical assays were used to screen for peptide sequences on the chip capable of binding to antibodies with both high affinity and specificity. With this, it was demonstrated that, for the very first time, thousands of μm-size spots of molecules may be generated within a small dimension, leading to potential miniaturization and high-throughput screenings of biological assays. The union of chemistry and semiconductor techniques laid the solid foundation for the revolution of DNA microarrays in biology. The following two years witnessed the generation of other types of arrays composed of peptidomimetic unnatural biopolymers, and oligonucleotides [45–51].

Surprisingly, this elegant technique has not advanced as much as expected for peptide synthesis because the focus mainly turned to synthetic oligonucleotide arrays. One reason for this could be the technical difficulties to construct arrays with 20 building blocks which need 20 special, photolabile protected amino acid derivatives and 20 different masks for each monomer elongation cycle (e.g., 20 × 10 = 200 masks for the synthesis of an array of decamer peptides). On the other side in the case of DNA microarrays, only 4 masks are needed for each coupling cycle. Furthermore, peptide synthesis in general is much less efficient than the oligonucleotide synthesis, making it extremely difficult to generate high-quality peptides on the glass slide.

Since this first report, a number of modified on-chip synthetic strategies have been reported. A maskless microarray setup was reported by Gao et al. to generate peptide and oligonucleotide arrays with the use of digital photolithography and photogenerated acids [52]. The approach used a conventional peptide synthesizer and a computer-controllable optical module, thus making it affordable for most research laboratories to access the powerful photolithographic techniques.

The most recent development related to photolithographic synthesis is in the generation of an array of peptoids (oligomers of *N*-substituted glycines) using a single photolabile synthon [37]. Kodadek et al. successfully inserted a light-dependent process into the generation of peptoids via the "submonomer" route. A four-step cycle was developed for the photolithographic synthesis of peptoids. First, glycolic acid protected with a light-sensitive MeNPOC group was coupled to an amine-modified surface. The hydroxyl group was unmasked by UV irradiation and then activated with tosyl chloride. Finally, the tosylate was displaced with a primary amine to complete the construction of a monomer unit. This chemistry should allow the spatially addressable synthesis of peptoids on an array by photolithography, because hydroxyl group unmasking, activation, and amine displacement will occur only at "addresses" that have been irradiated with UV light. Although recent advances in in situ synthesis are truly impressive, the chemistry of the in situ synthesized approach is more limited, particularly when photochemical reaction is a required synthetic step. As a result, only small peptides are used in the in situ synthesized arrays due to synthetic challenges on problems such as supports purity, stability, and so on.

#### 7.3.1.2 SPOT™ Synthesis

In 1990 a technique for simultaneous parallel chemical synthesis on membrane support, so-called SPOT synthesis, was presented by Frank et al. [16], which became a quite popular tool for studying numerous aspects of molecular recognition [53]. For several years, synthesis kits, membranes, and custom-made membrane-bound peptide arrays have been commercially available and have proven their reliability. The introduction of a semiautomated SPOT synthesizer in 1993 by the company ABIMED (Langenfeld, Germany) resulted in the creation of a number of opportunities for the facile and rapid synthesis of a large number of peptides on membrane or paper. The technique comprises the dispensing of a small volume of solutions containing Fmoc-amino acids and other coupling reagents to a designated spot on the membrane and subsequent deprotecting and coupling steps for the target molecule. The membrane is thoroughly washed and the peptides deprotected prior to the next coupling cycle. Many novel applications and assay principles concerning the use of peptide arrays for functional analysis of proteins have been developed based on this format.

The benefits of SPOT synthesis are (1) availability of well-established and convenient methods for the generation of arrays, (2) easy adaptability of the strategy to a wide range of assay and screening methods. This peptide array format has been modified and applied for the functional studies of different classes of proteins [54–56]. Nevertheless, the density of the spots obtained by this method is relatively low (e.g., 25 spots per $cm^2$) which limits the efficacy of this method for high-throughput applications.

## *7.3.2 Spotting Microarray Approach*

Complementary to the SPOT technology which describes the generation and screening of peptides as spots on membranes or paper, this approach provides another platform for epitope-mapping and performing enzymatic assays. With recent advances in microfabrication processes and surface modification technologies, the spotting array format has become a common tool for high-throughput screening these days. In contrast to the in situ synthesis methods, this approach makes use of the high-throughput spotting of presynthesized peptide products by an automatic arrayer onto a suitably derivatized solid surface. This approach is far more efficient because each compound needs to be synthesized only once, and multiple replicates can be produced simply by spotting.

Development of more efficient solid-phase synthesis such as new solid supports, linkers, and the like, novel peptide coupling chemistries, automated synthesis systems, and so on, have made acceleration of peptide library synthesis a lot easier. The peptide libraries can be prepared using different methods such as parallel synthesis, split and mix synthesis, and reagent mixture synthesis, among others. The choice of a suitable immobilization surface is also equally important. These are the fields where researchers focus on generating different chemical surfaces that allow efficient protein/peptide immobilzation by means of appropriate functional groups present on these biomolecules.

Conventional immobilization surfaces such as polystyrene, polyvinylidene fluoride, agarose thin film, nitrocellulose membranes, and so on were not chosen for a microarray format, primarily because these surfaces use noncovalent forces such as hydrophobic interactions for immobilization, resulting in the generation of low-density arrays of biomolecules that are randomly oriented on the surface. As a consequence, these surfaces often give rise to relatively low signal-to-noise ratios in downstream protein/peptide screening assays. Glass slides, however, present the ideal surface for microarray applications inasmuch as they are relatively inexpensive, have low intrinsic fluorescence, and also provide a homogeneous chemical surface suitable for immobilizing biomolecules at very high densities.

In order to be widely useful for the preparation of a variety of biochips, an immobilization method should have several characteristics. Firstly, the chemical reaction should occur rapidly at lower concentrations of reagents used. Secondly, the chemistry should require little postsynthetic modification of ligands before immobilization. Finally, the immobilization process should occur specifically in the presence of common functional groups present in peptides such as amines, thiols, carboxylic acids, alcohols, and so on. In addition, the reaction should have well-behaved kinetics and should be able to monitor easily with conventional spectroscopic methods. Several groups have reported immobilization chemistries that possess one or more of the above features (Fig. 7.3).

A promising approach from the Schreiber group led to the successful generation of a high-density microarray of proteins and peptides spotted with an automatic robotic arm [24]. This approach virtually made it possible for the potential immobilization and simultaneous screenings of tens of thousands of peptide substrates on a small glass surface. When compared to the SPOT technology and Fodor's techniques, Schreiber's approach was highlighted by its easy adoption to the generation of microarrays made of virtually any kind of biomolecules. To achieve beneficial features such as accomplishing the optimum spatial orientation to yield the maximum interaction between proteins and their ligands, minimal nonspecific protein absorptions to the surface of immobilized peptides, and so on, the authors introduced a molecular layer of BSA on the glass surface, followed by derivatization of the BSA with *N*-hydroxysuccimide (NHS). Here the spotting of the peptides allowed the covalent reaction between the reactive NHS group on the BSA surface and the nucleophilic groups in the peptide (–NH2, –SH, –OH, etc.). The limitation of the aforementioned strategy was the nonsite-specific immobilization of peptides on the glass surface. This could pose a critical issue if one needed to generate a highly dense array of peptides and, at the same time, retain their biological activities. Although unlike their counterparts, proteins, peptides do not typically possess well-defined three-dimensional structures, immobilization of peptides onto the glass surface with a correct orientation is still imperative for the peptides to interact effectively with their targeting proteins.

Falsey et al. [25] conceived a method using *N*-terminal cysteine-containing peptides immobilized on glyoxylic acid functionalized slides via oxime bond formation or thiazolidine ring formation. This versatile strategy allowed regiospecific covalent immobilization of the peptides on the glass slide. Commercially available slides were derivatized with APETS to form amine slides that were then converted to glyoxyl derivatized glass slides via the sequential steps of (1) coupling of Fmoc-Ser followed by deprotection and oxidation and (2) coupling with protected glycolic acid and deprotection. Although the above report is the first example of chemoselective ligation of peptide on a glass surface, the relative instability of the oxime bond as well as the unfavorably restricted orientation due to the thiazolidine ring formation virtually hindered the free interaction of the peptides with their target proteins and become the major impediments of the strategy.

Mrksich and colleagues used the classic Diels–Alder reaction to couple *N*-terminal cyclopentadiene-containing peptides onto quinone groups on the SAM

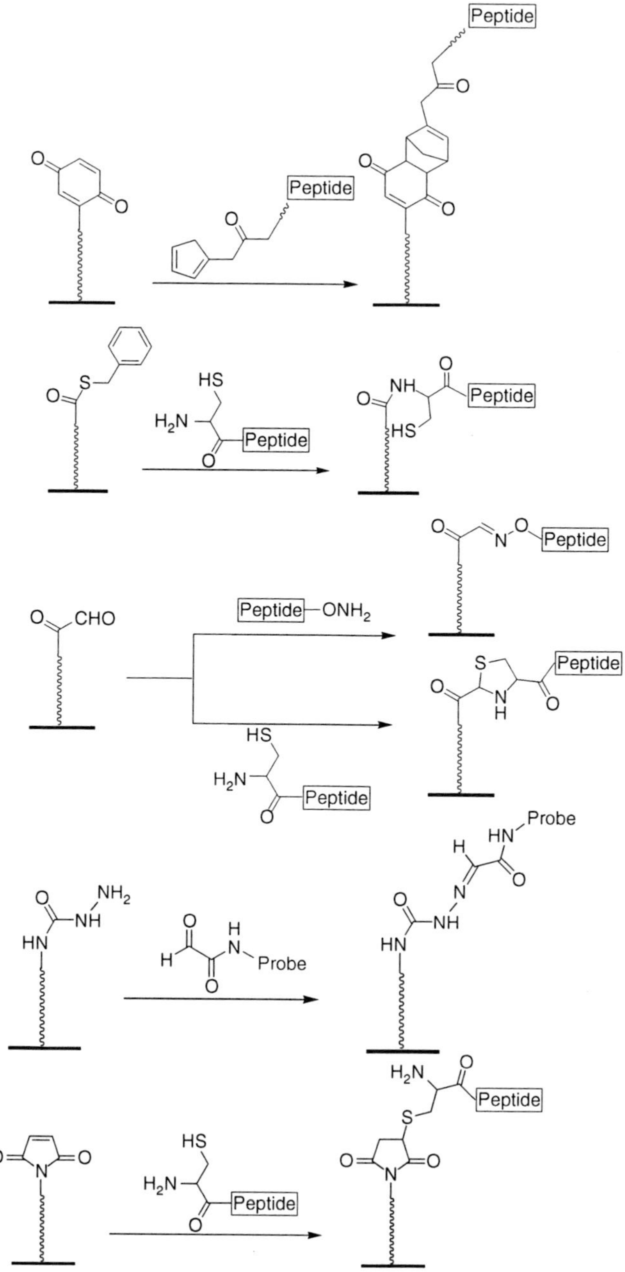


**Fig. 7.3** Chemoselective immobilization chemistries in peptide microarrays

surface [26]. Peptides conjugated to a cyclopentadiene linker were applied to a self-assembled monolayer of alkanethiols derivatized with benzoquinone groups. The Diels–Alder reaction occurs between the diene and quinone groups resulting in the rapid and selective immobilization of the peptides on the glass. This method has several advantages over the other existing methods as the self-assembled monolayers are inert toward nonspecific interactions with proteins thus eliminating the need of blocking. The method also provides a regular homogeneous environment for immobilized peptide ligands which makes them well suitable for quantitative assays.

However, this method requires the conjugation of peptides with an unnatural cyclopentadiene moiety, which is synthetically challenging and not easily accessible. The strategy was later modified by developing a class of maleimide-terminated self-assembled monolayers for the immobilization of peptides and carbohydrates [27]. The strategy makes use of the reaction between thiol and maleimide groups to immobilize ligands to self-assembled monolayers of alkanethiolates on gold. This approach does not require the postsynthetic modification of ligands as did their earlier strategy. SPR spectroscopy was used to measure the real-time interaction between immobilized peptides and carbohydrates with proteins.

By exploiting the same interaction, recently Kodadek et al. fabricated peptoid arrays to obtain "molecular fingerprints" against three fluorescently labeled proteins, namely GST, maltose-binding protein, and ubiquitin [57]. Octameric peptoids with C-terminal cysteines were immobilized on maleimide-functionalized glass slides and screened against the target proteins. A unique protein fingerprint was generated for each protein across the whole peptoid array. Their findings also revealed that the unique fingerprint of GST can be discerned when it arrayed in the presence of a large excess of other bacterial proteins. This report provides a valuable asset for identification of proteins through unique small-molecule fingerprinting and may have diagnostic use to identify cancer or other disease states.

To minimize the overwhelming shortcomings such as unstable peptide attachment and the "orientation and effective interaction" problem in peptide microarrays, we recently culminated two new approaches for site-specific immobilization of peptides on a glass plate [28, 29]. The crux and essence of the new strategies relies on the chemistries used, the native chemical ligation, and the biotin–avidin interactions. The first method makes use of a chemical ligation reaction between *N*-terminal cysteine residues engineered on the peptide substrates and the thioester moieties displayed on the array platform. The most noteworthy advantage of this method is its simplicity in that (1) *N*-terminal cysteine containing peptides can be easily synthesized using standard peptide synthesis protocols with no extra modifications needed, (2) a thioester-derivatized glass surface may be easily generated, and (3) the native chemical ligation reaction occurs with free peptides under physiological conditions with no external reagents needed.

Another attribute of this immobilization method is the exclusive formation of a highly stable, native peptide bond between the peptide and glass. As a proof-of-concept experiment, *N*-terminal-cysteine-containing kinase peptide substrates were immobilized onto a thioester-functionalized slide, and their activity was probed with the corresponding kinases, followed by successful detection with

FITC-labeled antiphosphotyrosine and antiphosphoserine [28, 29]. In order to minimize nonspecific bindings of the glass surface, a molecular layer of polyethylene glycol (PEG) was incorporated on the slide surface before attaching the thioester moieties. The use of the PEG layer was advantageous inasmuch as, unlike BSA or other macromolecular "cushions," it did not prevent the immobilized peptides from interacting with incoming targets. We subsequently used these peptide arrays to detect various enzymatic activities. Alanine-scanning, deletion, and positional-scanning peptide libraries of a kinase peptide substrate were site-specifically arrayed onto glass slides [30].

The second approach exploited biotin–avidin interaction, one of the strongest known noncovalent interactions (of $K_d = 10^{-}15\,M$) to immobilize *N*-terminally biotinylated peptides onto avidin functionalized surfaces. We immobilized *N*-terminally biotinylated peptides, which were chemically synthesized using solid-phase peptide chemistry, onto avidin-functionalized surfaces. Avidin is an extremely stable protein, making it an excellent candidate for slide derivatization and immobilization. The reaction with its natural ligand takes place almost instantaneously and efficiently, which considerably reduces the incubation times typically needed when compared to alternative immobilization methods. Avidin also acts as a molecular layer that minimizes nonspecific binding on the slide surface thereby eliminating the need of blocking procedures. We have successfully extended this approach also for the site-specific immobilization of biotinylated proteins in the generation of a highly versatile protein microarray [58].

Oliver et al. delineated the preparation and characterization of semicarbazide glass slides for the fabrication of peptide/oligonucleotide arrays using the site-specific α-oxosemicarbazide ligation in 2003. The glass slides are silanized using either a multistep procedure or a direct silanization with semicarbazide silane [34]. The functional density and homogeneity of the semicarbazide glass slides were optimized by analyzing the reactivity of the layer toward a synthetic glyoxylyl fluorescent probe. In summary they presented immobilization of a glyoxylyl peptide labeled with biotin on semicarbazide slides and subsequent detection with streptavidin or an antibiotin antibody.

Within the domain of peptide microarrays, Salisbury et al. recently reported an efficient strategy for the determination of protease substrate specificity [31]. A 361-member spatially addressable peptide library was synthesized and immobilized on aldehyde-derivatized glass slides. The peptides contained potential protease substrate sequences and were conjugated with fluorogenic moiety 7-amino-4carbamoylmethyl coumarin (ACC). Proteases would act to cleave certain substrate sequences preferentially, and upon release of the peptide, the ACC was unmasked thus generating a strong fluorescence which may be quantitated.

In a similar approach by our group (Fig. 7.4), a fluorogenic coumarin derivative has been used to generate a series of substrates for different classes of hydrolytic enzymes, and the resulting conjugates have been immobilized on a glass slide to generate a small molecule-based microarray capable of sensitive detection of different hydrolytic enzymes [32]. Each conjugate contains two different units: a fluorogenic moiety and an enzyme recognition head. The fluorogenic moiety serves as a sensitive reporter group that translates enzymatic activities into fluorescence readouts. It is a bifunctional coumarin derivative, containing a carboxyl group used as a handle for immobilization onto a glass surface, and an electron-donating group (phenolic or anilide group) serving as the site for conjugation to a potential enzyme

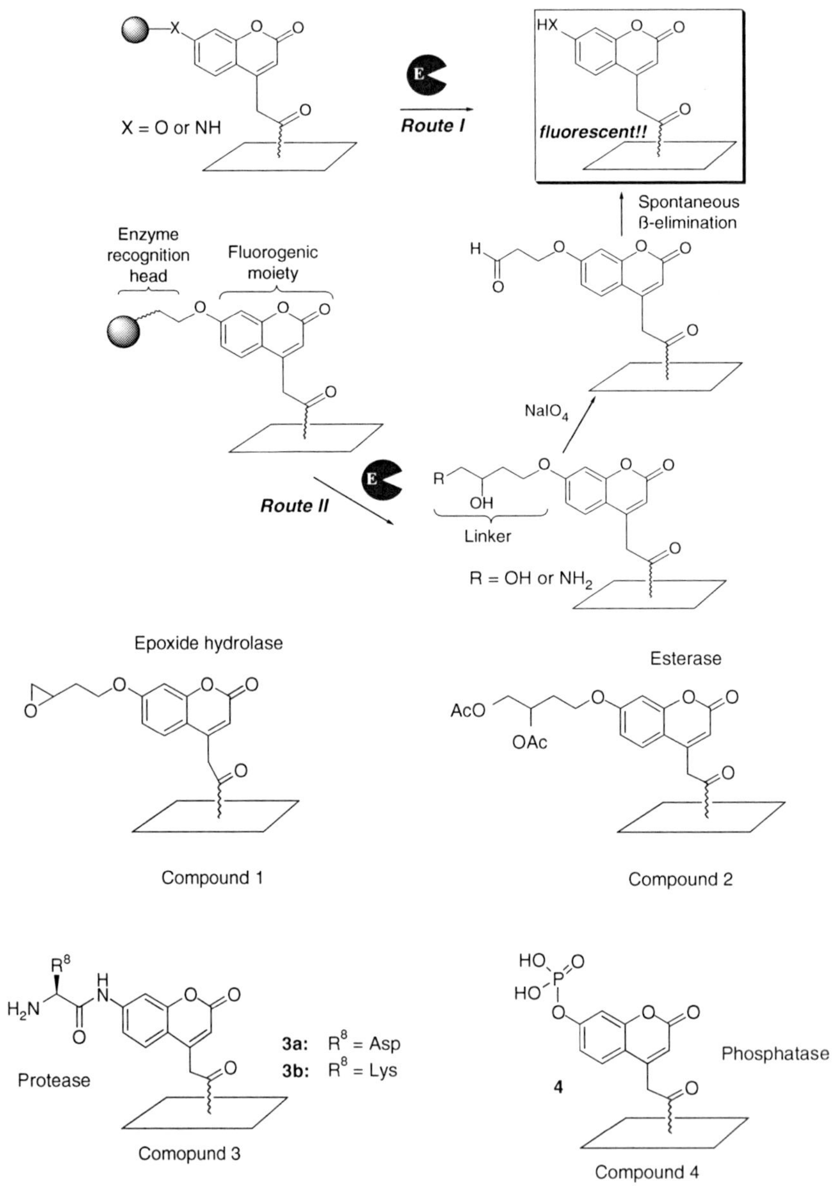


**Fig. 7.4** Strategies for detection of hydrolytic enzymes

substrate. The enzyme recognition head contains a unique chemical structure that serves as a potential enzyme substrate and may be fine-tuned to target different enzymes of choice.

Upon enzymatic cleavage, the release of the highly fluorescent coumarin on the surface of the glass slide renders it possible to detect the enzyme activity both quantitatively and specifically. It should be noted that our method was inspired by the extensive and elegant work carried out by Goddard et al., in which they showed similar approaches could be used in a microplate format to generate "fingerprints" of a variety of enzymes [59]. An elegant method for site-specific immobilization of peptides onto a microarray involves a traceless version of the Staudinger ligation [33]. An azido group incorporated into either a side chain or the main chain in a peptide reacted with a phosphinothioester-derivatized surface to form an amide bond. This reaction is one of the most rapid and high-yielding coupling reactions, occurring at room temperature in either aqueous or wet organic solvents. As the synthesis of azido-peptides is readily available, this strategy has been applied by others as well.

A recent report from Kiyonaka et al. described the development of supramolecular hydrogels that provide a semiwet environment to peptide/protein microarrays which is more compatible with enzyme assays [36]. Aqueous cavities created in the gel matrix act as a suitable semiwet reaction medium for enzymes, whereas the hydrophobic domains of the fiber are useful as a unique site for monitoring the reaction. This represents a novel class of peptide chips with a promising future. Recently Cheng et al. [38] introduced a new technique of facile synthesis of metal-chelating hexapeptide on chip that is very useful for preparing a specific metal binding peptide on chips. The technique can be useful for the purification of other histidine-tagged proteins or any other target containing a histidine tag.

Recently Marcella et al. investigated the suitability of a glass slide coated with a copolymer of DMA, NAS, and MAPS as a peptide array substrate [60]. The polymeric surface was used for peptide immobilization. Very recently Andresen et al. reported an alternate approach for manufacturing functional peptide microarrays. Their method relies on the solution-phase coupling of biotinylated synthetic peptides to NeutrAvidin (NA) and localized microdispensing of peptide-NA-complexes onto activated glass surfaces. Antibodies are captured in a sandwich manner between surface immobilized peptide probes and fluorescence-labeled secondary antibodies. By using spacer molecules of different type and length for NA-mediated peptide presentation, the authors demonstrate the significance of spacer length when compared to the direction of peptide immobilization for effective antibody affinity binding [61].

In contrast to the in situ synthesis method, the spotting microarray approach can generate many copies of the same chips with the aid of automated spotting techniques and efficient immobilization methods. Although the past few years witnessed an increasing number of reports in the field of the immobilization of peptides for microarrays, the search for more and newer methodologies is still of paramount importance.

## 7.4 Applications of Peptide-Based Microarray

Over the past few years, peptide-based microarrays have provided a powerful tool to study protein recognition and functions. Early research in peptide microarrays has primarily been focused in the high-throughput characterization of peptide–peptide and peptide–protein interactions. Lately, applications of the peptide microarray for probing enzymatic activity have been documented, such as in vitro characterization of kinases and proteases. Information regarding the catalytic activity and the substrate specificity of an enzyme can be extremely useful in the design of enzyme inhibitors, as well as providing insights into the enzyme's biological functions. Not surprisingly, the high-throughput screening property of the peptide array has thus far been widely utilized to profile the substrate specificity of different enzymes. The peptide-based microarray, due to its high throughput on screening thousands of enzyme–substrate reactions simultaneously, has been widely utilized to profile enzyme substrate specifity, mainly that of kinases and proteolytic enzymes.

### *7.4.1 Kinase Detection*

Most peptide arrays generated to date investigate kinases because phosphorylation of proteins by kinases is one of the most important mechanisms for the regulation of cell functions [62, 63]. Kinases are enzymes that regulate protein activity and function by phosphorylation of target proteins and are the second largest drug target group, after proteases. Kinases are classified according to the types of target residues that they are able to phosphorylate, namely serine, threonine, and tyrosine. It is important to note that the specificity of kinases is dependent upon their ability to recognize short sequences of amino acids flanking the phosphorylation site. Existing detection methods such as the one-bead–one-compound peptide libraries [20], phage-display peptide libraries [64], and peptide libraries using affinity-column selection [62] require time-consuming downstream work to identify the "hit" peptide sequence. Although the SPOT technology can alleviate this by generating peptide "macroarrays" to profile kinase substrates [54, 65], it does not produce many spots needed for high-throughput screening experiments. Peptide microarrays on the other hand, are the ideal platform, allowing simultaneous detection of multiple kinases in a spatially addressable fashion. In addition, each microarray-based assay requires only a small amount of the valuable enzyme. In direct contrast, other in vitro-based kinase assays usually need relatively large quantities of enzymes [66].

The earliest functional kinase assay in a peptide microarray format was achieved by MacBeath and Schreiber [24]. In their work, three different kinase substrates were immobilized on the BSA-NHS functionalized chips and incubated with kinase in the presence of [$\gamma^{33}$P]-ATP. After incubation, the slides were dipped into photographic emulsion and developed manually, and then the phosphorylation of the immobilized peptides could be visualized by automated light microscope. Later, Zhu et al. adopted the same strategy and successfully characterized the substrate specificity of over 100 yeast kinases using an array of peptides immobilized in microwells [67]. The group was able to identify as many as 27 proteins possessing novel tyrosine kinase activities.

In 2001, Falsey and coworkers demonstrated a proof-of-concept experiment to validate the successful phosphorylation of peptides immobilized on the microarray. These authors described the incorporation of $^{33}P$ into an immobilized peptide by $p60^{c\text{-}src}$ tyrosine kinase [25]. To date, this same strategy has been successfully applied to studies of substrate specificity of other kinases in a larger framework [68, 69]. A recent example is the identification of peptidic kinase using "phospho-site" collection arrays, which consist of libraries of peptides derived from annotated phosphorylation sites in human proteins [69]. Following synthesis by SPOT technology, the peptides were released from the cellulose membrane and immobilized chemoselectively via an *N*-terminus aminooxyacetyl moiety. Fluorescently tagged antibodies or autoradiography-based detection were used to profile different kinases including PKA, CK2, Abl-tyrosine kinase, and NEK-6. The method also allowed the study of key events in kinase recognition and regulation [54]. The phosphorylation of cytoplasmic domains of human membrane proteins by CK2 and the selectivity, subsite specificity, and cross-reactivity of generic antiphosphoamino antibodies have also been investigated using this method.

Another method devised by Houseman et al. smoothed the progress of rapid and quantitative evaluation of kinase activity [26]. Following incubation of a peptide array with the kinase, [$\gamma$-$^{32}P$]-ATP and varying amounts of the inhibitors quertin, tyrphostin A47, and PP1, the radioactivity incorporated into the peptides was measured and inhibition constants ($K_i$) were calculated. A direct determination of $K_i$ values was possible in this study inasmuch as the data acquired were a true reflection of the equilibrium binding between inhibitors and the kinase because, in an immobilized format, a large excess of the inhibitors relative to peptides precludes any competitive binding of the peptides to the enzyme. More recently Buss et al. [70] performed kinase substrate screening on the peptide arrays prepared by the SPOT synthesis method, and analyzed the specific peptide sequence containing phosphorylated serine by radiolabeling with [$^{32}P$] ATP method.

The major limitation of the radioisotope-based methods is the usage of hazardous radioactive reagents, and the long exposure time needed for sensitive detection of peptide phosphorylation. To overcome these limitations, we developed a fluorescence-based approach to detect the phosphorylation of kinases in a peptide microarray [28, 29]. For our experiment, we used fluorescently labeled antibodies that can specially recognize phosphorylated peptides (Fig. 7.5). A fluorescently tagged antiphosphoamino acid antibody was applied to the peptide array upon completion of the kinase reaction.

The production of quantifiable signals may be further used to compare between spots so as to efficiently identify the best peptide substrates for a given kinase. For a real-time detection of kinase activity in a microarray format, we explored the utility of the aforementioned fluorescence-based method in both concentration- and time-dependent detection of peptide phosphorylation on a chip: the fluorescence intensity was directly proportional to the concentration of the substrate, showing the feasibility for the determination of concentration-dependent kinase activity. The time-dependent kinase activity was also determined, indicating that the fluorescence intensity correlates well with kinase incubation time. To further expand the utility of microarray-based techniques in the profiling of kinase substrate specificity, we developed a combinatorial method to characterize kinase substrates (Fig. 7.5; [30].

Conventional microarrays heavily rely on the "one-spot–one-compound" method that permits the direct identification of positive hits. This, however, limits the diversity of compounds studied in arrays to what may be synthesized individually, thus severely imposing the constraints on the throughput. It is thus imperative

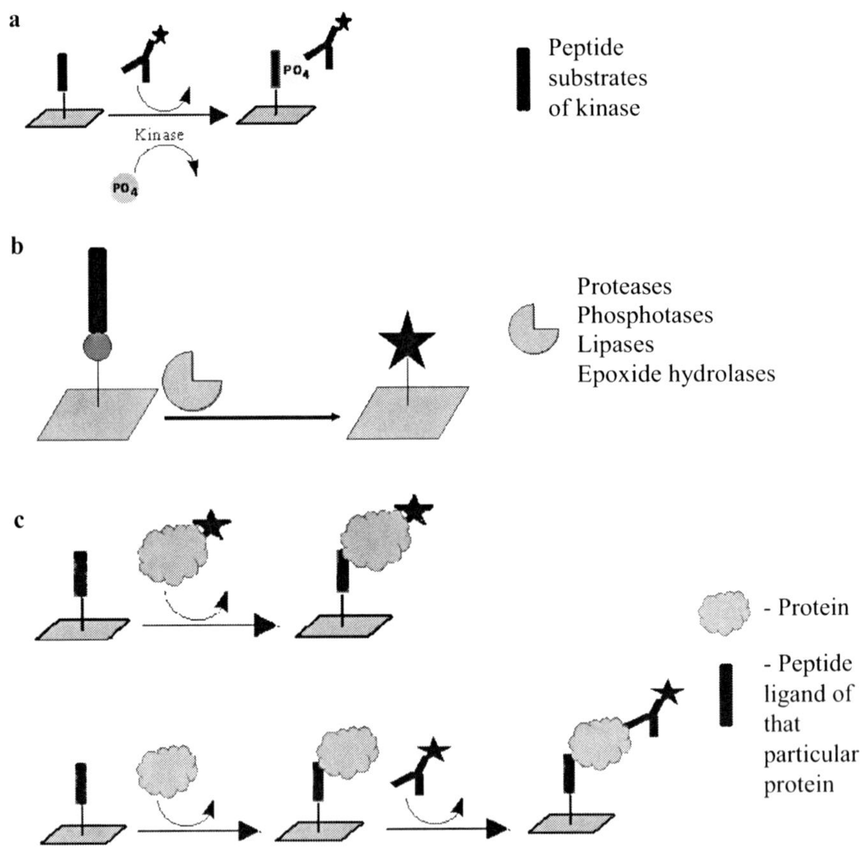


**Fig. 7.5** (a) Kinse substrate assay probed with FITC-labeled antibody. (b) Detection of enzymes using immobilized flourogenic peptide substrates. (c) Protein–peptide interaction mapping

that alternative strategies such as combinatorial peptide synthesis be suitably employed instead, with array-based methods to facilitate the development of a new generation of peptide arrays in order to study a greater diversity of compounds in a more efficient manner. We have shown that, by spotting libraries with various combinations of peptide sequences, it is possible to draw conclusions about positive hits or substrate specificity without generating large numbers of peptides sequences individually.

The augmentation of combinatorial strategies for rapid diversity production with the throughput attainable using microarrays provides for an even more rapid means of generating and applying peptide microarrays in high-throughput studies. When compared with similar methods developed in the past, our fluorescence-based method not only is safe, quantitative, and highly sensitive, but also requires a much shorter time for detection of phosphorylated amino acids and peptides on chip. Thus the newly developed strategy has significant advantages over all existing methods and it could offer a more competitive and compatible high-throughput screening of kinases in future proteomic scenario.

***Protocol IV. Detecting Kinases Using Peptide Substrate Arrays***

1. Apply 50 μl of the PKA (2U in 25 mM Tris pH 7.4, 15 mM $MgCl_2$, 1 mM DTT, 2 mM EGTA, 100 μM ATP) or p60c-src (2U in 25 mM pH 7.4, 35 mM $MgCl_2$, 7 mM $MnCl_2$, 0.5 mM EGTA, 100 μM ATP) enzyme solutions on glass slides with immobilized peptides.[a]
2. Incubate for different time periods depending on kinase activity. For p60c-src and PKA, we applied the kinases onto the slides for 5–6 hours in order to obtain good fluorescence before detection.
3. Probe slides with the corresponding labeled antibody for 1–2 hours. Antibody solutions are made in PBS with 1% BSA. Wash slides with PBST and water; dry and scan using the arrayWoRx™ (Applied Precision, WA) scanner.

More recently, Wang and colleagues developed a novel enzymatic detection system, in which phosphorylation of peptide substrates was detected by using biotinylated ATP in the kinase reaction, thus resulting in the biotinylation of target peptides in the microarray format [71]. The biotinylated peptides were then specifically labeled and detected with avidin-conjugated nanoparticles using a resonance light scattering detector.

[a] An *N*-terminal cysteine or biotin may be added to virtually any synthetic peptide. This makes it possible to apply peptides synthesized from various combinatorial strategies, such as positional-scanning libraries, onto an array format. Within reasonable lengths, under 10 amino acid residues, Fmoc synthesis yields sufficiently pure peptides that may be directly applied to an array without extensive purification (also dependent on the nature of the residues within the peptide and quality of synthesis).

### 7.4.2 Detection of Proteases and Other Hydrolytic Enzymes

Another emerging application of peptide-based microarrays is the substrate/inhibitor detection of hydrolases which include proteases and other types of hydrolytic enzymes, such as lipases and esterases among others. Proteases, enzymes that catalyze hydrolysis of peptide bonds, comprise approximately 2% of the encoded genes in organisms whose genomes have been decoded. These enzymes are critical to cell signaling, growth, and the cell cycle. Their activity is thus tightly regulated to ensure normal cell functions. Much work has been done with proteases, as they are probably the most ubiquitous enzymes in cells.

A number of approaches such as phage-displayed peptide libraries [72], positional-scanning libraries [73], and mixture-based oriented peptide libraries [74, 75] have obtained great success in the protease substrate specificity study. Other related strategies include fluorescence-quenched peptide substrate libraries [76] and end-labeled peptide substrate libraries [77]. However, these strategies do not possess the degree of throughput normally provided by microarray-based technologies. As a result, the complete specificity profile of a protease together with the kinetic evaluation of all of its potential substrates is not easily attainable. We recently explored a microarray-based strategy that utilizes mechanism-based suicide probes to detect different enzymatic activities in a microarray format [11, 78].

By using a solid support for peptide immobilization, Kiyonaka et al. studied protease activities using fluorogenic peptides trapped in a 3D supramolecular hydrogel [36]. Cleavage of a pentapeptide-DANSen conjugate by lysyl endopeptidase (LEP) was monitored as a function of the increase in fluorescence intensity of DANsen. The method was extended to profile other proteases such as V8 and chymotrypsin, as well as to screen for LEP inhibitors such as TLCK.

Gosalia and Diamond [79] recently described the use of a liquid-phase microarray that utilizes nanoliter sample volumes for protease screenings. For the purpose of array fabrication, glycerol droplets were spotted on glass slides to form individualized "reaction centers." Homogeneous enzyme assays were then assembled onto the array by simple aerosol deposition of reagents, thus doing away with the need for elaborate surface modifications that are common in other chip-based enzyme assays. Such fluid-phase reactions are advantageous as it is possible to tailor optimized reaction conditions at each individual position on the array. Using fluorogenic peptide substrates, the authors were able to detect multiple enzymes on a single array and demonstrated that the approach is amenable to enzymatic profiling and inhibition studies in a high throughput manner. A chromogenic assay for detecting hydrolases was described by Park et al. [80]. In this method, hydrolases were encapsulated in sol–gel microstructures to create the "solzyme" array and hydrolysis was monitored by the color change of a generic indicator, bromothymol blue. A panel of 20 different hydrolases was studied. Activities of the encapsulated enzymes were found to be consistent with those in solution-based assays. The array was also used to probe the enzyme specificity and inhibition using different peptide substrates.

The methods developed by Ellman et al. and subsequently by our group possess significant applications in catalomics. As described in the previous section of this chapter, Ellman's method was used to obtain quantitative measurements of protease activities on a microarray. A protease fingerprint can be thus obtained using a wide range of protease substrates. More recently, these authors have employed the strategy to the large-scale "fingerprinting" of proteases by using a 722-member library of fluorogenic protease substrates [81]. They were able to unveil the conservation of substrate specificity of thrombin across species. Using a similar approach, we have been able to efficiently detect different classes of enzymes including proteases, phosphatases, lipases, and epoxide hydrolases. This method has been described in previous sections of this chapter [32].

***Protocol III. Detecting Hydrolytic Enzymes Using Fluorogenic Substrate Arrays***

1. Immobilize compounds on amine functionalized slides.[a]
2. Treat each slide with the 50 μl of desired enzyme (1 mg/ml) in 20 mM borate buffer pH 8.8 containing 1 mM sodium periodate and 2 mg/ ml BSA (for esterases and epoxide hydrolase) or 50 mM Tris-HCl pH 8.0 (for alkaline phosphatase and tryspsin).[b]
3. Incubate the slides for 4–8 h, as required. Rinse slides with distilled water and dry under a stream of nitrogen or use a centrifuge with appropriate adapters. Leave slides to air, if necessary, until completely dry.
4. Scan the slides using an arrayWoRx™ (Applied Precision, WA) microarray scanner at $\lambda_{ex/em}$ = 360/457 nm to assess the release of the highly fluorescent coumarin.

Recently we have developed an alternative "nanodroplet" method for screening of protease inhibitors. Enzymes belonging to different classes were spotted in spatially addressable, segregated droplets on a glass slide coated with fluorogenic substrates [82]. Upon incubation, nanodroplets containing active enzymes showed up on the microarray as discrete fluorescent spots. The intensities of these spots were directly proportional to the relative activity of the spotted enzymes. The feasibility of our strategy for high-throughput identification of enzyme inhibitors was demonstrated by screening a 400-member peptide hydroxamate library against thermolysin [83]. Each member of the library had a hydroxamic acid "warhead" and an invariant isobutyl moiety at the P1′ position and diversity was generated by substituting P2′ and P3′ positions with combinations of all 20 natural amino acids. By printing preincubated nanodroplets of enzyme–inhibitor mixtures onto a protease-sensitive glass surface, we obtained the inhibitor fingerprint profiles

[a] The carboxyl moiety of the coumarin was first activated by NHS/DCC/DIEA followed by direct spotting on the slides.

[b] The presence of periodate and BSA in the reaction mix does not interfere with enzymatic activity but aids in the efficient oxidation sodium periodate oxidation of the 1, 2 diol/1, 2-aminoalcohol linker and the subsequent β-elimination.

for thermolysin in the terms of fluorescence intensity of the spots. Overall this strategy offers not only a rapid method for inhibitor profiling and discovery, but also a viable method for the chemical screening of huge combinatorial libraries against virtually any enzyme class.

### 7.4.3 *Other Applications of Peptide Arrays*

Knowledge about peptide/protein and protein/protein interactions is essential for a better understanding of important biological processes, as well as providing essential information that leads to drug discovery and development. For example, techniques that allow the accurate identification of specific peptide binding sequences of a protein could provide insights how enzyme/substrate, enzyme/inhibitor, antibody/antigen, and protein/protein interact. Zang et al. synthesized cyclic peptide libraries using a split and mix synthesis method to screen for peptide-based affinity ligands binding Streptavidin. They found that conformationally constrained cyclic peptide ligands bound 1000-fold more tightly than their linear analogues [84].

One-bead–one-peptide libraries also have been used for the screening of affinity ligands for proteins such as lime antibody [85], glycosylated hemoglobin [86], RNase S-Protein [87], and immunoglubulin G [88]. In 2004, Powell et al. described a protein–ligand binding assay method that is suitable for high-throughput screening applications (100,000 ligands per day) using a MALDI-TOF instrument [89].

Duburcq et al. [90] used a peptide/protein microarray to study pathogen infection in human lymphocytes. The authors immobilized peptides, as well as some proteins, on a semicarbazide glass slide using site-specific ligation reaction. Upon incubation of the resulting slide with human serum samples, they were able to sensitively detect specific antibody–antigen interactions. The microarray-based technique displayed high sensitivity and specificity for the detection of antibodies directed against different pathogens. In summary, they successfully proved that the peptide/protein microarray technology may provide an extremely useful platform for the clinical detection of human pathogens.

Keating and Newman used the peptide array platform and developed a strategy to study protein dimerization in a high-throughput manner [35]. Two thousand four hundred and one ($49^2$) interactions were studied in 49 individual experiments using a peptide microarray immobilized with each of the 49 peptides comprising the candidate coiled-coil strands of human basic-region leucine zipper (bZIP) transcription factors. Takahashi et al. developed a unique protein-detection system using a peptide microarray [91]. They generated a fluorescently labeled peptide loop library. Upon immobilization, the corresponding peptide array could be used to generate different "protein fingerprints." Recently, Rodriguez et al. [92] devised a novel approach called the oriented peptide array library (OPAL) approach that could be potentially used for the high-throughput proteomic analysis of protein–protein interactions. OPAL integrates the principles of both the oriented peptide libraries and array technologies. Hundreds of pools of oriented peptide libraries were synthesized as amino acid scan arrays. The authors demonstrated that these arrays can be used to map the specificities of a variety of interactions, including antibodies, protein domains such as Src homology 2 domains, and protein kinases.

There have been some reports on the discovery of catalytic peptides from peptide libraries. For example, Copeland et al. reported the selection of enantioselective acyl transfer catalysts from a pooled peptide library through a fluorescence-based activity assay [93]. Another similar example is the discovery of a Diels–Alder catalyst from a peptide library [94]. The high enantioselectivity and the enormous diversity of libraries are the merits of peptide catalysts. Sugars and their conjugates, for example, glycoproteins, mediate important cellular events such as cell–cell communication, cell adhesion, signal transduction, the attachment of microbes to cells during infection, and so forth. Recently glycopeptide libraries are emerging as α-galactosyl epitope mimetics and lectin ligands [95, 96].

Peptide nucleic acid (PNA), a DNA/RNA mimic that possesses a peptide-like backbone, has inspired the development of a variety of hybridization-based methods for detection, quantification, purification, and characterization of nucleic acids [97]. Microarrays based on PNA molecules have been on stage for almost one decade. Arrays of PNA oligomers were synthesized on a membrane using the SPOT technology and were used to investigate PNA/DNA interactions on a solid support [98]. Brandt et al. [99] developed PNA microarrays on glass slides. PNAs were synthesized by Fmoc-chemistry protocols and arrayed onto functionalized slides. After a series of postspotting modifications and validation experiments, hybridization experiments were carried out on the slides with fluorescently labeled, as well as unlabeled, oligonucleotides. Specific binding of the labeled DNA was detected by fluorescence scanner, whereas that of the unlabeled DNA was detected by the TOF-SIMS. In both conditions, the selective binding of DNA on the PNA microarray was observed with high selectivity and sensitivity.

Schultz and colleagues reported the development of a peptide nucleic acid (PNA)-tagged, small molecule array that in conjunction with the standard DNA microarray technology can be used to monitor the levels and activities of enzymes on a proteomic scale [100]. Using the split-pool combinatorial method, a small molecule library based on mechanism-based inhibitors of cysteine proteases was synthesized, with each member of the library having a PNA tag. The PNA tags did not alter the activity/selectivity of these small molecules and, at the same time served as molecular "barcodes" that could be used to decode the library by hybridization to a suitable DNA microarray. Recently, the PNA arrays have found wide application such as measuring protease activity in crude cell lysates, clinical blood samples, dust mite extracts, and so on [101, 102].

Cell-surface binding peptides are useful agents for cell–cell communications and for specific targeting of cancer cells. There have been a number of reports about the peptide-cell adhesion studies using peptide libraries and macroarray methods [103, 104]. However, the first report of using a peptide microarray to study peptide–cell interactions was reported by Lam et al. [25]. The authors showed for the first time that glass surface-immobilized peptides may be used to capture intact cells in a cell-adhesion assay, in which the platform could be used to determine not only the binding specificity of the peptide against different cell lines, but also the functional cell signaling of attached cells using in situ immunofluorescence techniques. Houseman et al. investigated the peptide–cell interaction on maleimide-functionalized chips and observed the cell adhesion with good specificity [26]. Stoevesandt et al. [105] recently examined the application of peptide microarrays in detecting signaling-dependent changes of molecular interactions. Recruitment of a protein into a complex upon stimulation of a cell leads to the masking of an otherwise exposed binding site. In cell lysates this masking can be detected by reduced binding to a microarray carrying a peptide that corresponds to the binding motif of the respective interaction domain.

## 7.5 Conclusions

Driven by the need for various high-throughput studies, the last few years have seen a major push towards the development of peptide microarray technologies that allow site-specific and stable immobilization of peptides on a variety of solid surfaces [105, 106]. With a peptide array, it is now possible to implement high-throughput screening assays that allow for the rapid identification, design, and selection of effective enzyme substrates/inhibitors, as well as potential drug candidates. In addition, such arrays may also be employed for the determination of ligand–receptor interactions, the assessment of antigen–antibody affinities, and the establishment of other similar interactions. Compared with other microarray-based technologies, peptide arrays are highly versatile, in that they provide easy access to a large number of molecular entities (e.g., peptides and peptidelike molecules) which are often natural ligands of most biological molecules, yet extremely robust and compatible with most biological assays. Despite the great progress made in only a few years, the field of peptide microarrays is still expanding at a remarkable pace. Together with other emerging technologies, the peptide microarray no doubt will be a major player in the postgenomic era and in the future of drug discovery.

## References

1. Venter, J.C. et al. (2001) The sequence of the human genome. *Science*, *291*, 1304–1351.
2. Rabilloud, T. (2002) Two-dimensional gel electrophoresis in proteomics: Old, old fashioned, but it still climbs up the mountains. *Proteomics*, 2, 3–10.
3. Gauss, C., Kalkum, M., Lowe, M., Lehrach, M.H., and Klose, J. (1999) Analysis of the mouse proteome. (I) Brain proteins: Separation by two-dimensional electrophoresis and identification by mass spectrometry and genetic variation. *Electrophoresis* 20, 575–600.

4. Unlu, M., Morgan, M.E., and Minden, J.S. (1997) Difference gel electrophoresis. A single gel method for detecting changes in protein extracts. *Electrophoresis* 18, 2071–2077.
5. Steinberg, T.H., Pretty, K., Berggren, K.N., Kemper, C., Jones, L., Diwu, Z., Haugland, R.P., and Pattonet, W.F. (2001) Rapid and simple single nanogram detection of glycoproteins in polyacrylamide gels and on electroblots. *Proteomics* 1, 841–855.
6. Washburn, M.P., Wolters, D., and Yates, J.R. (2001) Large-scale analysis of the yeast proteome by multidimensional protein identification technology. *Nat. Biotechnol.* 19, 242–247.
7. Gygi, S.P., Rist, B., Gerber, S.A., Turecek, F., Gelb, M.H., and Aebersold, R. (1999) Quantitative analysis of complex protein mixtures using isotope-coded affinity tags. *Nat. Biotechnol.* 17, 994–999.
8. Drews, J. (2000) Drug discovery: A historical perspective. *Science* 287, 1960–1964.
9. Hanash, S. (2003) Disease proteomics. *Nature* 422, 226–232.
10. Hu, Y., Uttamchandani, M., and Yao, S.Q. (2006) Microarray: A versatile platform for high-throughput functional proteomics. *Comb. Chem. High Throughput Screen*ing 9, 203–212.
11. Chen, G.Y.J., Uttamchandani, M., Lue, R.Y.P., Lesaicherre, M.L., and Yao, S.Q. (2003a) Array-based technologies and their applications in proteomics. *Curr. Top. Med. Chem. 3*, 705–724.
12. Reineke, U., Volkmer-Engert, R., and Schneider-Mergener, J. (2001) Applications of peptide arrays prepared by the SPOT-technology. *Curr. Opin. Biotechnol. 12*, 59–64.
13. Li, M. (2000) Applications of display technology in protein analysis. *Nat. Biotechno*l. *18*, 1251–1256.
14. Southern, E.M. (1988) Analysing polynucleotide sequences. Great Britain Patent Application GB 8810400.5.
15. Geysen, H.M., Melven, R.H., and Barteling, S.J. (1984) Use of peptide synthesis to probe viral antigen54s for epitopes to a resolution of a single amino acid. *Proc. Natl. Acad. Sci. U.S.A. 81*, 3998–4002.
16. Frank, R. (1992) Spot-synthesis: An easy technique for the positionally addressable, parallel chemical synthesis on a membrane support. *Tetrahedron 48*, 9217–9232.
17. Fodor, S.P.A., Read, J.L., Pirrung, M.C., Stryer, L., Lu, A.T., and Solas, D. (1991) Light-directed, spatially addressable parallel chemical synthesis. *Science 251*, 767–773.
18. www.affymetrix.com.
19. Lockhart, D.J. and Winzeler, E.A. (2000) Genomics, gene expression and DNA arrays. *Nature 405*, 827–836.
20. Lam, K.S., Salmon, S.E., Hersh, E.M., Hruby, V.J., Kazmierski, W.M., and Knapp, R.J. (1991) A new type of synthetic peptide library for identifying ligand-binding activity. *Nature* 354, 82–84.
21. Lam, K.S., Lebl, M., and Krchnak, V. (1997) The "one-bead-one-compound" combinatorial library method. *Chem. Rev*. 1997, *97*. 411–448.
22. Houghten, R.A. (1984) General method for the rapid solid-phase synthesis of large numbers of peptides: Specificity of antigenantibody interaction at the level of individual amino acids. *Proc. Natl. Acad. Sci. U.S.A.* 82, 5131–5135.
23. MacBeath, G., Koehler, A.N., and Schreiber, S.L. (1999) Printing small molecules as microarrays and detecting protein-ligand interactions en masse. *J. Am. Chem. Soc. 121*, 7967–7968.
24. Macbeath, G. and Schreiber, S.L. (2000) Printing proteins as microarrays for high-throughput function determination. *Science 289*, 1760–1763.
25. Falsey, J.R., Renil, R., Park, S., Li, S., and Lam, K.S. (2001) Peptide and small molecule microarray for high throughput cell adhesion and functional assays. *Bioconj. Chem. 12*, 346–353.
26. Houseman, B.T., Huh, J.H., Kron, S.J., and Mrksich, M. (2002) Peptide chips for the quantitative evaluation of protein kinase activity. *Nat. Biotechnol. 20*, 270–274.
27. Houseman, B.T., Gawalt, E.S., and Mrksich, M. (2004) Maleimide-functionalized self-assembled monolayers for the preparation of peptide and carbohydrate biochips *Langmuir 19*, 1522–1531.

28. Lesaicherre, M.L., Uttamchandani, M., Chen, G.Y.J., and Yao, S.Q. (2002a) Developing site-specific immobilization strategies of peptides in a microarray. *Bioorg. Med. Chem. Lett. 12*, 2079–2083.
29. Lesaicherre, M.L., Uttamchandani, M., Chen, G.Y.J., and Yao, S.Q. (2002b) Antibody-based fluorescence detection of kinase activity on a peptide array. *Bioorg. Med. Chem. Lett. 12*, 2085–2088.
30. Uttamchandani, M., Chan, E.W.S., Chen, G.Y.J., and Yao, S.Q. (2003) Combinatorial peptide microarrays for the rapid determination of kinase specificity. *Bioorg. Med. Chem. Lett.* 2003, *13*, 2997–3000.
31. Salisbury, C.M., Maly, D.J., and Ellman, J.A. (2002) Peptide microarrays for the determination of protease substrate specificity. *J. Am. Chem. Soc. 124*, 14868–14870.
32. Zhu, Q., Uttamchandani, M., Li, D.B., Lesaicherre, M.L., and Yao, S.Q. (2003) Enzymatic profiling system in a small-molecule microarray. *Org. Lett.* 2003, *5*, 1257–1260.
33. Soellner, M.B., Dickson, K.A., Nilsson, B.L., and Raines, R.T. (2003) site-specific protein immobilization by Staudinger ligation. *J. Am. Chem. Soc. 125*, 11790–11791.
34. Oliver, C., Hot, D., Huot, L., Oliver, N., El-Mahdi, O., Gouyette, C., Huynh-Dinh, T., Gras-Masse, H., Leomione, Y., and Melynk, O. (2003) α-Oxo semicarbazone peptide or oligodeoxynucleotide microarrays. *Bioconj. Chem. 14*, 430–439.
35. Newman, J.R.S. and Keating, A.E. (2003) Comprehensive identification of human bZIP interactions with coiled-coil arrays. *Science* 300, 2097–2101.
36. Kiyonaka, S., Sada, K., Yoshimura, I., Shinkai, S., Kato, N., and Hamachi, I. (2004) Semi-wet peptide/protein array using supramolecular hydrogel. *Nat. Materials 3*, 58–64.
37. Li, S., Bowerman, D., Marthandan, N., Klyza, S., Luebke, K.J., Garner, H.R., and Kodadek, T. (2004) Photolithographic synthesis of peptoids. *J. Am. Chem. Soc. 126*, 4088–4089.
38. Cheng, C.W., Lin, K.C., Pan, F.M., Sinchaikul, S., Wong, C.W., Su, W.C., Hsu, C.H., and Chen, S.T. (2004) Facile synthesis of metal-chelating peptides on chip for protein array. *Bioorg. Med. Chem. Lett. 14*, 1987–1990.
39. Usui, K., Tomizaki, K., Ohyama, T., Nokihara, K., and Mihara, H. (2006) A novel peptide microarray for protein detection and analysis utilizing a dry peptide array system. *Mol. BioSyst.* 2, 113–121.
40. Ofir et al. (2005) Versatile protein microarray based on carbohydrate-binding modules, *Proteomics*, 5, 1806–1816.
41. Cha, T., Guo, A., Zhu, X.Y. (2005) Enzymatic activity on a chip: The critical role of protein orientation. *Proteomics*, 5, 416–419.
42. Beyer, M., Felgenhauer, T., Bischoff, F.R., Breitling, F., and Stadler, V. (2006) A novel glass slide-based peptide array support with high functionality resisting non-specific protein adsorption. *Biomaterials* 27, 3505–3514.
43. Wegner, G.J., Wark, A.W., Lee, H.J., Codner, E., Saeki, T., Fang, S., and Corn, R.M. (2004) Real-time surface plasmon resonance imaging measurements for the multiplexed determination of protein adsorption/desorption kinetics and surface enzymatic reactions on peptide microarrays. *Anal. Chem.* 76, 5677–5684
44. Su, J., Bringer, M.R., Ismagilov, R.F., and Mrksich, M. (2005) Combining microfluidic networks and peptide arrays for multi-enzyme assays. *J. Am. Chem. Soc.* 127, 7280–7281.
45. Combimatrix.com.
46. Cho, C.Y., Moran, E.J., Cherry, S.R., Stephans, J.C., Fodor, S.P.A., Adams, C.L., Sundaram, A., Jacobs, J.W., and Schultz, P.G. (1993) An unnatural biopolymer. *Science* 261, 1303–1305.
47. Abell, A. (1999) Advances in amino acid mimetics and peptidomimetics, JAI Press, Greenwich, CT (USA).
48. Lipshutz, R.J. (1993) Likelihood DNA sequencing by hybridization. *J. Biomol. Struct. Dynamics 11*, 637–653.
49. Fodor, S.P.A., Rava, R.P., Huang, X.C., Pease, A.C., Holmes, C.P., and Adams, C.L. (1993) Multiplexed biochemical assays with biological chips. *Nature* 364, 555–556.

50. Sheldon, E.L., Briggs, J., Bryan, R., Cronin, M., Oval, M., McGall, G., Gentalen, E., Miyada, C.G., Masino, R., Bodlin, D., Pease, A., Solas, D., and Fodor, S.P.A. (1993) Matrix DNA hybridization. *Clin. Chem.* 39, 718–719.
51. Pease, A.C., Solas, D., Sullivan, E.J., Cronin, M.T., Holmes, C.P., and Fodor, S.P.A. (1994) Light-generated oligonucleotide arrays for rapid DNA sequence analysis. *Proc. Natl. Acad. Sci. U.S.A.* 91, 5022–5026.
52. Pellois, J.P., Zhou, X.C., Srivannavit, O., Zhou, T.C., Gulari, E., and Gao, X.L. (2002) Individually addressable parallel peptide synthesis on microchips. *Nat. Biotechnol. 20*, 922–926.
53. Frank, R. (2002) The SPOT-synthesis technique: Synthetic peptide arrays on membrane supports—Principles and applications. *J. Immunol. Methods 267*, 13–26 and references cited therein.
54. Dostmann, W.R.G., Taylor, M.S., Nickl, C.K., Brayden, J.E., Frank, R., and Tegge, W.J. (2000) Highly specific, membrane-permeant peptide blockers of cGMP-dependent protein kinase Iα inhibit NO-induced cerebral dilation. *Proc. Natl. Acad. Sci. U.S.A. 97*, 14772.
55. Toepert, F., Knaute, T., Guffler, S., Pires, J.R., Matzdorf, T., Oschkinat, H., and Schneider-Mergener, J. (2003) Combining SPOT synthesis and native peptide ligation to create large arrays of WW protein domains. J. *Angew. Chem. Intl. Ed. 42*, 1136–1140.
56. Bes, C., Briant-Longuet, L., Cerutti, M., Heitz, F., Troadec, S., Pugniere, M., Roquet, F., Molina, F., Casset, F., Bresson, D., Peraldi-Roux, S., Devauchelle, G., Devaux, C., Granier, C., and Chardes, T. (2003) Mapping the paratope of anti-CD4 recombinant fab 13B8.2 by combining parallel peptide synthesis and site-directed mutagenesis. *J. Biol. Chem. 278*, 14265–14273.
57. Muralidhar, R. and Kodadek, T. (2005) Protein "fingerprinting" in complex mixtures with peptoid microarrays. *Proc. Natl. Acad. Sci. U.S.A.* 102, 12672–12677.
58. Lue, R.Y.P., Chen, G.Y.J., Hu, Y., Zhu, Q., and Yao, S.Q. (2004) Versatile protein biotinylation strategies for potential high-throughput proteomics. *J. Am. Chem. Soc. 126*, 1055–1062.
59. Goddard, J.P. and Reymond, J.L. (2004) Recent advances in enzyme assays. *Trends Biotechnol.*, 22, 363–370.
60. Chiari, M., Cretich, M., Corti, A., Damin, F., Pirri, G., and Longhi, R. (2005) Peptide microarrays for the characterization of antigenic regions of human chromogranin A. *Proteomics* 5, 3600–3603.
61. Andresen, H., Grötzinger, C., Zarse, K., Kreuzer, O.J., Ehrentreich-Förster, E., and Bier, F.F. (2006) Functional peptide microarrays for specific and sensitive antibody diagnostics. *Proteomics* 6, 1376–1384.
62. Songyang, Z., Carraway III, K.L., Eck, M.J., Harrison, S.C., Feldman, R.A., Mohammadi, M., Schlessinger, J., Hubbard, S.R., Smith, D.P., Eng, E., Lorenzo, M.J., Ponder, B.A.J., Mayer, B.J., and Cantley, L.C. (1995) Catalytic specificity of protein-tyrosine kinases is critical for selective signaling. *Nature 373*, 536–539.
63. Pawson, T. and Scott, J.D. (1997) Signaling through scaffold, anchoring, and adaptor proteins. *Science* 278, 2075–2080.
64. Caserine, G. (1992) Peptide display on filamentous phage capsids: A new powerful tool to study protein—ligand interaction. *FEBS Lett. 307*, 66–70.
65. Tegge, W., Frank, R., Hofmann, F., and Dostmann, R.G. (1995) Determination of cyclic nucleotide-dependent protein kinase substrate specificity by the use of peptide libraries on cellulose paper. *Biochemistry* 34, 10569–10577.
66. Sills M.A., Weiss, D., Pham, Q., Schweitzer, R., Wu, X., and Wu, J.Z.J. (2002) Comparison of assay technologies for a tyrosine kinase assay generates different results in high throughput screening. *J. Biomol. Screening 7*, 191–214.
67. Zhu, H., Klemic, J.F., Chang, S., Bertone, P., Casamayor, A., Klemic, K.G., Smith, D., Gerstein, M., Reed, M.A., and Snyder, M. (2000) Analysis of yeast protein kinases using protein chips. *Nat. Genet. 26*,283–289.

68. Rychlewski, L., Kschischo, M., Dong, L., Schutkowski, M., and Reimer, U. (2004) Target specificity analysis of the Abl kinase using peptide microarray data. *J. Mol. Biol.* 336, 307–311.
69. Schutkowski, M., Reimer, U., Panse, S., Dong, L., Lizcano, M., Alessi, D.R., and Schneider-Mergener, J. (2004) High-content peptide microarrays for deciphering kinase specificity and biology. *Angew. Chem. Int. Ed.* 43, 2671–2674.
70. Buss, H., Dörrie, A., Schmitz, M.L., Frank, R., Livingstone, M., Resch, K., and Kracht, M. (2004) Phosphorylation of serine 468 by GSK-3 beta negatively regulates basal p65 NF-kappa B activity. *J. Biol. Chem.* 279, 49571–49574.
71. Wang, Z., Lee, J., Cossins, A.R., and Brust, M. (2005) Microarray-based detection of protein binding and functionality by gold nanoparticle probes. *Anal. Chem.* 77, 5770–5774.
72. Deng, S. (2000) Substrate specificity of human collagenase 3 assessed using a phage-displayed peptide library. *J. Biol. Chem. 275*, 31422–31427.
73. Nazif, T. and Bogyo, M. (2001) Global analysis of proteasomal substrate specificity using positional-scanning libraries of covalent inhibitors. *Proc. Natl. Acad. Sci. U.S.A. 98*, 2967–2972.
74. Benjamin, E.T., Huang, L.L., Piro, E.T., and Cantley, L.C. (2001) Determination of protease cleavage site motifs using mixture-based oriented peptide libraries. *Nat. Biotechnol. 19*, 661–667.
75. Benjamin, E.T. and Cantley, L.C. (2004) Using peptide libraries to identify optimal cleavage motifs for proteolytic enzymes. *Methods* 32, 398–405.
76. Meldal, M., Svendsen, I., Breddam, K., and Auzanneau, F.I. (1994) Portion-mixing peptide libraries of quenched fluorogenic substrates for complete subsite mapping of endoprotease specificity. *Proc. Natl. Acad. Sci. U.S.A.* 1994, *91*, 3314–3318.
77. Leon, S., Quarrell, R., and Lowe, G. (1998) Evaluation of resins for on-bead screening: A study of papain and chymotrypsin specificity using pega-bound combinatorial peptide libraries. *Bioorg. Med. Chem. Lett. 8*, 2997–3002.
78. Chen, G.Y.J., Uttamchandani, M., Zhu, Q., Wang, G. and Yao, S.Q. (2003b) Developing a strategy for activity-based detection of enzymes in a protein microarray. *Chem. Bio. Chem.* 2003, *4*, 336–339.
79. Gosalia, D.N. and Diamond, S.L. (2003) Printing chemical libraries on microarrays for fluid phase nanoliter reactions. *Proc. Natl. Acad. Sci. U.S.A.* 100, 8721–8726.
80. Park, C.B. and Clark, D.S. (2002) Sol-gel encapsulated enzyme arrays for high-throughput screening of biocatalytic activity. *Biotechnol. Bioeng.* 78, 229–235.
81. Gosalia, D.N., Salisbury, C.M., Maly, D.J., Ellman, J.A., and Diamond, S.L. (2005) Profiling serine protease substrate specificity with solution phase fluorogenic peptide microarrays. *Proteomics* 5, 1292–1298.
82. Uttamchandani, M., Huang, X., Chen, G.Y.J., and Yao, S.Q. (2005) Nanodroplet profiling of enzymatic activities in a microarray. *Bioorg. Med. Chem. Lett. 15*, 2447–2451.
83. Wang. J., Uttamchandani, M., and Yao, S.Q. (2006) Activity-based high-throughput profiling of metalloprotease inhibitors using small molecule microarrays. *Chem. Commun.* 717–719.
84. Zang, X., Yu, Z., and Chu, Y.-H. (1998) Tight-binding streptavidin ligands from a cyclic peptide library. *Bioorg. Med. Chem. Lett.* 8, 2327–2332.
85. Yu, Z., Tu, J., and Chu, Y.-H. (1997) Confirmation of crossreactivity between lyme antibody H9724 and human heat shock protein 60 by a combinatorial approach. *Anal. Chem.* 69, 4515–4518.
86. Chen, B., Bestetti, G., Day, R.M., and Turner, A.P.F. (1998) The synthesis and screening of a combinatorial peptide library for affinity ligands for glycosylated haemoglobin. *Biosens. Bioelectron.* 13, 779–785.
87. Barnes, C.A.S. and Clemmer, D.E. (2001) Assessment of purity and screening of peptide libraries by nested ion mobility- TOFMS: identification of RNase S-protein binders. *Anal. Chem.* 73, 424–433.
88. Verdoliva, A., Marasco, D., De Capua, A., Saporito, A., Bellofiore, P., Manfredi, V., Fattorusso, R., Pedone, C., and Ruvo, M. (2005) A new ligand for immunoglobulin G subdomains by screening of a synthetic peptide library. *Chem. Bio. Chem.* 6, 1242–1253.

89. Powell, K.D. and Fitzgerald, M.C. (2004) High-throughput screening assay for the tunable selection of protein ligands. *J. Comb. Chem.* 6, 262–269.
90. Duburcq, X., Olivier, C., Malingue, F., Desmet, R., Bouzidi, A., Zhou, F., Auriault, C., Gras-Masse, H., and Melnyk, O. (2004) Peptide-protein microarrays for the simultaneous detection of pathogen infections. *Biocon. Chem. 15*, 307–316.
91. Takahashi, M., Nokihara, K., and Mihara, H. (2000) Construction of a protein-detection system using a loop peptide library with a fluorescence label. *Chem. Biol. 10*, 53–60.
92. Rodriguez, M., Li, S. S.–C., Harper, J.W., and Songyang, Z. (2004) An oriented peptide array library (OPAL) strategy to study protein-protein interactions. *J. Biol. Chem.* 279, 8802–8807.
93. Copeland, G.T. and Miller, S.J. (2001) Selection of enantioselective acyl transfer catalysts from a pooled peptide library through a fluorescence-based activity assay: An approach to kinetic resolution of secondary alcohols of broad structural scope. *J. Am. Chem. Soc.* 123, 6496–6502.
94. Lingard, I., Bhalay, G., and Bradley, M. (2003) Dyad beads and the combinatorial discovery of catalysts. *Chem. Commun.* 2310–2311.
95. Xian, M., Fatima, Z., Zhang, W., Fang, J., Li, H., Pei, D., Loo, J., Stevenson, T., and Wang, P.G. (2004) Identification of α-galactosyl epitope mimetics through rapid generation and screening of C-linked glycopeptide library. *J. Comb. Chem.* 6, 126–134.
96. Ying, L., Liu, R., Zhang, J., Lam, K., Lebrilla, C.B., and Gervay- Hague, J. (2005) A topologically segregated one-bead-one-compound combinatorial glycopeptide library for identification of lectin ligands. *J. Comb. Chem.* 7, 372–384.
97. Nielsen, P.E. (2001) Peptide nucleic acid: A versatile tool in genetic diagnostics and molecular biology. *Curr. Opin. Biotechnol. 12*, 16–20.
98. Weiler, J., Gausepohl, H., Hauser, N., Jensen, O.N., and Hoheisel, J.D. (1997) Hybridisation based DNA screening on peptide nucleic acid (PNA) oligomer arrays. *Nucleic Acids Res. 25*, 2792–2799.
99. Brandt, O., Feldner, J., Stephan, A., Schroè, M., Arlinghaus, H.F., Hoheisel, J.D., and Jacob, A. (2003) PNA microarrays for hybridisation of unlabelled DNA samples. *Nucleic Acids Res.* 2003, *31*, e119.
100. Winssinger, N., Harris, J.L., Backes, B.J., and Schultz, P.G. (2001) From split-pool libraries to spatially addressable microarrays and its application to functional proteomic profiling. *Angew. Chem. Intl. Ed.* 40, 3152–3155.
101. Winssinger, N., Ficarro, S., Schultz, P.G., and Harris, J.L. (2002) Profiling protein function with small molecule microarrays. *Proc. Natl. Acad. Sci. U.S.A.* 99, 11139–11144.
102. Winssinger, N., Damoiseaux, R., Tully, D.C., Geierstanger, B.H., Burdick, K., and Harris, J.L. (2004) PNA-encoded protease substrate microarrays. *Chem. Biol.* 11, 1351–1360.
103. Pennington, M.E., Lam, K.S., and Cress, A.E. (1996) The use of a combinatorial library method to isolate human tumor cell adhesion peptides. *Mol. Divers.* 2, 19–28.
104. Otvos, L.J., Pease, A.M., Bokonyi, K., Giles-Davies, W., Rogers, M.E., Hintz, P.A., Hoffmann, R., and Ertl, H.C.J. (2000) In situ stimulation of a T helper cell hybridoma with a cellulose-bound peptide antigen. *J. Immunol. Methods* 233, 95–105.
105. Stoevesandt, O., Elbs, M., Köhler, K., Lellouch, A.C., Fischer, R., André, T., and Brock, R. (2005) Peptide microarrays for the detection of molecular interactions in cellular signal transduction. *Proteomics* 5, 2010–2017.
106. Panse, S., Dong, L., Burian, A., Carus, R., Schutkowski, M., Reimer, U., and Schneider Mergener, J. (2004) Profiling of generic anti-phosphopeptide antibodies and kinases with peptide microarrays using radioactive and fluorescence-based assays. *Mol. Divers.* 8, 291–299.
107. Yeo, S.Y.D., Panicker, R.C., Tan, L.P., and Yao, S.Q. (2004) Strategies for immobilization of biomolecules in a microarray. *Comb. Chem. High Throughput Screening* 7, 213–221.

# Chapter 8
# Protein Microarrays for the Detection of Biothreats

**Amy E. Herr**

**Abstract** Although protein microarrays have proven to be an important tool in proteomics research, the technology is emerging as useful for public health and defense applications. Recent progress in the measurement and characterization of biothreat agents is reviewed in this chapter. Details concerning validation of various protein microarray formats, from contact-printed sandwich assays to supported lipid bilayers, are presented. The reviewed technologies have important implications for in vitro characterization of toxin–ligand interactions, serotyping of bacteria, screening of potential biothreat inhibitors, and as core components of biosensors, among others, research and engineering applications.

## 8.1 Introduction

Both public health and defense concerns necessitate the development and availability of analytical technologies capable of ready detection and reliable identification of biothreats. For the purposes of this discussion, biothreats consist of pathogens, viruses, and toxins. Examples of naturally occurring biothreats include emerging infectious diseases, infections that often appear unexpectedly and have the potential to spread through a population or increase in severity of each incidence. Recent outbreaks include a highly contagious African viral hemorrhagic fever (Ebola) in 1974, hantavirus pulmonary syndrome in the southwestern United States in 1993, and severe acute respiratory syndrome (SARS) in 2003 [1]. Biothreats with relevance to defense concerns include ricin, anthrax, and botulinum.

Table 8.1 summarizes biothreats of interest to the U.S. National Institutes of Allergy and Infectious Diseases, an institute of the National Institutes of Health. Historically, identification of microbes and viruses has relied upon bacterial culture and viral replication in a host, respectively, whereas identification of toxins employed

---

A.E. Herr
Department of Bioengieering; University of California, Berkeley
e-mail: aeh@berkeley.edu

K. Dill et al. (eds.), *Microarrays: Preparation, Microfluidics, Detection Methods, and Biological Applications*, 

**Table 8.1** U.S. National Institutes of Allergy and Infectious Diseases (NIAID) Priority Pathogens

| |
|---|
| **Category A** |
| *Bacillus anthracis* (anthrax) |
| *Clostridium botulinum* |
| *Yersinia pestis* |
| *Variola major* (smallpox) and other pox viruses |
| *Francisella tularensis* (tularemia) |
| Viral hemorrhagic fevers |
| Arenaviruses (LCM; Junin, Machupo, Guanarito viruses; Lassa fever); Bunyaviruses (Hantaviruses, Rift Valley fever) |
| Flaviruses (Dengue) |
| Filoviruses (Ebola, Marburg) |
| **Category B** |
| *Burkholderia pseudomallei* |
| *Coxiella burnetii* (Q fever) |
| *Brucella* species (brucellosis) |
| *Burkholderia mallei* (glanders) |
| Ricin toxin (from *Ricinus communis*) |
| Epsilon toxin of *Clostridium perfringens* |
| *Staphylococcus* enterotoxin B |
| Typhus fever (*Rickettsia prowazekii*) |
| Food and waterborne pathogens (bacteria, viruses, protozoa) |
| **Category C** |
| Tickborne hemorrhagic fever viruses (Crimean-Congo hemorrhagic fever) |
| Tickborne encephalitis viruses |
| Yellow fever |
| Multidrug-resistant TB |
| Influenza |
| Other rickettsias |
| Rabies |
| Severe acute respiratory syndrome-associated coronavirus (SARS-CoV) |
| Antimicrobial resistance |

Source: Accessed on September 18, 2006 at: http://www3.niaid.nih.gov/Biodefense/bandc_priority.htm.

biological assays [2, 3]. Consequently, these conventional means for identification of biothreats are not rapid enough to facilitate quick responses from public health agencies in a large-scale event associated with a biothreat.

Ideal biothreat surveillance necessitates the measurement of multiple analytes quickly in a single sample with little to no sample preprocessing. In reality, multiplexed quantitative measurement of analytes is difficult and samples are typically complex, being of biological or environmental origin. That said, recent engineering advancements in the development of high-throughput systems, as well as improvements regarding the affinity, specificity, and availability of molecular recognition components underpin the development and probable success of next-generation analytical methods [4]. Incorporation of advanced sensing and measurement techniques in deployable instruments would significantly improve the odds of reduced casualties associated with either a pandemic or biodefense situation. Usage scenarios ranging from clinical diagnostics to food safety, and from industrial applications to molecular medicine and biodefense would benefit from the advantages afforded by early detection of such an event and efficient triage of victims.

Nucleic acids-based methods such as real-time polymerase chain reaction or PCR [5] and DNA microarrays [6, 7] have been instrumental in identifying pathogens based on genetic signatures, as well as allowing characterization of postexposure biological response to toxins [8]. To augment DNA-based techniques, researchers are developing multiplexed immunoaffinity-based methods for biothreat screening relevant to public health laboratories and field deployment. Immunoaffinity-based methods provide substantial advantages for multiplexed identification of biothreats and are, arguably, one of the most important technologies aside from PCR-based methods and DNA microarrays [9]. Immunoaffinity-based methods may be even more important than genomic methods when developing systems for field use. This chapter details the role protein microarray methods play in development of increasingly sophisticated analytical platforms for detection, measurement, or identification of biothreats.

## 8.2 Molecular Recognition

As is true in any immunoassay method, protein microarrays typically rely on antigen binding to antibodies as a means to detect. Protein microarrays are a plausible assay solution if (1) a high affinity exists between the antigen in question and the capture or detection antibodies and (2) capture or detection antibodies for a particular antigen are available. The former point determining the effectiveness of the microarray and the latter point establishing the degree of multiplexing available for a particular assay.

A review by Iqbal and colleagues describes advances made in molecular recognition technologies for detection of biothreats [4]. A 2003 article by Andreotti and co-workers presents a detailed review of available means for molecular recognition of antigens using immunoassay methods [1]. As mentioned by these authors, binding of recognition molecules to protein targets requires that the detection probe recognize and have sufficient affinity for specific domains on the protein target.

Polyclonal and monoclonal antibodies are often used as both capture and detection probes in protein microarrays. Polyclonal antibodies may provide sufficient assay sensitivity, but may not enable the specificity necessary to measure the presence of specific antigens. Use of polyclonal and monoclonal antibodies in combination or use of monoclonal antibodies alone can surmount the specificity requirements, sometimes at the expense of sensitivity. In certain cases, neither monoclonal nor polyclonal antibodies with the required biological characteristics are available.

Recently, recognition molecules such as recombinant antibodies produced via phage display, random peptides, and aptamers have become available and grown in importance. Molecules such as these may play a significant role in the future of protein microarray, and biosensor, development [10]. Regardless of the recognition component identity, improvements in the molecular recognition functions (i.e., affinity, specificity) and the availability of the molecule will determine the success or failure of the particular detection technology.

## 8.3 Contact-Printed Protein Microarrays

### *8.3.1 Direct and Competitive Immunoassays for Toxin Measurement*

In work from our own laboratory, Rucker et al. report on the development of antibody-based microarray techniques for the multiplexed detection of protein toxins in solution [11]. Details regarding assay development and fabrication of a six-element monoclonal antibody microarray for cholera toxin β-subunit, diphtheria toxin, anthrax lethal factor and protective antigen, *Staphylococcus aureus* enterotoxin B, and tetanus toxin C fragment were presented. Samples included both a model buffer system and bovine serum samples. We reported on assessment of two detection schemes. Namely, a direct assay (fluorescently labeled toxins were detected directly by immobilized capture antibodies) and a competition assay (unlabeled toxins employed as reporters for the quantification of native toxin in solution).

In the direct assays, six unique fluorescently labeled toxins were exposed to arrayed antibodies as a means to determine the strength of the antibody–toxin interaction in both buffer and diluted bovine serum samples. Fluorescence measured at each array element was correlated with known labeled toxin concentration to yield binding information (e.g., Langmuir isotherms, affinity constants). Both dissociation constants and limits of detection for the antibody microarray were determined. Limits of detection for the direct binding assays ranged from 14 to 704 ng/mL, depending on the analyte. The dilute bovine serum sample matrix allowed investigation of cross-reactivity between the arrayed antibodies or spiked toxin and background sample constituents.

We also reported on experiments directed toward the ultimate goal of detecting and identifying unlabeled toxins at low concentration. To this end, a competition assay was designed for the detection and characterization of unlabeled toxins in solution. Competition assays have been reported previously for the detection of toxins using immobilized gangliosides (membrane-embedded receptors that are negatively charged ceramide-based glycolipids with one or more sialic residues recognized by several bacterial toxins when infecting a host cell) and G protein-coupled receptors, whereas others have used a competition assay for serum-profiling experiments 2000a.

A significant advantage of the competition assay over reported profiling assays was the minimal sample preparation required. The competition assay obviated the need to fluorescently label native proteins in the sample of interest. In our study, the 50% inhibition constants for the competition between fluorescently labeled reporter toxin and unlabeled toxin were characterized for all six analytes in buffer and diluted bovine serum. Both the calculated inhibition constants for the binding of the unlabeled toxin to the immobilized antibodies and the calculated detection limits using this competition assay for native toxin detection were reported.

Dose–response curves and detection limits were established for both assay formats. Although the sensitivity of the direct assay was superior to that of the competition assay (limits ranging from 24 to 5300 ng/mL), detection limits for unmodified toxins in the competition assay were comparable to values reported previously for sandwich-format immunoassays of antibodies arrayed on planar substrates. As a demonstration of the potential of the competition assay for unlabeled toxin detection, a straightforward multiplexed assay was demonstrated for the differentiation and identification of both native *S. aureus* enterotoxin B and tetanus toxin C fragment in spiked dilute serum samples.

### 8.3.2 Sandwich Immunoassays for Measurement of Pathogenic Bacteria

Although analytical methods such as plate culture, enzyme-linked immunosorbent assasys (ELISA), and PCR have been used to detect bacteria, multiplexed methods appropriate for screening of bacterial pathogens in biological samples are needed. Gehring and coauthors [15] have recently reported on a microarray-based method for the detection of *Escherichia coli* O157:H7. Both biotinylated and antibodies bound to biotinylated protein G were used as capture moieties. Biotinylated capture antibodies for *E. coli* O157:H7 were contact printed onto streptavidin-coated microarray slides. Printed slides were blocked by static incubation with 100 μL of PBS plus 1% BSA (*w/v*) for 1 h at room temperature.

After washing and drying steps, 100 μL of bacterial solution was added to each array and incubated (1 h, room temperature) to allow bacterial capture. After a second set of washing and drying steps, 100 μL of solution consisting of fluorescently labeled reporter antibodies in PBS with 0.5% BSA (*w/v*) was added to each slide, incubated for 1 h at room temperature, washed, dried, and then scanned for fluorescence. Fluorescence detection yielded a linear detection range from $3.0 \times 10^6$ to $9.0 \times 10^7$ cells/mL, with an apparent limit of detection at $3.0 \times 10^6$ cells/mL. Below $10^6$ cells/mL, protein G-bound antibody did not produce a measurable signal.

### 8.3.3 Serotyping

Cai and coworkers report on the development of an antibody microarray that makes use of simultaneous analysis of multiple antigens inherent in protein microarray methods to perform *Salmonella* serotyping [16]. The somatic (O) and flagellar (H) antigens present on *Salmonella* bacteria result in over 2500 serovars (strains). The work reported by Cai et al. makes important improvements to current serotyping methods; namely, the described microarray allows concurrent detection of multiple antibody–antigen interactions, requires small reagent volumes, and requires significantly less time than current three-day serotyping durations. The *Salmonella* serotyping array was designed to detect and identify the 20 most clinically relevant serovars, strains that represent more than 80% of *Salmonella* isolates collected in Canada.

The array reported by Cai et al. was composed of 35 antibodies (11 O factor antisera, 9 H phase antisera, 13 H factor antisera, 2 O multigroup antisera) in an 8 × 15 array. Antibodies from polyclonal rabbit antisera were contact printed on commercially available epoxide-functionalized glass slides. Optimal antibody concentrations ranged from 1 to 5 mg/mL.

After printing, the authors blocked unreacted epoxide groups and applied fluorescently labeled *Salmonella* cells to the array. The cells were labeled with Eosin Y and Cy3 by incubating the cells with the dyes. For operator safety, cells were inactivated by a brief heating step. Simultaneous detection of O serogroups and H phase 1 and phase 2 antigens was accomplished using the assay. Evaluation of the array consisted of screening 117 *Salmonella* strains, covering the 20 target serovars, and 73 strains, covering 38 nontarget serovars. Correct identification of 74% of the target strains was made using the serotyping microarray.

## 8.4 Membrane Microarrays

Contact-printed supported membranes exhibiting characteristics of a physiologically fluid environment provide a microarray format especially relevant for studying cellular processes. Especially pertinent to the study of signal transduction pathways, physiological fluidity of the supported membrane allows development of high-throughput biomimetic assays, thus allowing the study of protein–receptor interactions on time scales relevant to in vivo questions. Membrane fluidity is largely dependent upon the preparation of the lipid bilayer and the characteristics of solid substrate supporting the membrane. The fluidity of molecules inserted in planar lipid bilayers facilitates dimerization and other higher-order interactions necessary for biological signaling events. Membranes also allow the study of toxin interactions with gangliosides. Gangliosides are components of the cell membrane and act to modulate cell signal transduction events, including those relevant to toxin signaling pathways.

### *8.4.1 Gangliosides Immobilized on Optical Waveguides*

A nonantibody-based flow assay utilizing ganglioside GM1 immobilized on the surface of optical waveguides was reported for rapid detection of cholera toxin [12, 13]. Direct and sandwich assays for cholera using a flow-patterning method (described in more detail in Section 8.6) to immobilize GM1 at discrete locations on a planar optical waveguide were described. As the authors note, interest in development of an identification and detection system for cholera stems from the difficulty associated with distinguishing cholera from other causes of acute diarrhea. In the case of cholera, mortality approaches 50% if left untreated, yet, when properly treated, cholera mortality rates can be as low as 1%. Conventional culture-based diagnosis can take 8 to 24 hours and may not be effective, as only a single serotype of Vibrio cholerae (O1) causes disease (WHO, 1994).

Patterning of gangliosides relied upon the hydrophobic interaction between octade-cyltrichlorosilane (OTS)-silanized surfaces and ganglioside reagents. Several buffer and solvent systems were investigated in an effort to optimize ganglioside immobilization. Slides were treated with a long-chain silane to allow hydrophobic interactions between the surface and the ganglioside capture molecule. OTS was chosen due to the chain length of the sphingosine (C18). Ganglioside-based ELISAs for cholera toxin and other toxins have been reported [17–20].

Direct assays involved measurement of binding of fluorescently labeled cholera toxin to immobilized GM1. In a manner similar to ELISAs, sandwich assays using fluorescently labeled monoclonal and polyclonal detection antibodies were also developed. The sandwich assays enabled limits of detection as low as 40 ng/mL in the case of polyclonal detection antibodies and 5 ug/mL with monoclonal detection antibodies. The detection sensitivity of the reported system for GM1 binding of cholera toxin (40 ng/mL) compared well with other reported ganglioside-based assays [21–23].

To take advantage of the multiplexing ability of the presented sensor platform, the authors explored binding of cholera toxin to a number of different glycolipids (GD1b, Gb3, Gb4, and GT1b). As was the case with GM1, each glycolipid was flow-patterned onto a silanized slide. Negligible binding to GT1b was observed, however, the authors did observe dose-dependent binding of cholera toxin to GD1b, Gb3, and Gb4. As the authors discuss, the ligand–receptor binding results from the array sensor are in disagreement with published reports. Factors such as different experimental conditions, shear fluid forces, toxin concentration ranges, orientation, and density of receptor molecules patterned on the planar surface are all discussed as possible sources of deviation in the measured ligand to receptor binding.

### *8.4.2 Ganglioside Microarrays for Toxin Detection*

In the early 2000s, researchers at Corning reported on microarrays of lipids containing gangliosides for detection of bacterial toxins and screening of potential inhibitors, two important applications of membrane microarrays [24–26]. A typical bacterial toxin is composed of two domains: the *A* or *activating* domain is involved in intracellular enzymatic activity and the *B* or *binding* domain is involved in binding to the cell membrane. The surface of host cells displays several types of molecules that act as toxin-binding sites [27]. The authors provide a brief overview of surface receptors important as potential bacterial toxin inhibitors (cholesterol, carbohydrate derivatized lipids (sphingoglycolipids), and gangliosides, in particular).

At Corning, Fang and colleagues [24–26] described development of membrane microarrays as a tool to study the interaction of toxins (cholera and tetanus) with carbohydrates in near-native environments. The work also investigated screening of possible inhibitory compounds using the developed membrane microarray. Supported lipids are of interest as these features allow both immobilization and confinement of the probe ligand and the associated lipids, while maintaining lateral movement of individual molecules within the arrayed lipid microspot. Recent studies have shown that ideal supported membrane surfaces used in microarrays should (i) resist physical desorption and (ii) exhibit long-range lateral fluidity.

As background information, in 1999 Boxer and coworkers reported that supported lipids patterned on bare-glass desorbed as the glass substrate was drawn through an air–water interface [28]. In 2000, Mrksich and others used self-assembled monolayers to provide carbohydrate ligands covalently attached to the surface making studies of biomolecular recognition feasible [29]. The lack of lateral fluidity in the system precluded the biomimetic display of carbohydrate ligands, as is necessary in cases where ligand clustering is important [30]. Through use of supported lipids, Fang and coworkers have combined advantages inherent to microarray analysis (throughput, efficiency) with near-native environments to carbohydrate-mediated recognition.

As described in Fang et al. [24–26] the supported lipid system reported utilized surfaces derivatized with γ-aminopropylsilane (GAPS) to provide a substrate resistant to desorption and yet supportive of lateral fluidity (with a mobile fraction of ~0.5). The fabrication protocol reported relies upon quill-printing of DLPC (1 mg/mL) in the absence and presence of 4 mol% ganglioside (GM1 or GT1b) in 20 mM phosphate buffer (pH 7.4) on the GAPS-coated slide. Quill pins were dipped into individual wells of a 384 microtiter plate to pick up each lipid solution followed by a 1 h incubation period in a humidity chamber. Repeated dipping of slides printed with fluorescently labeled lipid revealed robust performance of the described system. The lateral fluidity of supported lipids was tested by traditional fluorescence recovery after photobleaching experiments. The authors report that microarrays of G protein-coupled receptors exhibiting ligand binding affinities and specificities consistent with the literature [24–26].

To demonstrate detection of toxins in a competitive assay, arrays were incubated with 20 μL of a solution consisting of fluorescently labeled toxin (0.031–2 nM) and varying concentrations (0–100 nM) of unlabeled toxin. Fluorescence signal was measured at the completion of the assay using a microarray scanner. The authors report detection limits of 10 nM for the unlabeled cholera toxin and 50 nM for the unlabeled tetanus toxin. For labeled toxins, the detection limits were measured at 0.03 nM for FITC-labeled cholera toxin and 0.06 nM for FITC-labeled tetanus toxins.

### 8.4.3 *Polymer Lift-Off Technique for Lipid Patterning*

Moran-Mirabal and colleagues present a characterization of bacterial toxins (cholera toxin B-subunit and tetanus toxin C-fragment) binding to micron-sized lipid domains patterned onto planar substrates and within microfluidic channels [31]. The authors detail an innovative polymer lift-off technique used to fabricate ganglioside-populated (GM1, GT1B) supported lipid bilayers.

In brief, the patterning procedure used to deposit supported lipid bilayers on planar substrates consisted of an initial deposition of a polymer film on a silicon substrate by vapor deposition. Photoresist is then spun on the coated silicon substrate and exposed with ultraviolet (UV) light. The photoresist is developed and removed from UV exposed areas. The resulting exposed parylene is etched via oxygen plasma, revealing well-defined micron-sized features of exposed silicon. The substrates are oxidized and incubated with lipid vesicles, resulting in vesicle fusion and the formation of lipid bilayers.

Finally, the polymer is peeled off and ganglioside-populated lipid bilayer patterns remain. The bacterial toxin assays involve subsequent incubation of the patterned surface with aqueous samples of interest. An adaptation of the procedure is described for the fabrication of multiple lipid bilayer elements in a microfluidic channel, as is especially relevant to biosensor development. The authors point out that the lift-off technique eliminates the need for etched barriers to contain the bilayers, as the polymer defines the bilayer structures with micron-scale resolution.

Total internal reflection fluorescence microscopy allowed extraction of apparent binding constants and suggested that protein binding to the ganglioside receptors was influenced by the microenvironment lipid bilayer and the underlying substrate. Characterization of the binding did enable estimates of the limits of detection at down to 100 pM for cholera toxin B-subunit and 10 nM for tetanus toxin C-fragment. Arrays of lipid domains having different compositions were demonstrated on a single microfluidic device and enabled segregation and selective binding from a composite mixture of the two toxins, as determined by epifluorescence microscopy.

### 8.4.4 *Lipid Bilayers on Nonglass Substrates*

Although not in a multiplexed format, Phillips and Cheng [32] report on the development of heterogeneous assays for cholera toxin using supported lipid bilayers in poly (dimethylsiloxane) (PDMS) microfluidic channels. The heterogeneous immunoassay developed could be readily adapted to a microarray format, possibly using contact printing. The technique revealed the ability to quickly and specifically detect cholera toxin using the cell surface receptor GM1 integrated into a supported lipid bilayer.

The assay provided a dynamic range spanning three orders of magnitude and a detection sensitivity of 8 fmol of cholera toxin when performed under flow conditions. In addition, the supported lipid bilayer had good fluidity, as measured by fluorescence recovery after photobleaching, exhibited minimal nonspecific protein adsorption, and was robust under flow conditions. The work highlighted advantages associated with using a microfluidic fluid delivery approach. Furthermore, the work demonstrates the potential for use of alternate materials for planar substrates, in this case PDMS.

Building on the study described above, Phillips and coworkers recently detailed development of a lipid bilayer-based sensor for detection of cholera toxin in environmental water samples [33]. The work addresses the design challenge of providing sufficient mechanical stability in the lipid bilayer while retaining lateral fluidity. In the previous work, the authors show that oxidized PDMS exhibits hydrophobic recovery over short periods of time, whereas surfaces covered with phosphatidylcholine membranes maintain hydrophilic properties.

As it is important to robust sensor development, the authors report in this recent work that PDMS surfaces treated with vesicles from cationic lipids (i.e., ethylphosphocholine, DOPC+) exhibit exceptionally strong resistance to air-dry damage. Fluorescence recovery after photobleaching measurements and protein adsorption studies conducted by the authors reveal that the mobile fraction of PC membranes decreased by nearly 10% after drying/rehydration cycles. Membrane fluidity was reduced as well, with the lateral diffusion coefficient decreasing by close to 30% after drying/rehydration of the PC membranes. The DOPC+ membranes developed in the study reveal little to no reduction in either mobile fraction or diffusivity.

As an example application, the authors packaged the DOPC+ membrane (GM1/DOPC+ membranes) in a PDMS flow-based immunoassay. A detection limit of 250 amol for cholera toxin was obtained from on-chip calibration curves. Cholera toxin spiked into river water samples revealed similar response and sensitivity. The group has recently reported on use of a similar system for immunosensing of SEB in milk [34].

### *8.4.5 Microfabrication as a Means to Corral Lipid Bilayers*

Yamazaki and co-workers report a technology for fabricating multiplexed, high information-content cell membrane microarrays as a tool for high throughput biological assays [35]. The authors use the membrane microarrays to validate the approach in the study of ligand/receptor binding and interactions with live intact cells. The authors report on three classes of interactions. Specifically, the group assesses the interaction between the cholera toxin B-subunit and the membrane ganglioside GM1. The interaction of the pentameric cholera toxin B-subunit with membrane gangliosides is hypothesized to be multivalent and involve up to five GM1 receptors [36] under membrane conditions with sufficient fluidity to allow assembly of the ligand/receptor complex.

A second system of interest is the display of membrane components that are important drug targets for treating diseases. The authors present results on displaying LPS (lipopolysaccharide) from gram-negative bacteria and the mammalian proteins ICAM-1 and I-Ek to show relevance to conditions ranging from septic shock to autoimmune dysfunction.

Finally, a third system reported by Yamazaki [35] and coworkers investigates T-cell adhesion and activation on membrane microarrays displaying proteins of immunological importance. Two proteins important in mediating the adhesion and activation of T cells on antigen-presenting cells, ICAM-1 and I-Ek, were investigated using the reported methodology.

A method that combines lipid biochemistry with microfabrication methods common to the semiconductor industry and robotic handling was developed to generate high-density arrays of membranes contained within discrete "corrals" on planar substrates. Glass wafers with chrome barriers were, in some parts of the study, used to define the membrane microarray elements. To fabricate the planar supported bilayers, small unilamellar vesicles were fused onto the exposed glass regions using a microarrayer capable of programmable aspiration and dispensing.

The cholera toxin B-subunit/GM1 study substantiated the validity of the reported fabrication approach through generation of dose–response curves based on varying concentrations of membrane-incorporated GM1 and aqueous cholera toxin B-subunit. FRAP measurements were used to assess the membrane fluidity. After incorporation of lipid A, the principal endotoxic moiety of LPS, in lipid bilayers, the authors measured an approximately fourfold increase in immunoreactivity as compared with membranes not containing lipid A.

The study of T-cell adhesion revealed that monoclonal antibodies specific for the membrane-arrayed surface receptors (ICAM-1, I-Ek) nearly eliminated T-cell adhesion. The authors incorporated the ICAM-1 or I-Ek ectodomains in supported lipid bilayers (through glycosylphosphatidylinositol tethering), and examined these membrane-incorporated proteins for their ability to capture murine T cells. Results from the study suggested that each target protein, alone or in combination, mediated specific adhesion of T cells. Furthermore, the authors measured spatial clustering of membrane-incorporated I-Ek molecules as an indicator of T-cell activation. The authors have shown that incorporation of transmembrane proteins into the arrayed membranes holds promise for biomimetic studies of ligand/surface receptor interactions in both industrial and research settings.

## 8.5 Hydrogel Microarrays

Detection of biothreats using arrayed proteins and antibodies is an important potential application of microarrays, as discussed later in this chapter, however, protein microarrays have been demonstrated for characterization of biothreats. Two recent examples of the use of microarrays for measuring ligand binding properties and serotyping biological strains are presented in this section. Both studies make use of hydrogel spots as supports for immobilization and as nanoliter reaction volumes for reactions between samples of interest and immobilized compounds. The studies presented in this section, as well as those reported elsewhere for nonbiothreat applications, highlight several advantages afforded by hydrogel-based microarrays [37–42]. Briefly, these advantages include: an increased immobilization capacity as compared to planar glass surfaces, a stable substrate amenable to covalent attachment of proteins, and minimal background fluorescence [43]. A further advantage is of particular importance to cell surface receptor–protein interactions, in that the three-dimensional (3D) gel is thought to mimic a solution-phase environment.

As an example of hydrogel-based protein microarrays, take the synthetic 3D immobilization strategy developed by Charles and co-workers which is based on a thin film crosslinked with bis (sulfosuccinimidyl) suberate under acidic conditions [44]. The work makes use of studies conducted by several groups exploring the use of hydrogels or polyacrylamide gel pads as a means to create 3D spots on planar surfaces.

In the methods developed by Charles and colleagues, pendant NHS-ester reacts with amide moieties within a hydrogel film and a secondary NHS-ester group reacts with the primary amine on SEB under neutral conditions. The protocol produces NHS as a side product for the formation of a stable bond between the protein and the 3D hydrogel. Binding measurements made using the 3D hydrogel microarray to study immobilized SEB binding with Cy5-labeled anti-SEB revealed significant differences in background fluorescence between the 3D and the 2D substrates. Fluorescence signal from the 3D hydrogels was threefold higher than that of planar glass surfaces when immobilized SEB was at concentrations greater than 10 μg/mL.

The authors conclude that binding epitopes on hydrogel-immobilized SEB were more accessible for antibody binding than epitopes of SEB printed on the planar glass surface. The hydrogel-based SEB microarray exhibited a linear detection range from 0.1 to 30 μg/mL and demonstrated low background signal. Citing previous work on hydrogel-based platforms, the authors conclude that femtomolar (pg/mL) detection sensitivities should be possible.

## 8.6 Sensor Technologies

Significant progress regarding detection of chemical agents has been made over the last few decades; that said, chemical agents are more readily detected than biothreats, as victims of biological agents typically display a delayed physiological response owing to incubation periods on the order of up to several days. As Ivnitski and coauthors describe [5], the potential threat from biological agents has sharpened owing to advances in molecular biology, genetic engineering, and the engineering of efficient delivery and dispersion systems, including increased civilian air travel.

Sensors capable of environmental monitoring as a means to detect the presence of biothreat agents, as well as portable systems capable of monitoring physiological conditions to detect and diagnose illness are two important components of biothreat surveillance. An ideal surveillance network would rely upon fast analysis for detection of the presence of biothreats, acting in concert with more sophisticated analytical sensor systems that would enable identification of the biothreat present.

In a clinical setting, timely identification is also critical, but difficult to implement with generally accepted diagnostic indicators commonly relied upon by public health and hospital-based clinical labs. Clinical indicators are largely based on physiological observables related to a patient's general health state (e.g., reported flulike symptoms). Unfortunately, the initial clinical signs and symptoms of many agents are nonspecific and similar to those observed from common infections. Improvements in clinical diagnostic technologies would fill this gap. Such a diagnostic system would be desirable if it were a sensitive, specific, inexpensive, easy-to use system that could rapidly and accurately identify toxins. Additional capability regarding the ability to detect the presence of a variety of possible biothreats in a single sample is of special interest.

Portable multiplexed tools are viewed as one means to achieve these goals. Although chemical sensing systems are more advanced than those developed for detection of biothreats, technological advances arising from miniaturization (e.g., microarrays, microfluidics, microelectromechanical systems) have accelerated the development of portable inexpensive sensing [5]. Ivnitski and co-authors present a review of DNA-based detection and identification of biothreat agents. Commercial systems, as well as those developed by national laboratories and academic groups are discussed. Further technological innovation regarding automation of sample collection and preparation are required before real-time environmental monitoring (water treatment plants, agriculture, food products) and clinical diagnosis fulfill their potential.

### 8.6.1 Flow-Patterned Protein Array Biosensors

Sensor development relying upon immunological recognition of proteins, bacteria, and viruses at the Naval Research Laboratory (NRL) is one nascent example of such a system, as summarized in review articles by the NRL authors [45, 46]. A summary of select applications is given in Table 8.2. The platform developed over the last decade at the NRL is designed to provide field-portable instrumentation for use in military and civilian settings.

Immobilization of capture probes on planar substrates has been developed as a detection cartridge for integration into a fully automated, user-friendly instrument. Major performance requirements for such instrumentation include: rapid analysis times (assays complete in roughly ten minutes), concurrent analysis of multiple samples for the presence of multiple analytes, and a small form factor amenable to

**Table 8.2** Summary of NRL Array Platform Assays and Sensitivity Estimates

| Category | Toxin | Limit of Detection |
|---|---|---|
| Protein toxins | SEB | 0.5 ng/mL |
| | Cholera toxin | 1.6 ng/mL |
| | Botulinum toxoid A | 40 ng/mL |
| | Botulinum toxoid B | 200 ng/mL |
| | Ricin | 8 ng/mL |
| Protein allergens | Ovalbumin | 0.025 ng/mL |
| Physiological markers | Y. Pestis F1 | 25 ng/mL |
| | D-dimer | 25–50 ng/mL |
| Gram-negative bacteria | E. herbicola | $10^4$ cfu/mL |
| | B. abortus (killed) | $3 \times 10^3$ cfu/mL |
| | F. tularensis LVS | $10^5$ cfu/mL |
| | Salmonella | $8 \times 10^4$ cfu/mL |
| | | $8 \times 10^3$ cfu/g excreta |
| Gram-positive bacteria | B. globigii | $10^5$ cfu/mL |
| | B. anthracis | $10^3$ cfu/mL |
| Virus | MS2 | $10^7$ pfu/mL |

Source: Adapted from information available from the *World Technology Evaluation Center (WTEC) Workshop on International R&D in Biosensing* held on May 13, 2003.

portable operation. The sensor is designed for end-use by minimally trained personnel in a field setting. The following sections detail protocols and results of select studies performed with the NRL platform.

Rowe and coworkers reported on a fluidic method for immobilization of capture probes in a small array patterned on a planar waveguide [47]. Biotin-labeled capture molecules were immobilized on NeutrAvidin-coated slides, as described in Bhatia et al. [48]. A flow-patterning process developed by the authors at NRL was used to pattern capture antibodies. A standard sandwich assay format was designed to rely on fluorescence reporting of analyte concentration in a sample. Excitation light was incident on the end of the waveguide to generate an evanescent excitation of the surface-bound fluorescent detection antibodies. Evanescent wave techniques enable rapid response time and relative insensitivity to complex biological matrices. Relying upon detection of fluorescent signal eliminates confounding signal arising from nonspecific adsorption. Furthermore, the narrow penetration depth (100–200 nm) of evanescent waves allows detection of events occurring on the surface only, with little signal from the bulk solution. Real-time measurements of turbid or inhomogeneous samples are possible. A two-dimensional graded index of the refraction lens array allowed imaging of the surface using a CCD camera.

The two-step flow process utilized two fluid modules: (1) a molded PDMS flow chamber module consisting of six channels was used to pattern six vertical lines of biotinylated capture antibody on the waveguide surface, and (2) a second PDMS assay module consisting of six flow channels to introduce sample and detection antibody over the patterned waveguide. After patterning, a washing procedure was implemented and patterned slides were either used immediately or treated with a storage preparation buffer (15 mM sodium phosphate/0.1 M trehalose supplemented with 10 mg/mL BSA) for 15 min and then stored at 4°C for at least three weeks.

The protocol for sample analysis consisted of flowing fluid over the patterned surface using the PDMS assay module. The assay module also consisted of six flow channels. The module was placed on the patterned planar waveguide, but perpendicular to the patterned vertical stripes allowing probing of immobilized capture antibodies at locations where patterned lines intersected with assay module channels. The sample was incubated in the channels for 15 min under flowing conditions (0.3 mL/min). After PBST washing, a second assay step introduced fluorescently labeled detection antibody that was recirculated through the channels at 0.3 mL/min for a 15 min period.

As a demonstration of the instrument for detection and identification of multiple analytes, the authors report on a study focused on measurement of Staphyloccal enterotoxin B (SEB), F1 antigen from *Yersinia pestis*, and a marker of sepsis and thrombotic disorders, D-dimer, spiked into minimally pretreated biological sample matrices at clinically relevant concentrations [49]. SEB causes food poisoning, nausea, vomiting, and diarrhea when ingested [50]. Inhalation of aerosolized SEB can be life threatening and may lead to hypotension, respiratory distress, shock, and death [51].

The glycoprotein F1 antigen is a major component of the outer membrane of *Y. pestis*. F1 antigen is the etiologic agent of plague and is secreted by *Y. pestis* only upon invasion of a mammalian host. F1 antigen is routinely detected as a means to diagnose plague [52]. DNA hybridization, PCR, and ELISA are more rapid than bacteriological and serological tests, however, these assays are not rapid enough to be used as diagnostic tools and are used mainly as confirmatory tests. Although d-dimer is a normal component of blood in healthy individuals, high concentrations are indicative of disseminated intravascular coagulation, pulmonary embolism, myocardial infarction, and deep venous thrombosis [53], as well as sepsis and infection [54].

Biological samples analyzed with the immunosensor included serum, nasal swabs, and saliva. In the serum analyses, blood samples were collected from healthy volunteers and from ICU patients clinically suspected as having sepsis. Platelet-poor plasma was prepared from heparinized whole blood, centrifuged at 3000 g, with serum obtained from clotted whole blood by centrifugation at 2000 g for 10 min. Prior to spiking with analytes of interest, the serum was diluted 1:1 with buffer. Nasal fluid samples were collected from healthy volunteers by simply swiping the interior of the nasal cavity using two cotton-tipped swabs per nostril. After collection, the swabs were placed in 4 mL of a PBS/0.05% triton x-100 buffer and incubated for 15 min. Swabs were removed and the fluid was retained.

Animal studies have shown that SEB concentrations can reach 500 ng/mL in serum and 1–10 ng/mL in urine for up to 24 h postexposure. The described immunosensor provided limits of detection at 1 ng/mL in buffer and spiked nasal fluid. The authors highlight that analysis of nasal swabs is potentially important as nasal fluid is a preferred means to diagnose exposure to aerosolized infectious agents. The immunosensor was not able to detect physiological levels of SEB in spiked serum, saliva, or urine. F1 antigen has been detected in clots and serum at the high nanogram per mL to low microgram/mL range two to three days after exposure [55].

The described NRL immunosensor platform yielded detection of F1 antigen at 25 ng/mL in buffer, spiked serum, urine, nasal swabs, and saliva. The authors assessed the sensitivity of the platform for d-dimer in buffer, plasma, and whole blood samples. Normal healthy individuals typically have between 25 ng/mL and 150 ng/mL levels of d-dimer in their blood, whereas patients with sepsis, myocardial infarction, or thrombotic disorders can have levels of d-dimer greater than 125 ng/mL. The described immunosensor detected concentrations of d-dimer greater than 50 ng/mL in buffer, plasma, and diluted whole blood. The authors did note an apparent inhibitory effect of plasma on antibody binding to D-dimer. The platform was used to assay for all three analytes individually, as well as both F1 and D-dimer in plasma.

The NRL array biosensor has also been employed to analyze 126 blind samples for the presence of bacterial, viral, and protein analytes [49]. Specifically, the authors investigated limits of detection and assay throughput regarding analysis of the nonpathogenic gram-positive sporulating soil bacteria Bacillus globigii, the small RNA bacteriophage MS2, and the toxin SEB. Single analyte assays were run in parallel with the analysis of a mixture of the three analytes, thus demonstrating the sensor's capability to detect multiple species in a single assay. Sensitivity limits of the 14 minute Bacillus globigii and MS2 assays approached those of ELISA, with limits of detection for Bacillus globigii and MS2 reported to be $10^5$ cfu/mL, $10^7$ pfu/mL, respectively. The array sensor had a 10 ng/mL limit of detection for SEB, a factor of tenfold less sensitivity than ELISA. The authors attributed the poorer SEB sensitivity to use of polyclonal, not monoclonal, antibodies. Analysis of the 126 samples yielded a 0% false negative rate. False positives were present at 0.8%, the same level reported via ELISA.

Detection of toxins, toxoids, and killed or nonpathogenic (vaccine) strains of pathogenic bacteria has also been demonstrated on the array flow sensor [13]. Again, a sandwich format was employed in conjunction with fluorescence detection. An automated version of the array flow sensor was introduced. In the reported study, improvements regarding the optical coupling of the waveguide and flow cell assembly to the detector were made by incorporating a patterned reflective silver-based cladding to optically decouple the waveguide and the flow cell. Detection of bacterial analytes (B. anthracis (Stern), Francisella tularensis LVS, Brucella abortus) and toxins (Botulinum toxoids A and B, S. aureus enterotoxin B (SEB), ricin, and cholera toxin) was reported. Although the sensitivity of the semiautomated system was compatible with immunosensors, the sensitivity was dependent on the type of antibody used (i.e., monoclonal vs. polyclonal). An automated version of the array biosensor suffered from fivefold less sensitivity, mainly arising from higher variation in background signal attributed by the authors to fluorescence arising from materials used in the flow cell.

### 8.6.2 Carbohydrate-Based Flow Biosensor

More recently, the NRL array biosensor provided an experimental platform to assess the binding interactions between sugars and protein toxins, as well as bacterial cells [56]. Carbohydrate derivatives were of interest to the authors, as a large number of bacterial toxins (including choler toxin, *Escherichia coli* heat-labile enterotoxin, shigalike toxins, pertussis toxin, botulinum toxin, and tetanus toxin), viruses, and bacteria target carbohydrates on the cell surface as a means to attach and, ultimately, enter a cell. Recent carbohydrate-based sensors have employed gangliosides as receptors for protein toxins [12, 13, 22, 24–26, 57, 58], thus limiting information regarding specific carbohydrates involved in key protein toxin–carbohydrate interactions.

In order to characterize and detect specific protein–carbohydrate interactions, a model array consisting of immobilized *N*-acetyl galactosamine (GalNAc) and *N*-acetylneuraminic acid (Neu5Ac) derivatives on a planar waveguide surface was developed. GalNAc and Neu5Ac were used as receptors to assay for *Salmonella typhimurium*, *Listeria monocytogenes*, *Escherichia coli*, and staphylococcal enterotoxin B (SEB), cholera toxin, and tetanus toxin. To immobilize the model receptors on maleimide-modified glass slides, the sugars were converted to anomeric thiophenyl glycosides containing para-hydroxyl thiophenol. An acid linker was added and coupled to a thioacetate terminating linker. The monosaccharide array was constructed using a flow pattering approach similar to that described previously, albeit with a 12-channel PDMS patterning template. Fluorescently labeled bacterial cells and protein toxins were used to probe the array.

The authors observed no binding of *Salmonella typhimurium*, *Listeria monocytogenes*, *Escherichia coli*, and staphylococcal enterotoxin B to either GalNAc or Neu5Ac. Measurable binding of both cholera toxin and tetanus toxin to GalNAc and Neu5Ac was observed. The authors conclude that the results illustrate the semiselective binding behavior of the Neu5Ac and GalNAc derivatives toward cholera and tetanus toxin, as supported by previous studies. Results of dose–response studies with cholera and tetanus toxins showed a carbohydrate density dependence on the observed semiselective binding of these two toxins to the carbohydrates. Detection limits for both toxins were 100 ng/ml and the assays were complete in 65 minutes.

### 8.6.3 Microarray Cartridge System for Malaria Detection

The emergence of antimicrobial-resistant strains of pathogens such as malaria and tuberculosis, in conjunction with increased global travel and trade, has made centuries-old diseases important biothreats. According to the Centers for Disease Control and Prevention, malaria afflicts an estimated 300 to 500 million new victims each year. With funding from the U.S. Special Operations Command, Wave 80

Biosciences, Inc. has developed a field-portable, quantitative biosensor for detection of antigens associated with malaria in whole blood [59].

As described in Laser et al., the system provides detection and quantification of malarial antigens using low-density microarrays in a cartridge format. A 20 microliter whole blood sample (lysed and labeled within the cartridge) passes over a 12 spot antigen/antibody microarray in a flow chamber 120 microns deep × 3 mm wide. The malarial antigens HRP-II (associated with *Plasmodium falciparum*) and aldolase (a pan-malarial antigen relevant to *Plasmodium vivax* and other species) are detected and quantified with dynamic range greater than two orders of magnitude through variations in solid-phase probe surface concentration. With a double-sandwich assay configuration, the optical detection limit is below 100 ng per milliliter. The malaria cartridges run on a battery-powered handheld instrument with external volume (package size) less than 0.5 liters.

## 8.7 Future Directions

Protein microarrays, whether based on conventional recognition moieties such as antibodies or on newer affinity capture probes such as aptamers, are already playing a strengthening role in the identification, detection, and ultimate measurement of biothreats. Developed for basic science at the bench or robustly packaged as deployable sensors, the information available from such multiplexed systems promises to complement that available through genomic studies in the laboratory and DNA-based detection systems in the field.

Early protein microarrays were built upon concepts that borrowed from DNA microarrays and conventional immunoassays, ELISAs in particular. Recent protein microarrays for the measurement of biothreat agents have become more complex and, arguably, more adaptable. One recent example is a system based upon nanoscale glassification of gold substrates for surface plasmon resonance of supported lipid membranes, thus allowing glass surface-based assay techniques to be readily adapted for label-free SPR analysis without the need to rely on thiol-based materials [60]. Flexible detection schemes, as well as less stringent surface requirements [61], will indelibly lead to more versatile sensor platforms.

Although this chapter has focused primarily on laboratory-based protein microarrays, two commercially sponsored protein microarray systems were launched. Invitrogen, Inc. was contracted by the U.S. Department of Defense to develop Invitrogen's ProtoArray™ protein microarray technology to both detect and analyze agents such as hemorrhagic fever viruses, poxviruses, *Bacillus anthracis*, smallpox, and *Yersinia pestis*. The goal of the work is to improve the military's ability to detect dangerous bacteria and viruses in air, food, and water.

Invitrogen also demonstrated their PathAlert™ system in public venues (2006 Winter Olympics, 2006 Commonwealth Games, Technical Readiness Assessment at the U.S. Army's Dugway Proving Grounds) and reported the system as a sensitive, specific molecular approach for detecting agents such as anthrax and plague. The

ProtoArray technology is proprietary, however, Invitrogen describes the protein microarray as capable of analyzing entire pathogen and yeast proteomes or up to 5000 human proteins in a single experiment. Furthermore, the ProtoArray technology has been designed to analyze proteins, antibodies, and small molecules in a single array. Prior to 2006, Invitrogen has used the ProtoArray platform for cancer and autoimmune disease protein biomarker discovery. Protein microarray technologies developed for important areas such as biomarker discovery and proteomics form a core technology base that is beginning to prove essential to solving defense and public health biothreat concerns.

## References


1. Andreotti, P.E., Ludwig, G.V., Peruski, A.H., Tuite, J.J., Morse, S.S., and Peruski, L.F. (2003) Immunoassay of infectious agents. *BioTechniques*, 35, 850–859.
2. Feng, P. (1997) Impact of molecular biology on the detection of foodborne pathogens. *Mol. Biotechnol.*, 7, 267.
3. vanderZee, H. and Huis in't Veld, J.H. (1997) Rapid and alternative screening methods for microbiological analysis. *J. AOAC Int.*, 80, 934–940.
4. Iqbal, S.S., Mayo, M.W., Bruno, J.G., Bronk, B.V., Batt, C.A., and Chambers, J.P. (2000) Review of molecular recognition technologies for detection of biological threat agents. *Biosens. Bioelectron.*, 15, 549.
5. Ivnitski, D., O'Neil, D.J., Gattuso, A., Schlicht, R., Calidonna, M., and Fisher, R. (2003) Nucleic acid approaches for detection and identification of biological warfare and infectious disease agents. *BioTechniques*, 35, 862.
6. Schena, M. (2003) Microarrays as toxin sensors. *Pharmacogenomics J.*, 3, 125.
7. Stears, R.L., Martinsky, T., and Schena, M. (2003) Trends in microarray analysis. *Nature Med.*, 9, 140.
8. Stenger, D.A., Andreadis, J.D., Vora, G.J., and Pancrazio, J.J. (2002) Potential applications of DNA microarrays in biodefense-related diagnostics. *Curr. Opin. Biotechnol.*, 13, 208–212.
9. Cirino, N., Musser, K., and Egan, C. (2004) Multiplex diagnostic platforms for detection of biothreat agents. *Expert. Rev. Mol. Diagn.*, 4, 841–857.
10. O'Sullivan, C.K. (2002) Aptasensors — the future of biosensing? *Anal. Bioanal. Chem.*, 372, 44–48.
11. Rucker, V.C., Havenstrite, K.L., and Herr, A.E. (2005) Antibody microarrays for native toxin detection. *Anal. Biochem.*, 339, 262–270.
12. Rowe-Taitt, C.A., Cras, J.J., Patterson, J.P., Golden, J.P., and Ligler, F.S. (2000a) A ganglioside-based assay for cholera toxin using an array biosensor. *Anal. Biochem.*, 281, 123–133.
13. Rowe-Taitt, C.A., Golden, J.P., Feldstein, M.J., Cras, J.J., Hoffman, K.E., and Ligler, F.S. (2000b) Array biosensor for detection of biohazards. *Biosens. Bioelectron.*, 14, 785–794.
14. Barry, R., Diggle, T., Terrett, J., and Soloviev, M. (2003) Competitive assay formats for high-throughput affinity arrays. *J. Biomol. Screen.*, 8, 257–263.
15. Gehring, A.G., Albin, D.M., Bhunia, A.K., Reed, S.A., Tu, S.-I., and Uknalis, J. (2006) Antibody microarray detection of Escherichia coli O157:H7: Quantification, assay limitations, and capture efficiency. *Anal. Chem.*, 78, 6601–6607.
16. Cai, H.Y., Lu, L., Muckle, C.A., Prescott, J.F., and Chen, S. (2005) Development of a novel protein microarray method for serotyping Salmonella enterica strains. *J. Clin. Microbiol.*, 43, 3427–3430.
17. Holmgren, J. (1981) Actions of cholera toxin and the prevention and treatment of cholera. *Nature*, 292, 413–417.

18. Holmes, R.K. and Twiddy, E.M. (1983) Characterization of monoclonal antibodies that react with unique and cross-reacting determinants of cholera enterotoxin and its subunits. *Infect. Immun.*, 42, 914–923.
19. Svennerholm, A.M., Wikstrom, M., Lindblad, M., and Holmgren, J. (1986) Monoclonal antibodies against Escherichia coli heat-stable toxin (STa) and their use in a diagnostic ST ganglioside GM1-enzyme-linked immunosorbent assay. *J. Clin. Microbiol.*, 24, 585–590.
20. Dubey, R.S., Lindblad, M., and Holmgren, J. (1990) Purification of El Tor cholera enterotoxins and comparisons with classical toxin. *J. Gen. Microbiol.*, 136, 1839–1847.
21. Holmgren, J., Elwing, H., Fredman, P., and Svennerholm, L. (1980) Immunoassays based on plastic-adsorbed gangliosides. *Adv. Exp. Med. Biol.*, 125, 339–348.
22. Charych, D., Cheng, Q., Reichert, A., Kuziemko, G., Stroh, M., Nagy, J.O., Spevak, W., and Stevens, R.C. (1996) A 'litmus test' for molecular recognition using artificial membranes. *Chem. Biol.*, 3, 113–120.
23. Athanassopoulou, N., Davies, R.J., Edwards, P.R., Yeung, D., and Maule, C.H. (1999) Cholera toxin and GM1: A model membrane study with IAsys. *Biochem. Soc. Trans.*, 27, 340–343.
24. Fang, Y., Frutos, A.G., and Lahiri, J. (2002a) G-protein-coupled receptor microarrays. *J. ChemBio-Chem.*, 3, 987–989.
25. Fang, Y., Frutos, A.G., and Lahiri, J. (2002b) Ganglioside microarrays for toxin detection. *Langmuir*, 19(5), 1500–1505, 2003.
26. Fang, Y., Frutos, A.G., and Lahiri, J. (2002c) Membrane protein microarrays. *J. Am. Chem. Soc.*, 124, 2394–2395.
27. Rappuoli, R. and Montecucco, C. (Eds.) (1997) *Guidebook to Protein Toxins and Their Use in Cell Biology*, Oxford University Press, Oxford.
28. Cremer, P.S., and Boxer, S.G. (1999) Formation and spreading of lipid bilayers on planar glass supports. *J. Phys. Chem. B*, 103, 2554–2559.
29. Mrksich, M. (2000) A surface chemistry approach to studying cell adhesion. *Chem. Soc. Rev.*, 29, 267–273.
30. Song, X., Shi, J., and Swanson, B. (2000) Flow cytometry-based biosensor for detection of multivalent proteins. *Anal. Biochem.*, 284, 35–41.
31. Moran-Mirabal, J.M., Edel, J.B., Meyer, G.D., Throckmorton, D., Singh, A.K., and Craighead, H.G. (2005) Micrometer-sized supported lipid bilayer arrays for bacterial toxin binding studies through total internal reflection fluorescence microscopy. *Biophys. J.*, 89, 296.
32. Phillips, K.S. and Cheng, Q. (2005) Microfluidic immunoassay for bacterial toxins with supported phospholipid bilayer membranes on poly (dimethylsiloxane). *Anal. Chem.*, 77, 327–334.
33. Phillips, K.S., Dong, Y., Carter, D., and Cheng, Q. (2005) Stable and fluid ethylphosphocholine membranes in a poly(dimethylsiloxane) microsensor for toxin detection in flooded waters. *Anal. Chem.*, 77, 2960–2965.
34. Dong, Y., Phillips, K.S., and Cheng, Q. (2006) Immunosensing of Staphylococcus enterotoxin B (SEB) in milk with PDMS microfluidic systems using reinforced supported bilayer membranes (r-SBMs). *Lab Chip*, 6, 675–681.
35. Yamazaki, V., Sirenko, O., Schafer, R.J., Nguyen, L., Gutsmann, T., Brade, L., and Groves, J.T. (2005) Cell membrane array fabrication and assay technology. *BMC Biotechnol.*, 16, 18–29.
36. Schon, A. and Freire, E. (1989) Thermodynamics of intersubunit interactions in cholera toxin upon binding to the oligosaccharide portion of its cell surface receptor, ganglioside GM1. *Biochemistry*, 28, 5019–5024.
37. Parinov, S., Barsky, V., Yershov, G., Kirillov, E., Timofeev, E., Belgovskiy, A., and Mirzabekov, A. (1996) DNA sequencing by hybridization to microchip octa- and decanucleotides extended by stacked pentanucleotides. *Nucleic Acids Res*, 1, 2998–3004.
38. Guschin, D., Yershov, G., Zaslavsky, A., Gemmell, A., Shick, V., Proudnikov, D., Arenkov, P., and Mirzabekov, A. (1997) Manual manufacturing of oligonucleotide, DNA, and protein microchips. *Anal. Biochem.*, 1, 203–211.

39. Arenkov, P., Kukhtin, A., Gemmell, A., Voloshchuk, S., Chupeeva, V., and Mirzabekov, A. (2000) Protein microchips: Use for immunoassay and enzymatic reactions. *Anal. Biochem.*, 15, 123–131.
40. Revzin, A., Russell, R.J., Yadavalli, V.K., Koh, W.G., Deister, C., Mellott, M.B., and Pishko, M.V. (2001) Fabrication of poly(ethylene glycol) hydrogel microstructures using photolithography. *Langmuir*, 17, 5440–5447.
41. Angenendt, P., Glokler, J., Murphy, D., Lehrach, H., and Cahill, D.J. (2002) Toward optimized antibody microarrays: A comparison of current microarray support materials. *Anal. Biochem.*, 309, 253–260.
42. Stevens, P.W., Wang, C.H., and Kelso, D.M. (2003) Immobilized particle arrays: coalescence of planar- and suspension-array technologies. *Anal. Chem.*, 75, 1141–1146.
43. Rubina, A.Y., Dementieva, E.I., Stomakhin, A.A., Darii, E.L., Pan'kov, S.V., Barsky, V.E., Ivanov, S.M., Konovalova, E.V., and Mirzabekov, A.D. (2003) Hydrogel-based protein microchips: Manufacturing, properties, and applications. *Biotechniques*, 34, 1008–1022.
44. Charles, P.T., Taitt, C.R., Goldman, E.R., Rangasammy, J.G., and Stenger, D.A. (2004) Immobilization strategy and characterization of hydrogel-based thin films for interrogation of ligand binding with staphylococcal enterotoxin B (SEB) in a protein microarray format. *Langmuir*, 20, 270–272.
45. Ligler, F.S., Taitt, C.R., Shriver-Lake, L.C., Sapsford, K.E., Shubin, Y., and Golden, J.P. (2003) Array biosensor for detection of toxins. *Anal Bioanal Chem.*, 377, 469–477.
46. Golden, J., Shriver-Lake, L., Sapsford, K., and Ligler, F.S. (2005) A "do-it-yourself" array biosensor. *Methods*, 37, 65–72.
47. Rowe, C.A., Scruggs, S.B., Feldstein, M.J., Golden, J.P., and Ligler, F.S. (1999) An array immunosensor for simultaneous detection of clinical analytes. *Anal. Chem.*, 15, 433–439.
48. Bhatia, S.K., Shriver-Lake, L.C., Prior, K.J., Georger, J.H., Calvert, J.M., Bredehorst, R., and Ligler, F.S. (1989) Use of thiol-terminal silanes and heterobifunctional crosslinkers for immobilization of antibodies on silica surfaces. *Anal. Biochem.*, 178, 408–413.
49. Rowe, C.A., Tender, L.M., Feldstein, M.J., Golden, J.P., Scruggs, S.B., MacCraith, B.D., Cras, J.J., and Ligler, F.S. (1999) Array biosensor for simultaneous identification of bacterial, viral, and protein analytes. *Anal. Chem.*, 1, 3846–3852.
50. Hughes, J.M. and Tauxe, R.V. (1990) *Principles and Practice of Infectious Diseases*, 3rd ed. G.L. Mandell, R.G. Douglas, and J.E. Bennett (Eds.), New York: Churchill Livingstone.
51. Mattix, M.E., Hunt, R.E., Wilhelmsen, C.L., Johnson, A.J., and Baze, W.B. (1995) Aerosolized staphylococcal enterotoxin B-induced pulmonary lesions in rhesus monkeys (Macaca mulatta). *Toxicol. Pathol.*, 23, 262–268.
52. Drobkov, V.I., Abdullin, T.G., and Darmov, I.V. (1991) The use of immunoenzyme analysis in the laboratory diagnosis of plague. *Zh. Mikrobiol. Epidemiol. Immunobiol.*, 10, 40–42.
53. Eisenberg, P.R., Sherman, L.A., Perez, J., and Jaffe, A.S. (1987) Relationship between elevated plasma levels of crosslinked fibrin degradation products (XL-FDP) and the clinical presentation of patients with myocardial infarction. *Thromb. Res.*, 46, 109–120.
54. Deitcher, S.R. and Eisenberg, P.R. (1993) Elevated concentrations of cross-linked fibrin degradation products in plasma. An early marker of gram-negative bacteremia. *Chest*, 103, 1107–1112.
55. Shepherd, A.J., Hummitzsch, D.E., Leman, P.A., Swanepoel, R., and Searle, L.A. (1986) Comparative tests for detection of plague antigen and antibody in experimentally infected wild rodents. *J. Clin. Microbiol.*, 24, 1075–1078.
56. Ngundi, M.M., Taitt, C.R., McMurry, S.A., Kahne, D., and Ligler, F.S. (2006) Detection of bacterial toxins with monosaccharide arrays. *Biosens. Bioelectron.*, 21, 1195–1201.
57. Song, X. and Swanson, B.I. (1999) Direct, ultrasensitive, and selective optical detection of protein toxins using multivalent interactions. *Anal. Chem.*, 71, 2097–2107.
58. Singh, A.K., Harrison, S.H., and Schoeniger, J.S. (2000) Gangliosides as receptors for biological toxins: Development of sensitive fluoroimmunoassays using ganglioside-bearing liposomes. *Anal. Chem.*, 72, 6019.

59. Laser, D.J., Hardham, C.T., and Kim, J. (2006) Flow-chamber microarray detection of malarial antigens In: *Solid-State Sensors, Actuators, and Microsystems Workshop*, Spangler, L. and Kenny, T.W. (Eds.). Transducers Research Foundation, Hilton Head Island, SC, pp. 408–409.
60. Phillips, K.S., Han, J.H., Martinez, M., Wang, Z., Carter, D., and Cheng, Q. (2006) Nanoscale glassification of gold substrates for surface plasmon resonance analysis of protein toxins with supported lipid membranes. *Anal. Chem.*, 78, 596–603.
61. Rucker, V.C., Havenstrite, K.L., Simmons, B.A., Sickafoose, S.M., Herr, A.E., and Shediac, R. (2005) Functional antibody immobilization on 3-dimensional polymeric surfaces generated by reactive ion etching. *Langmuir*, 21, 7621.

Kigutha, H., 1994. "*Guidelines for cholera control*: *By World Health Organisation* (Ed.) 61 pp., ISBN 92-4-154449-X," Health Policy, Elsevier, vol. 27(2), pages 197-197, February.

# Chapter 9
# Photo-Generation of Carbohydrate Microarrays

**Gregory T. Carroll, Denong Wang, Nicholas J. Turro, and Jeffrey T. Koberstein**

**Abstract** The unparalleled structural diversity of carbohydrates among biological molecules has been recognized for decades. Recent studies have highlighted carbohydrate signaling roles in many important biological processes, such as fertilization, embryonic development, cell differentiation and cell–cell communication, blood coagulation, inflammation, chemotaxis, as well as host recognition and immune responses to microbial pathogens. In this chapter, we summarize recent progress in the establishment of carbohydrate-based microarrays and the application of these technologies in exploring the biological information content in carbohydrates. A newly established photochemical platform of carbohydrate microarrays serves as a model for a focused discussion.

## 9.1 Introduction

The human genome project has shown that about 30,000 genes are available for constructing the human proteome. However, the number of genes revealed by whole genome sequencing does not set the upper limit of the repertoire of proteins. Protein posttranslational modifications, especially glycsylation, further diversify the available repertoire of functional proteins in a living organism. Protein glycosylation results in attachment of carbohydrate moieties at certain sites of a newly synthesized protein by either *N*-glycosylation or *O*-glycosylation. About 50% of the proteins in eukaryotes

G.T. Carroll
Department of Chemistry, Columbia University, 3000 Broadway, MC 3157, 10027, New York, NY

D. Wang (✉)
Stanford Tumor Glycome Laboratory, Department of Genetics, Stanford University School of Medicine, Beckman Center B006, 94305, Stanford, CA
e-mail: dwang1@stanford.edu

N.J. Turro and J.T. Koberstein
Department of Chemical Engineering, Columbia University, 500 West 120th Street, 10027, New York, NY 10027

K. Dill et al. (eds.), *Microarrays: Preparation, Microfluidics, Detection Methods, and Biological Applications*,

are found to be glycosylated [1]. Virtually all mammalian cells and micro-organisms are decorated with characteristic carbohydrate moieties. Exploring the biological information content in carbohydrates represents one of the current focuses of postgenomic research and technology development.

Carbohydrates are structurally suitable for the generation of diversity. Analogous to nucleotides and amino acids, monosaccharides are the building blocks of sugar chains. As in DNA and proteins, a carbohydrate's structural diversity comes from the composition of its repeat units and the sequence of these units. In mammals, there are nine common monosaccharide repeat units. A unique source of diversity to carbohydrates is the possibility to form branched structures because linkages can occur at multiple locations on a monosaccharide. As discussed below, the location of the linkage can be an important parameter in molecular recognition. For example, the flu virus recognizes the same monosaccharide units in both humans and birds; however, strains specific for birds recognize a linkage at a different location than those specific for humans (see below). In addition, the configuration at the anomeric position is another source of diversity.

Monosaccharides are linked together by forming glycosidic bonds between an anomeric OH on a monosaccharide and any OH (or other reactive functional group) on another monosaccharide. A pyranoside (six-membered ring monosaccharide) can bond to another pyranoside at five different positions with each linkage being in an $\alpha$ or $\beta$ orientation. The potential to create differential linkages allows monosaccharides to form a more diverse array of oligomers in comparison to oligonucleotides and amino acids. For example, any fully hydroxylated pyranoside can form 11 possible disaccharides. A given amino acid can form only one dipeptide. A trimer of the nine common sugar residues found in mammals could potentially give rise to 119,736 different biological "expressions." A trimer of the 20 amino acids or four oligonucleotides can form only 8000 and 64 "expressions" in comparison. It should be noted that bacteria contain over 100 types of monosaccharides, often containing functional groups not found in mammalian monosaccharides, making the potential number of sugar structures found in nature well beyond the number found in mammals.

The "marriage" of carbohydrates with other biomolecules produces a large repertoire of carbohydrate-containing biomolecules with a variety of hybrid structures, called glycoconjugates. The two major classes of glycoconjugates, glycolipids and glycoproteins, are present in many living organisms. Each of these can be further subdivided. Glycoproteins bearing amino sugars, called glycosaminoglycans, are categorized as proteoglycans. Peptidoglycans are glycosaminoglycans cross-linked by peptides. Glycopeptides are oligosaccharides bound to oligopeptides and are either degradation products or are chemically synthesized. Glycosphingolipids are glycolipids in which an oligosaccharide is covalently attached to the lipid sphinganine. Lipopolysaccharides are glycolipids in which the sugar is a polymeric carbohydrate derivative.

Appending carbohydrates to lipids and proteins serves various roles. It allows carbohydrates to be anchored to the surface of biological membranes and can change the physical properties of the molecules to which they are conjugated. For example, carbohydrates attached to glycoproteins in deep-sea fish are thought to disrupt the formation of ice crystals allowing the fish to survive at temperatures below 0° C.

In addition, glycoconjugation can endow a molecule with a saccharide molecular marker. For example, sialic acid residues found on an erythrocyte *N*-linked glycoprotein indicates whether the blood cell is "new" or "old" [2, 3] as depicted in Fig. 9.1a. If the glycoprotein loses its sialic acid residue, a galactose residue is exposed and recognized by a liver surface protein called the asialoglycoprotein receptor (*asialo* meaning "without sialic acid"). When enough galactose residues are exposed, multiple interactions bind the "old" erythrocyte to the liver. The "old" erythrocyte is then absorbed into the liver for degradation. However, the native

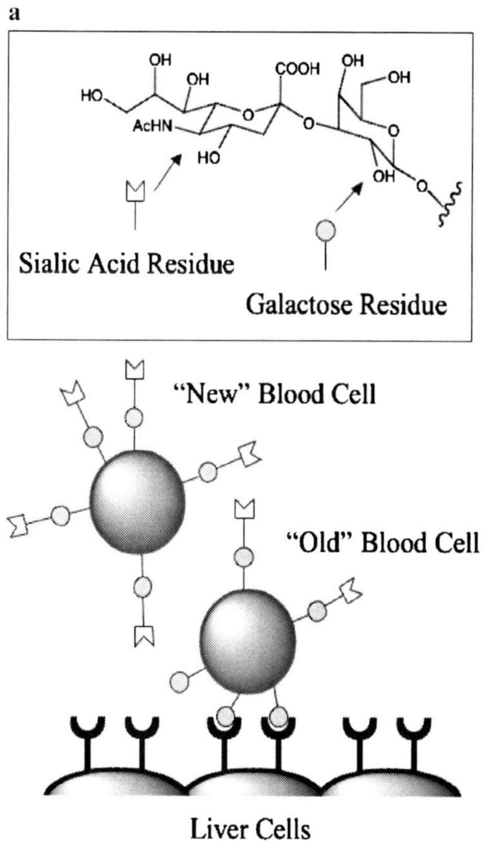


**Fig. 9.1** Characteristic carbohydrate moieties serve as markers for biological recognition. (a) Sialic acid residues found on erythrocytes provide a molecular marker to indicate whether the cell is "new" or "old" [3]. Loss of sialic acid exposes a galactose residue that is recognized by the asialoglycoprotein receptor in mammalian liver [4]. When multiple galactose residues are exposed multivalent interactions result in a tight binding to the liver cells

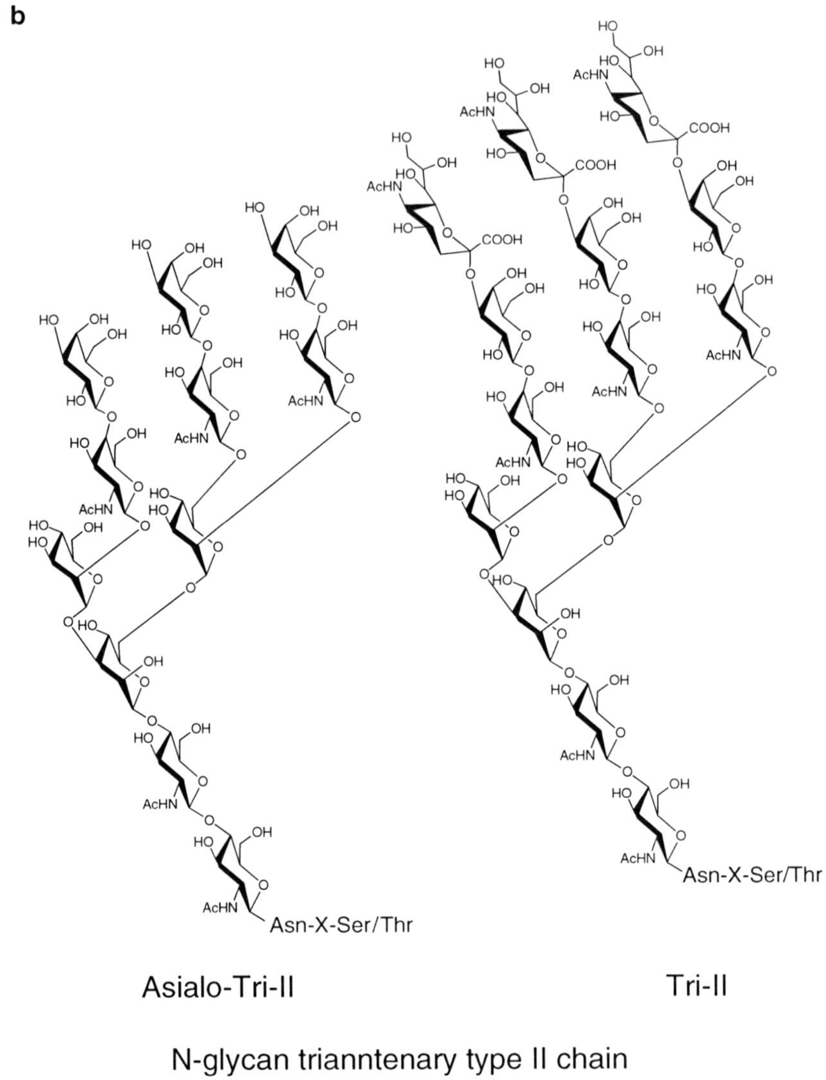


**Fig. 9.1** (continued) (b) *N*-glycan type II chains [Galβ(1 → 4)GlcNAc] in the triantennary cluster configurations with (Tri-II) and without sialic acid terminal residues (asialo-Tri-II). The asialo-Tri-II sugar moieties but not the Tri-II structures are specifically targeted by a horse-neutralization antibody of SARS-CoV [5]

structures of cellular *N*-glycans are much more complex than the schematics in Fig. 9.1a. The sugar moieties recognized by the liver asialoglycoprotein receptor are in triantennary and multiantennary configurations without sialic acid caps [4]; Fig. 9.1b).

Recent studies have demonstrated that carbohydrate-mediated molecular recognition plays key roles in many important biological processes. These include cellular events, such as fertilization [6, 7], embryonic development [8], and cell differentiation and cell–cell communication [9–11]. In addition, important molecular processes such as blood coagulation [12, 13], inflammation [3, 14], and chemotaxis [15], as well as host recognition and immune responses to microbial pathogens [16–18] are mediated or regulated by carbohydrate biosignals.

The carbohydrate-mediated processes mentioned above rely on the presence of appropriate biological molecules that are able to "decode" the biosignals of carbohydrate moieties. There are at least two classes of biological molecules in living organisms that have evolved to play such roles. They are known as anticarbohydrate antibodies produced by immune cells (B lymphocytes; [18]) and lectins of nonimmune origins [19].

The carbohydrate moieties that accommodate binding and are complementary to the combining-sites of lectins and/or antibodies are termed glycoepitopes. It was estimated that there are about 500 endogenous glycoepitopes in mammals [20]. However, this estimation did not consider the repertoires of the "hybrid" structures that are generated by protein posttranslational modification, including both *N*- and *O*-glycosylation. Furthermore, the conformational diversity of carbohydrates substantially increases the repertoire of carbohydrate-based antigenic determinants or glycoepitopes [21–23].

## 9.2 Carbohydrate Microarrays as Essential Tools in the Postgenomics Era

To meet the technical challenges posed by the structural diversity and functional complexity of carbohydrates, substantial efforts have been made to establish different platforms of carbohydrate microarrays [24–31]. The following examples demonstrate that these relatively nascent technologies have been utilized to explore the mysteries of life shrouded in the structure of carbohydrates.

A carbohydrate microarray printed on nitrocellulose-coated glass slides has been applied to explore the potential immunogenic sugar moieties expressed by a previously unrecognized viral pathogen, SARS-CoV [5]. The strategy established in the SARS virus investigation is likely applicable for other microbial pathogens. It involves three steps of experimental investigation. In the first step, a carbohydrate microarray is used to characterize antibody responses to an infectious agent or antigen preparation in order to recognize the disease- or pathogen-associated anticarbohydrate antibody specificities. The second step focuses on identifying lectins and/or antibodies that are specific for the glycoepitopes that are recognized by the pathogen-elicited antibodies. This provides specific structural probes to enable the third step of investigation, that is, to identify the glycoepitopes in the candidate pathogens using specific lectins or antibodies identified in steps 1 and 2.

The rationale for this approach is that if SARS-CoV expressed antigenic carbohydrate structures, then immunizing animals using the whole virus-based vaccines would elicit antibodies specific for these structures. In addition, if SARS-CoV displayed a carbohydrate structure that mimicked host cellular glycans, then vaccinated animals may develop antibodies with autoimmune reactivity to their corresponding cellular glycans. By characterizing the SARS-CoV neutralizing antibodies elicited by an inactivated SARS-CoV vaccine, autoantibody reactivity specific for the carbohydrate moieties of an abundant human serum glycoprotein, asialo-orosomucoid (ASOR), was detected [5]. This "chip hit" provides important clues for the selection of specific immunologic probes to further examine whether SARS-CoV expresses antigenic structures that imitate the host glycan. Given that lectin PHA-L is specific for glycoepitopes Tri-II or mII of ASOR (See Fig. 9.1b for an asialo-Tri-II structure of *N*-glycans), this lectin was applied as a structural probe to examine whether SARS-CoV expresses the PHA-L reactive antigenic structure. The results demonstrated that glycoepitopes Tri-II or mII of ASOR are highly expressed by SARS-CoV-infected cells and by the viral particles.

Another study [32] involved the dendritic cell receptor, DC-SIGN, and the endothelial cell receptor, DC-SIGNR, both of which play an important role in pathogen recognition. The ligand-binding properties of these receptors were elucidated using a glycan array fabricated with biotinylated mono- and oligosaccharides immobilized on streptavidin-coated wells. Screening the arrays with extracellular domains of the two cell receptors showed that in addition to the ligands that DC-SIGNR binds, DC-SIGN binds glycans that contain terminal fucose residues. Almost all the carbohydrates screened that bound to the CRD of DC-SIGN had branched terminal structures. Bulky and charged sialic acid residues were found to prevent binding.

Glycan arrays have also been used to investigate the interaction of the gp120 glycoprotein of HIV-1 [29], which interacts with CD4 of human T cells. The binding of gp120 to CD4 initiates events that subsequently allow the gp41 glycoprotein of HIV-1 to insert into the host cell membrane. Understanding how HIV carbohydrates interact with binding partners is expected to aid in developing agents to prevent HIV entry. A microarray of natural and modified glycoproteins, as well as neoglycoproteins were used to reveal the binding profiles of the following four gp120-binding molecules: DC-SIGN, a monoclonal antibody 2G12, cyanovirin-N (CVN), and scytovirin.

The dependence of carbohydrate moieties in their binding to the gp120 was studied by chemically modifying the protein ovalbumin with a high mannose oligosaccharide found on gp120. The four gp120-binding molecules only bind to ovalbumin when the high mannose oligosaccharide is present, indicating that their reactivity with gp120 is mainly dictated by the oligosaccharide residues as opposed to the polypeptide backbone. The binding profiles of the four proteins were investigated in further detail. A microarray composed of the high mannose oligosaccharide and components of the oligosaccharide revealed that both 2G12 and CVN bind to terminal Man$\alpha$1–2Man linkages whereas scytovirin requires an additional underlying $\alpha$1–6 trimannoside moiety. DC-SIGN was found to bind to all components investigated.

Aside from gp120 studies, the microarray revealed the novel finding that gp41, which expresses high-mannose oligosaccharides and is known to bind CVN and scytovirin, also binds both DC-SIGN and 2G12. The microarray allows for a rapid screening of fine structural details within a carbohydrate to ascertain the motifs responsible for molecular recognition. Such details will be beneficial in developing prophylaxis agents that prevent or inhibit HIV from infecting a host's cells.

A fourth pathogen-related application [33] involves the influenza A virus that has recently received much public attention due to several recent cases of the avian virus infecting humans. Aside from foreboding commentary of the emergence of a pandemic strain in the near future, virulent episodes from 1918 that claimed more than 50 million lives has left a lasting interest in understanding the immunogenic details of this virus. Influenza A infection is initiated by binding of hemagglutinin (HA), an antigenic protein found on the virus' coat, to carbohydrates on the surface of the host's epithelial cells. HA recognizes sialic acid terminated glycans and its linkage to galactose residues (see Fig. 9.1 for sialic acid terminated glycans in the Tri-II sugar chain configuration). The receptor specificity of various serotypes has been studied using cell-based assays. HA variants adapted to humans recognize an α2–6 linkage whereas strains specific for birds recognize an α2–3 linkage. Recently, researchers have applied carbohydrate microarrays to study the virus in a cell-independent assay [33].

Microarrays are expected to reduce complications involved in the cell studies. Using a combination of genomic sequence analysis and a glycan array displaying 200 carbohydrates, factors determining the specificity of influenza A for birds and mammals were investigated. By screening a variety of HAs and probing their interaction with a glycan array containing sugars with sialic acids attached via α2–3, α2–6, and α2–8 linkages among other glycans, specific mutations were shown to control the specificity of the HA for a given linkage as anticipated by previous studies. In addition, binding specificity was correlated to other fine structural motifs including charge, size, sulfation, fucosylation, and sialylation showing that the microarray can be used to identify different strains of the virus based on a fingerprint of the specificity.

The microarray could distinguish between the binding specificity of two different strains of human HAs, 18NY and 18SC (named after 1918 pandemic strains found in New York and South Carolina). 18SC HA recognizes only α2–6 linkages whereas the 18NY additionally recognizes α2–3 linkages. In the avian 18NY strain, a Glu190Asp mutation confers α2–6 recognition, resulting in virulence towards humans. The strain becomes more suited to α2–6 recognition after a second mutation, Gly225Asp. Thus, only two mutations are required to cross the major species barrier. This study shows that glycan arrays can be used to rapidly screen specificity profiles of pathogens and to predict the emergence of human pathogenic strains.

Another application of carbohydrate microarrays includes the study of carbohydrate-processing enzymes. Nature utilizes enzymes to synthesize carbohydrates. Understanding the specificity of a given enzyme upon carbohydrate modification, as well as identifying inhibitors of the enzyme aids in treating diseases that rely on carbohydrate interactions is important. For example, NB-DNJ is an inhibitor that prevents a glucosidase enzyme from constructing appropriate carbohydrates on gp120, a glycoprotein of HIV mentioned above, reducing the virus' ability to bind to leucocytes. The specificity of enzymes that modify carbohydrates can be probed by treating a carbohydrate array with a given enzyme. Modified carbohydrates can then be revealed by lectin interactions. In addition, potential inhibitors of the enzyme activity can be screened by mixing selected candidates with the enzyme prior to treating the array. Carbohydrates stabilized on a microtiter plate through hydrophobic interactions have been used to reveal inhibitors of fucosyltransferase, an enzyme responsible for transferring a fucose residue from GDP-fucose to a sialyl-lactosamine to form sialyl-Lewis$^x$, a tetrasaccharide involved in the inflammatory response in mammalian tissue [34]. By incubating various candidate inhibitors with the enzyme prior to immersing the microarray, four inhibitors with nanomolar $K_i$ values were discovered. The authors indicate that the method is 70% more cost-effective than a previously applied coupled-enzyme assay method.

## 9.3 Construction of Carbohydrate Microarrays

Developing new methods to fabricate carbohydrate microarrays has been an ongoing topic since 2002 when the first microarrays were reported [28]. Four important requirements involved in creating a functional microarray include (1) the ability to immobilize biological molecules on a flat substrate; (2) the immobilized molecules retain their biorecognition properties; (3) the sensitivity to detect a broad range of specificities; and (4) the ability to incorporate high-throughput equipment in creating the array. Both high- and low-density microarrays have been described. Low-density arrays are created on microtiter plates. High-density arrays are created on glass, metallic, and polymer surfaces. The latter can contain tens of thousands of sugars on one microscope slide for a large-scale characterization of saccharides and their receptors and has a unique advantage in exploring unknown carbohydrate targets and their potential receptors and antibodies. The former is suitable for a more focused biomedical application.

Fig. 9.2 shows the schematics of four approaches for immobilization of saccharides on a chip substrate: (1) noncovalent and nonspecific, (2) noncovalent and specific, (3) covalent and nonspecific, and (4) covalent and specific. Covalent immobilization links sugars on a surface by forming covalent bonds to the substrate. It ensures a stable immobilization of the saccharides regardless of their physicochemical properties. The ability to control the specificity of immobilization provides that a given face of the sugar will not be inactivated through chemical derivatization or burying at the substrate interface. Given that carbohydrates are structurally diverse and that key sugar moieties for biological recognition are frequently unknown to begin with, it remains technically challenging and practically

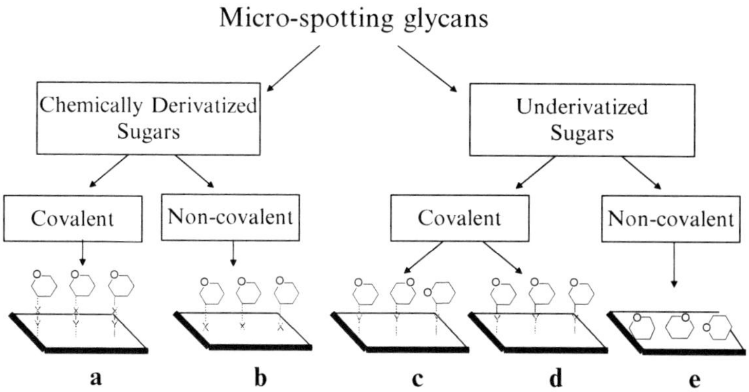


**Fig. 9.2** Microspotted sugars can be stabilized on a chip in a variety of ways [35]. Chemically derivatized carbohydrates can be site-specifically immobilized on a substrate derivatized with appropriate functional groups (a). Carbohydrates derivatized with hydrophobic groups such as long hydrocarbon tails adsorb to hydrophobic substrates. Additionally, biotinylated carbohydrates can be stabilized on streptavidin-coated surfaces (b). Photochemical methods can be used to covalently immobilize underivatized carbohydrates to surfaces containing photoactive groups such as aromatic carbonyls or diazo compounds (c). Underivatized carbohydrates can be site-specifically immobilized on hydrazide or amino-oxy derivatized surfaces (d). Underivatized carbohydrates can be physically adsorbed onto polymer surfaces such as nitrocellulose or oxidized polystyrene. The interaction of immobilized carbohydrates with various biological species can be probed using fluorescence as a detection method

difficult to achieve a generally useful method for covalent and site-specific saccharide immobilization for a broad range of applications.

A number of noncovalent methods for presenting carbohydrates on a surface have been developed and are currently in use by many investigators. These include the use of native polysaccharides, glycoproteins, and glycolipids, as well as the use of chemically derivatized carbohydrates [23–25, 28, 36–38]. Hydrophobic effects are utilized by chemically derivatizing the carbohydrates with a hydrophobic group such as a lipid and spotting onto a hydrophobic surface.

Lipid-linked oligosaccharides have been immobilized on commercially available nitrocellulose-coated surfaces [25]. Similarly, sugars derivatized with long tail hydrocarbon chains of 13–15 carbon atoms have been immobilized on microtiter plates [24, 29]. Carbohydrates derivatized with $C_8F_{17}$ fluorocarbon tails can be stabilized on glass slides containing a fluoroalkylsilane coating [40]. In addition, carbohydrates can be conjugated to bovine serum albumin and subsequently immobilized on an appropriate substrate [28, 41, 42]. Another strategy of carbohydrate presentation involves immobilizing biotinylated carbohydrates on streptavidin-coated wells [43] or glass slides [44].

The use of noncovalent immobilization in carbohydrate microarrays is analogous to bioassays that are based on the noncovalent immobilization of a biomolecule on a solid phase. These include the Southern blot for DNA hybridization, Northern blot for mRNA detection, and Western blot and ELISA assays for monitoring protein and/or carbohydrate-based biomarkers. However, the efficacy of saccharide immobilization by noncovalent bonding is under the influence of the physicochemical properties of the given molecules. The stability of immobilization and exposure of the desired glycoepitopes or antigenic determinants must be examined for each saccharide on the substrate [5, 28, 31, 37, 38].

A number of methods for covalent and site-specific saccharide immobilization have also been developed. For example, maleimide-linked carbohydrates have been attached to thiol-coated surfaces, and vice versa, by formation of a thioether linkage [27, 29, 45]. Other surface linking reactions include Diels–Alder reaction [26], dipolar cycloaddition [46], amine-*N*-hydroxy succinimide (NHS) coupling [47, 48], *p*-aminophenyl-cyanuric chloride coupling [49], dimethylaminopyridine-NHS coupling [50], and the attachment of thiol-derivatized carbohydrates to gold films [51, 52]. In most of these procedures, both a chemically derivatized surface and derivatized carbohydrate are needed.

Methods that allow for arrays to be created from underivatized sugars are important because this avoids altering the native structure of bioactive carbohydrates and is less time-consuming for array construction. A few methods for immobilizing underivatized sugars have been demonstrated. Nitrocellulose-coated glass chips have been used to create polysaccharide microarrays. Although this is the most convenient method presently available, it can only be used for high molecular weight polysaccharides unless lipids are attached (see above), in which case oligosaccharides can be immobilized. Similarly, oxidized black polystyrene substrates immobilize underivatized sugars. The black substrate has the advantage of giving a high signal-to-noise ratio.

Recent methods allow for underivatized sugars to be covalently bound to a surface. Hydrazide-coated glass slides have been shown to react with underivatized mono-, oligo-, and polysaccharides in a site-specific manner [53]. Similarly, oligosaccharide microarrays have been prepared on hydrazide-coated gold slides [54]. In addition, underivatized carbohydrates have been bound to amino [55, 56], aminooxy [53], aminooxyacety [57], and phenylboronic acid-coated [58] slides. Note that in some cases the surface reactions reported were not used to fabricate microarrays. Two other methods for covalently immobilizing underivatized sugars involve derivatizing a surface with photoactive groups that can form covalent bonds to a wide variety of molecules after irradiation with UV light. These methods are discussed below.

## 9.4 Photons as General Reagents for Covalent Coupling of Carbohydrates on a Chip

When an appropriate functional group within a molecule absorbs a photon, the physical and chemical characteristics of the molecule can be dramatically altered [59]. Fig. 9.3 illustrates the effects of light absorption on the electronic structure of a molecule when the photon is in the UV/visible region of the electromagnetic spectrum. Electrons in the ground electronic state, $S_0$, are promoted to a higher energy level, $S_1$, resulting in an excited singlet state (1). The excited molecule can relax to the ground state by emitting light (fluorescence) (2) or giving off heat (internal conversion) (3). The excited molecule can also undergo a chemical reaction (4) that is not possible in the ground state. In addition, the electron can flip its spin, a process called intersystem crossing (5), resulting in an excited triplet state. From the triplet state the electron can undergo phosphorescence (emission of light from the triplet state) (6), internal conversion (7), or chemical reaction (8).

Exploiting the photoactive nature of various chromophores (chemical units within a molecule that absorb light of a given wavelength) has allowed chemists to form covalent bonds between chemicals that are otherwise nonreactive. This feature has allowed underivatized carbohydrates to be immobilized on surfaces bearing appropriate chromophores. In addition, the use of photons as "traceless" and "weightless" reagents is a convenient, clean, and inexpensive method for fabricating microscale devices.

A variety of photochemical reactions has been performed on carbohydrates [60]. Our emphasis is focused on reactions that involve covalent bond formation between a carbohydrate and another molecule. H-abstraction from a C–H group by a photochemically excited state of carbonyl compounds is a common primary photochemical process. In this reaction, an excited carbonyl compound abstracts a hydrogen atom from a suitable C–H donor to form a pair of radicals. The susceptibility of a hydrogen atom to abstraction will depend on the C–H bond strength and the stability of the resulting radicals. The more substituted a carbon atom, the greater will be the stability of a radical forming at that carbon atom. In addition, inductive and resonance effects will influence the bond strength and stability of the resulting radical.

Studies on model compounds have shown that H-abstraction occurs preferentially at the anomeric center [60], however, when actual carbohydrates were studied, ESR results indicated that H-abstraction occurs preferentially at C–H bonds on $C_1$–$C_4$ with varying efficiency [61, 62]. Certain substituents can affect the selectivity

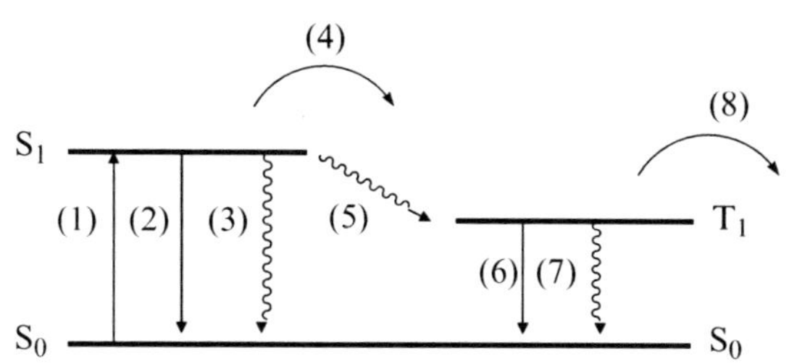


**Fig. 9.3** The possible photophysical and photochemical events that can occur upon absorption of a photon by a molecule: (1) singlet–singlet absorption of a photon; (2) fluorescence; (3) internal conversion; (4) reaction from the excited singlet state; (5) intersystem crossing to the triplet state; (6) phosphorescence; (7) internal conversion; and (8) reaction from the triplet state

of the reaction. For example, in D-galacturonic acid hydrogen abstraction occurs predominantly at $C_5$ due to a combination of steric and stereoelectronic effects from an adjacent carboxyl group [63].

It should be noted that electron-donating substituents such as amines can favor electron transfer as opposed to H-abstraction. The rate for electron transfer is at least an order of magnitude faster than hydrogen abstraction, so sugars such as glycosaminoglycans are expected to favor this pathway. The possibility for electron transfer to occur will depend on the excited state reduction potential of the chromophore and the oxidation potential of the carbohydrate. When proton transfer follows electron transfer, the resulting radicals can recombine to form a covalent bond. Other possible reactions include back transfer, disproportionation, and various rearrangements that are pH-dependent [63].

Photogeneration of carbenes and nitrenes provides another potential method to form covalent bonds to carbohydrates. Irradiation of diazo and azide compounds results in the loss of $N_2$ and subsequent formation of carbene and nitrene intermediates. These can undergo a variety of reactions that result in covalent bond formation including insertion into sigma and pi bonds, addition of a nucleophile or electrophile and hydrogen abstraction.

Aziridine derivatized polysaccharides have been used to cast films on a surface that react with underivatized sugars after irradiation with UV light [64]. Upon absorption of a photon, the aziridine group loses $N_2$ to form a highly reactive nitrene. The nitrene presumably reacts with spotted sugars to form a covalent bond. Only polysaccharides were investigated using this surface. Similarly, diazirine derivatized mono- and disaccharides have been synthesized and photochemically immobilized on diamond [65] and poly (styrene) [66] films by photogenerating a carbene that reacts with the film.

A recently reported photochemical method employs a self-assembled monolayer [67] on a glass chip presenting phthalimide chromophores at the air–monolayer interface as shown in Fig. 9.4 [31]. The phthalimide chromophore is known to undergo a variety of photochemical reactions [68, 69]. In the presence of a carbohydrate the most plausible primary process is hydrogen abstraction from a C–H group.

Recombination of the resulting radicals results in a covalent bond. The mechanism is illustrated in Fig. 9.5. After absorption of a photon, the excited phthalimide abstracts a hydrogen atom from a nearby molecule. This creates radical centers on both molecules. The radicals can then recombine to form a covalent bond. Tethering a monolayer of phthalimide-derivatized silanes to a glass surface, SAM **1** (Fig. 9.4), provides a platform for covalently immobilizing sugars on chips. The phthalimide molecule is covalently bound to the glass through a condensation reaction between a trimethoxy silane and Si–OH groups at the glass surface. Bond formation to the substrate as well as van der Waals interactions between the long alkyl chains provide a driving force for the molecules to self-assemble into oriented clusters one molecule thick at the surface. In general the clusters are collectively described as a monolayer of molecules.

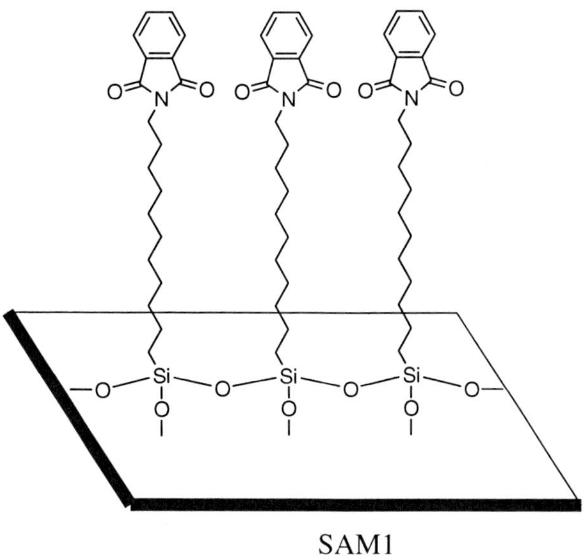


**Fig. 9.4** A self-assembled monolayer containing phthalimide endgroups that become reactive when irradiated with UV light

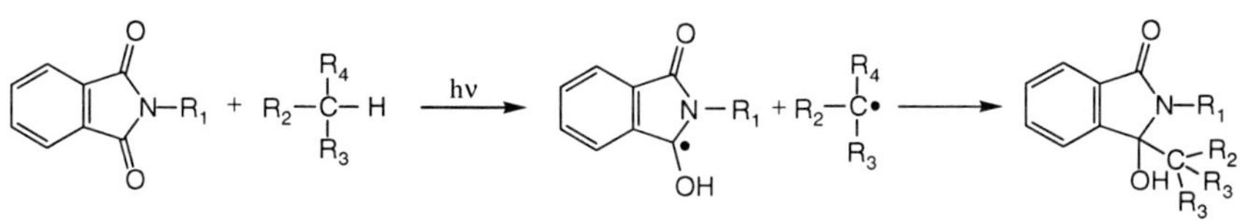


**Fig. 9.5** A phthalimide derivative can undergo a photochemical hydrogen abstraction reaction followed by recombination to form a covalent bond

The "two-dimensional" nature of the surface guarantees that the photo-reactive group will present itself at the solid–air interface where the carbohydrates will be adsorbed. This is in contrast to polymeric three-dimensional coatings where van der Waals interactions will dictate which groups migrate to the interface. A key advantage of this photoactive surface is that sugars of all sizes should in principle be stabilized on the substrate because a covalent bond is expected to form. A simple modification of this surface that involves mixing a hydrophilic molecule into the monolayer (discussed below) allows for the use of high-throughput equipment to create microarrays of underivatized poly-, oligo-, and monosaccharides. Microarrays fabricated in this way have proven successful in elucidating fundamental information concerning the immunogenic properties of oligosaccharides found on pathogen glycoproteins.

A first step in pursuing any new methodology of microarray fabrication involves testing the interfacial chemistry between the carbohydrate and surface. In order to test the photochemical reactivity of SAM **1** towards sugars, films of polysaccharides were spin-coated onto SAM **1**. Ellipsometry, contact angle, and fluorescence measurements were used to demonstrate that polysaccharides can be photochemically immobilized on this surface. In order to show the versatility of the method, a photolithographic patterning experiment was used to show that poly-, di-, and monosaccharides could be immobilized on the surface. Irradiation of homogeneous films of dextran polysaccharides, sucrose, and glucose through a photomask results in stabilized hydrophilic patterns that can be visualized by condensing water onto the surface. This demonstrates that sugar films composed of carbohydrates of any size can be immobilized on a phthalimide monolayer.

In order to be functional at the highest level, immobilized sugars must be accessible to assayed lectins and other such molecules and they must preserve their ability to react with cellular receptors or antibodies of defined specificities. Ideally, the surface must be suitable for patterning by way of conventional robotic spotters used to create microarrays because this allows for an automated production of potentially thousands of different carbohydrate spots on a single chip [38]. In order to perform a reaction at a surface, the thermodynamic properties of the system must favor adsorption of the reacting molecules at the interface. This is pertinent to constructing a microarray in that in order to use a robotic spotter, the surface must favor adsorption of a glycan solution onto the substrate. Otherwise, very little material will leave the pin of the spotter.

It was found that a monolayer consisting of only phthalimide end-groups was inappropriate for spotting. Even when a noticeable amount of material was spotted on SAM **1** or a benzophenone-terminated monolayer (another class of aromatic carbonyls that participates in H-abstraction when irradiated), the surfaces were unable to retain a detectable amount of carbohydrates after irradiation. This is in contrast to the film studies in which a pure monolayer was sufficient to immobilize sugars. The discrepancy is probably a result of the film's sampling a much larger area of the surface in comparison to a spot of approximately 200 μm, making immobilized sugars easier to detect. Also, the mechanical effects of spin-coating may press the sugars into gaps in the monolayer allowing the excited carbonyl more access to the sugar.

In order to make the surface more amenable for spotting, trimethoxyaminopropylsilane was mixed into the surface (PAM) as shown in Fig. 9.6. A ratio of 5:1 amine:phthalimide was found to give a reliable surface for spotting and immobilization. The amine acts as bait to pull the sugars onto the surface through a favorable hydrophilic interaction. The hydrophilic gaps are also expected to put the carbohydrate in a more favorable location for the phthalimide to abstract a hydrogen atom from the sugar. The structure of the surface is probably more complicated than the simple picture shown in Fig. 9.6. If a monolayer is formed, the phthalimide most likely will try to tilt over the amines to reduce the interfacial tension. In addition, it is possible that like molecules cluster or even phase-separate within the mixed monolayer, or those multilayers, oligomers, or polymers form, but these phenomena have not yet been investigated in this system. Regardless of the

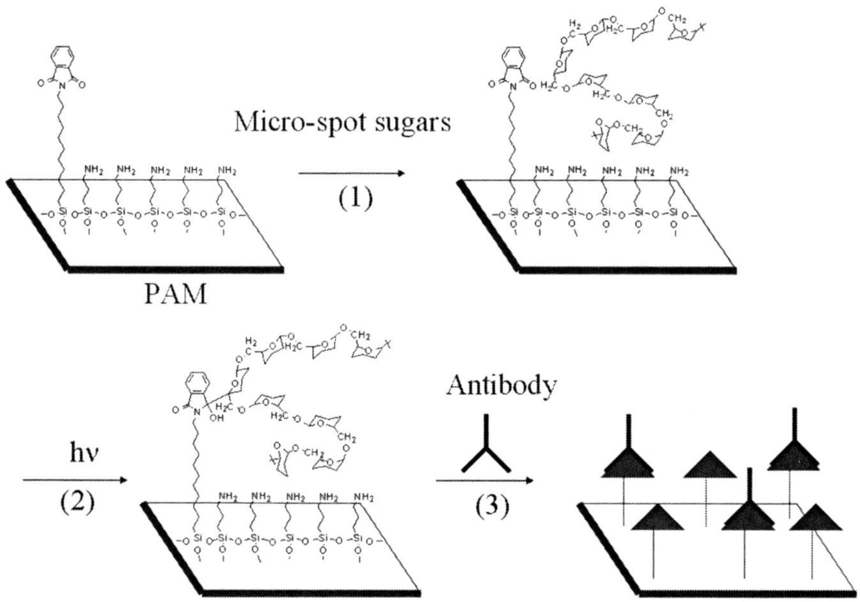


**Fig. 9.6** A mixed monolayer containing phthalimide and amine-terminated molecules provides a photoactive surface appropriate for spotting. After sugars are spotted, irradiation with UV light binds the sugars to the surface presumably through covalent bond formation. Antibodies can recognize the corresponding epitopes on the immobilized carbohydrates (triangles)

surface structure, mixing amines into the surface makes the substrate functional in a high-throughput microarray application.

In order to understand the utility of a given method it is critical to screen a variety of sugars with known specificities towards various lectins, antibodies, or other substances that interact with carbohydrates. Steric interactions and the specificity of the immobilization could potentially inhibit recognition. Immobilized polysaccharide dextran antigens were screened against antidextran antibodies. The antibodies were able to recognize the antigenic determinants of photoimmobilized polysaccharides. A more important application of PAM is the ability to assay underivatized oligosaccharides because nitrocellulose-coated surfaces are unable to hold such small molecules.

Photoimmobilized saccharides containing 3–7 glucose and mannose residues were found to recognize the lectin concanavalin A (con A), however, as the size of the oligosaccharide decreased the intensity of the fluorescence signal decreased. Photoimmobilized glucose and mannose monosaccharides were unable to recognize con A. This is most likely due to increased steric hindrance as the size of the sugar decreases. The close proximity of the sugar to the monolayer makes the epitope inaccessible to the protein. In addition, the expected nonspecific nature of the reaction can bury the $C_3$, $C_4$, and $C_5$ hydroxyl groups of the monosaccharide that are required for binding the lectin. As a sugar decreases in size, the number of biologically active epitopes will decrease along with the probability that the epitope displays itself at the surface. At least some active faces of the sugar are expected to be present at the surface.

Characterizing the immunogenic properties of carbohydrate structures on the surface of pathogens could lead to the development of improved vaccines, drugs, sensors, and diagnostic methods. PAM was recently used to identify immunogenic moieties on the surface of *Bacillus anthracis* spores, rodlike gram-positive bacteria responsible for anthrax infection [70]. Among the various proteins found on the exosporium, the outermost surface of *B. anthracis*, BclA (*Bacillus* collagen-like protein of *anthracis*) is the most prominent. BclA is a glycoprotein containing two types of *O*-linked oligosaccharides: a 324 Da disaccharide and a 715 Da tetrasaccharide. The structure of the tetrasaccharide is shown in Fig. 9.7.

The terminal amide-containing residue was given the name anthrose. The trisaccharide attached to anthrose is made up of rhamnopyranosyl units. The anomeric configuration of the rhamnopyranosyl residue attached to the glycoprotein is unknown. The presence of a given sugar on the surface of the exosporium does not guarantee that the sugar takes part in eliciting an immune response. In order to determine the immunogenic properties of the tetrasaccharide, the α and β conformers of the tetrasaccharide, components of the tetrasaccharide and additional sugars were photoimmobilized on PAM. Incubation of the microarray with

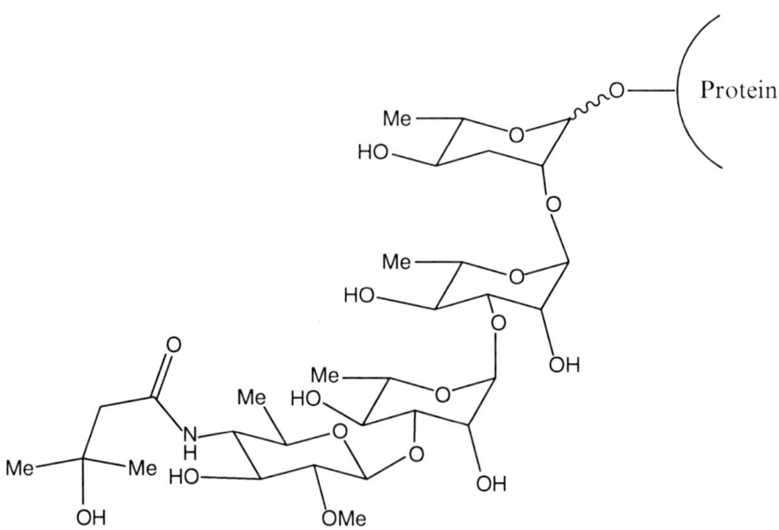


**Fig. 9.7** A tetrasaccharide found on the exosporium of *B. anthracis* spores [71]. The terminal monosaccharide residue has been given the name anthrose. Anthrose is attached to a trisaccharide component made up of rhamnopyranosyl residues. The orientation of the anomeric center of the residue attached to the protein is unknown

antibodies elicited by anthrax spore immunization demonstrated that the anthrose-containing tetrasaccharides are specifically recognized by the antibody.

Inhibition assays were also conducted with the microarray. The anthrose monosaccharide was found to inhibit the antibody from binding to the tetrasaccharide. Thus, the terminal anthrose residue, together with the trisaccharide containing rhamnopyranosyl units form a highly specific immunogenic sugar moiety of *B. anthracis* spores. The experimental approach is expected to allow for the high-throughput screening of the saccharide structures found on any pathogen to identify their key antigenic structures.

In summary, we have described a photochemical strategy that allows for carbohydrates to be immobilized on surfaces without chemical modification. This technique offers a clean and simple method to immobilize carbohydrates on a glass chip. A key advantage is that the carbohydrates do not need to be derivatized with a specific functional group for covalent immobilization on a chip surface. However, an intrinsic weakness of this method is that the immobilized saccharides are expected to lack a specific orientation. In a given microspot, the active part of an unknown percentage of the immobilized sugars can get buried at the interface if the photochemical reaction targets a C–H group on the epitope of the sugar. Smaller sugars are expected to be more hindered by this inasmuch as they contain a lesser amount of epitopes.

The photochemical method could be adapted to incorporate reactions mentioned above that result in selective immobilization of carbohydrates. Rather than directly spot carbohydrates onto the photoactive surface, a polymeric scaffold functionalized with a pertinent reactive group (hydrazide groups for underivatized sugars, e.g.) could be photoimmobilized onto the surface before spotting. The polymeric thin film will provide a thicker and more mobile layer of functional groups, potentially increasing the amount of carbohydrates adsorbed and immobilized per spot. In addition, the flexibility of a surface-bound macromolecule in comparison to a small molecule in a monolayer may increase the accessibility of immobilized carbohydrates to lectins, particularly if the assay conditions swell the polymer chain. This carbohydrate microarray platform provides a versatile tool for carbohydrate research. Its potential in biomedical applications is yet to be further explored.

## References


1. Apweiler, R., H. Hermjakob, et al. (1999). On the frequency of protein glycosylation, as deduced from analysis of the SWISS-PROT database. *Biochim. Biophys. Acta, Gen. Subj.* **1473**(1): 4–8.
2. Davis, B.G.F. and J. Antony (2002). *Carbohydrate Chemistry*. Oxford, Oxford University Press.
3. Lindhorst, T.K. (2003). *Essentials of Carbohydrate Chemistry and Biochemistry*. Weinheim, Wiley-VCH.
4. Schwartz, A.L., S.E. Fridovich, et al. (1981). Characterization of the asialoglycoprotein receptor in a continuous hepatoma line. *J. Biol. Chem.* **256**(17): 8878–8881.
5. Wang, D., and J. Lu (2004). Glycan arrays lead to the discovery of autoimmunogenic activity of SARS-CoV. *Physiol. Genomics* **18**(2): 245–248.

6. Rosati, F., A. Capone, et al. (2000). Sperm-egg interaction at fertilization: glycans as recognition signals. *Int. J. Dev. Biol.* **44**(6): 609–618
7. Focarelli, R., G.B. La Sala, et al. (2001). Carbohydrate-mediated sperm-egg interaction and species specificity: A clue from the Unio elongatulus model. *Cells Tissues Organs* **168**(1–2): 76–81.
8. Crocker, P.R. and T. Feizi (1996). Carbohydrate recognition systems: functional triads in cell-cell interactions. *Curr. Opin. Struct. Biol.* **6**(5): 679–691.
9. Ziska, S.E., and E.J. Henderson (1988). Cell surface oligosaccharides participate in cohesion during aggregation of Dictyostelium discoideum. *Proc. Natl. Acad. Sci. USA* **85**(3): 817–821.
10. Geijtenbeek, T.B., D.S. Kwon, et al. (2000). DC-SIGN, a dendritic cell-specific HIV-1-binding protein that enhances trans-infection of T cells. *Cell* **100**(5): 587–597.
11. Geijtenbeek, T.B., R. Torensma, et al. (2000). Identification of DC-SIGN, a novel dendritic cell-specific ICAM-3 receptor that supports primary immune responses. *Cell* **100**(5): 575–585.
12. Lindahl, U., L. Thunberg, et al. (1984). Extension and structural variability of the antithrombin-binding sequence in heparin. *J. Biol. Chem.* **259**(20): 12368–12376.
13. Capila, I. and R.J. Linhardt (2002). Heparin-protein interactions. *Angew. Chem. Int. Ed. Engl.* **41**(3): 391–412.
14. Kansas, G.S. (1996). Selectins and their ligands: Current concepts and controversies. *Blood* **88**(9): 3259–3287.
15. Gardinali, M., P. Padalino, et al. (1992). Complement activation and polymorphonuclear neutrophil leukocyte elastase in sepsis. Correlation with severity of disease. *Arch. Surg.* **127**(10): 1219–1224.
16. Karlsson, K.A., J. Angstrom, et al. (1992). Microbial interaction with animal cell surface carbohydrates. *APMIS. Suppl.* **27**: 71–83.
17. Feizi, T., and R.W. Loveless (1996). Carbohydrate recognition by Mycoplasma pneumoniae and pathologic consequences. *Am. J. Respir. Crit. Care. Med.* **154**(4 Pt 2): S133–6.
18. Wang, D. and E.A. Kabat (1996). Carbohydrate antigens (polysaccharides). *Structure of Antigens*. M.H.V.V. Regenmortal (Ed.). Boca Raton, FL, CRC Press. **3**: 247–276.
19. Lee, Y.C., and R.T. Lee (1995). Carbohydrate-protein interactions: Basis of glycobiology. *Acc. Chem. Res.* **28**(8): 321–327.
20. Drickamer, K. and M.E. Taylor (2002). Glycan arrays for functional glycomics. *GenomeBiol.* **3**(12): No pp given.
21. Mammen, M., S.-K. Chio, et al. (1998). Polyvalent interactions in biological systems: Implications for design and use of multivalent ligands and inhibitors. *Angew. Chem. Int. Ed.* **37**(20): 2755–2794.
22. Wang, D. and E.A. Kabat (1998). Antibodies, specificity. *Encyclopedia of Immunology*. Delves and Roitt (Eds.). London, Academic Press: 148–154.
23. Wang, D. (2004). Carbohydrate antigens. *Encyclopedia of Molecular Cell Biology and Molecular Medicine*. R.A. Meyers (Ed.), Weinheim, Wiley-VCH. **II:** 277–301.
24. Fazio, F., M.C. Bryan, et al. (2002). Synthesis of sugar arrays in microtiter plate. *J. Am. Chem. Soc.* **124**(48): 14397–14402.
25. Fukui, S., T. Feizi, et al. (2002). Oligosaccharide microarrays for high-throughput detection and specificity assignments of carbohydrate-protein interactions. *Nature Biotechnol.* **20**(10): 1011–1017.
26. Houseman, B.T. and M. Mrksich (2002). Carbohydrate arrays for the evaluation of protein binding and enzymatic modification. *Chem. Biol.* **9**(4): 443–454.
27. Park, S., and I. Shin (2002). Fabrication of carbohydrate chips for studying protein-carbohydrate interactions. *Angew. Chem. Int. Ed. Engl.* **41**(17): 3180–3182.
28. Wang, D., S. Liu, et al. (2002). Carbohydrate microarrays for the recognition of cross-reactive molecular markers of microbes and host cells. *Nature Biotechnol.* **20**(3): 275–281.
29. Adams, E.W., D.M. Ratner, et al. (2004). Oligosaccharide and glycoprotein microarrays as tools in HIV glycobiology glycan-dependent gp120/protein interactions. *Chem. Biol.* **11**(6): 875–881.

30. Willats, W.G.T. (2005). Microarrays for the high-throughput analysis of protein-carbohydrate interactions. *Protein Microarrays* 57–69.
31. Carroll, G.T., D. Wang, et al. (2006). Photochemical micropatterning of carbohydrates on a surface. *Langmuir* **22**(6): 2899–2905.
32. Guo, Y., H. Feinberg, et al. (2004). Structural basis for distinct ligand-binding and targeting properties of the receptors DC-SIGN and DC-SIGNR. *Nat. Struct. Mol. Biol.* **11**(7): 591–598.
33. Stevens, J., O. Blixt, et al. (2006). Glycan microarray analysis of the hemagglutinins from modern and pandemic influenza viruses reveals different receptor specificities. *J. Mol. Biol.* **355**(5): 1143–1155.
34. Bryan, M.C., L.V. Lee, et al. (2004). High-throughput identification of fucosyltransferase inhibitors using carbohydrate microarrays. Bioorg. Med. Chem. Lett. **14**(12): 3185–3188.
35. Shin, I., S. Park, et al. (2005). Carbohydrate microarrays: An advanced technology for functional studies of glycans. *Chem. Euro. J.* **11**(10): 2894–2901.
36. Willats, W.G., S.E. Rasmussen, et al. (2002). Sugar-coated microarrays: A novel slide surface for the high-throughput analysis of glycans. *Proteomics* **2**(12): 1666–1671.
37. Wang, D. (2003). Carbohydrate microarrays. *Proteomics*. **3**(11): 2167–2175.
38. Wang, R., S. Liu, et al. (2005). A practical protocol for carbohydrate microarrays. *Methods in Molecular Biology*. Totowa, NJ, **310**: 241–252.
39. Bryan, M.C., O. Plettenburg, et al. (2002). Saccharide display on microtiter plates. *Chem. Biol.* **9**(6): 713–720.
40. Ko, K.-S., F.A. Jaipuri, et al. (2005). Fluorous-based carbohydrate microarrays. *J. Am. Chem. Soc.* **127**(38): 13162–13163.
41. Manimala, J.C., Z. Li, et al. (2005). Carbohydrate array analysis of anti-Tn antibodies and lectins reveals unexpected specificities: implications for diagnostic and vaccine development. *Chembiochem*. **6**(12): 2229–2241.
42. Manimala, J.C., T.A. Roach, et al. (2006). High-throughput carbohydrate microarray analysis of 24 lectins. *Angew. Chem. Int. Ed. Engl.* **45**(22): 3607–3610.
43. Shao, M.C. (1992). The use of streptavidin-biotinylglycans as a tool for characterization of oligosaccharide-binding specificity of lectin. *Anal. Biochem.* **205**(1): 77–82.
44. Bochner, B.S., R.A. Alvarez, et al. (2005). Glycan array screening reveals a candidate ligand for Siglec-8. *J. Biol. Chem.* **280**(6): 4307–4312.
45. Park, S., M.-R. Lee, et al. (2004). Carbohydrate chips for studying high-throughput carbohydrate-protein interactions. *J. Am. Chem. Soc.* **126**(15): 4812–4819.
46. Bryan, M.C., F. Fazio, et al. (2004). Covalent display of oligosaccharide arrays in microtiter plates. *J. Amer. Chem. Soc.* **126**(28): 8640–8641.
47. Blixt, O., S. Head, et al. (2004). Printed covalent glycan array for ligand profiling of diverse glycan binding proteins. *Proc. Natl. Acad. Sci. USA* **101**(49): 17033–17038.
48. de Paz, J.L., C. Noti, et al. (2006). Microarrays of synthetic heparin oligosaccharides. *J. Am. Chem. Soc.* **128**(9): 2766–2767.
49. Schwarz, M., L. Spector, et al. (2003). A new kind of carbohydrate array, its use for profiling antiglycan antibodies, and the discovery of a novel human cellulose-binding antibody. *Glycobiology* **13**(11): 749–754.
50. Xia, B., Z.S. Kawar, et al. (2005). Versatile fluorescent derivatization of glycans for glycomic analysis. *Nature Meth.* **2**(11): 845–850.
51. Fritz, M.C., G. Hähner, et al. (1996). Self-assembled hexasaccharides: Surface characterization of thiol-terminated sugars adsorbed on a gold surface. *Langmuir* **12**(25): 6074–6082.
52. Revell, D.J., J.R. Knight, et al. (1998). Self-assembled carbohydrate monolayers: Formation and surface selective molecular recognition. *Langmuir* **14**(16): 4517–4524.
53. Lee, M.-R. and I. Shin (2005). Facile preparation of carbohydrate microarrays by site-specific, covalent immobilization of unmodified carbohydrates on hydrazide-coated glass slides. *Organic Lett.* **7**(19): 4269–4272.
54. Zhi, Z.-l., A. K. Powell, et al. (2006). Fabrication of carbohydrate microarrays on gold surfaces: Direct attachment of nonderivatized oligosaccharides to hydrazide-modified self-assembled monolayers. *Anal. Chem.* **78**(14): 4786–4793.

55. Yates, E.A., M.O. Jones, et al. (2003). Microwave enhanced reaction of carbohydrates with amino-derivatized labels and glass surfaces. *J. Mater. Chem.* **13**(9): 2061–2063.
56. Peramo, A., A. Albritton, et al. (2006). Deposition of patterned glycosaminoglycans on silanized glass surfaces. *Langmuir* **22**(7): 3228–3234.
57. Zhou, X. and J. Zhou (2006). Oligosaccharide microarrays fabricated on aminooxyacetyl functionalized glass surface for characterization of carbohydrate-protein interaction. *Biosens. Bioelectron.* **21**(8): 1451–1458.
58. Takahashi, S. and J. Anzai (2005). Phenylboronic acid monolayer-modified electrodes sensitive to sugars. *Langmuir* **21**(11): 5102–5107.
59. Turro, N.J. (1991). *Modern Molecular Photochemistry*. Sausalito, CA, University Science Books.
60. Binkley, E.R. and R.W. Binkley (1998). *Carbohydrate Photochemistry*. In: *ACS Monogr.*, 1998; 191.
61. Madden, K.P., and R.W. Fessenden (1982). ESR study of the attack of photolytically produced hydroxyl radicals on a-methyl-D-glucopyranoside in aqueous solution. *J. Am. Chem. Soc.* **104**(9): 2578–2581.
62. Shkrob, I.A., M.C. Depew, et al. (1993). Time-resolved, electron-spin resonance study of radical species derived from naturally occurring carbohydrates. *Chem. Phys. Lett.* **202**(1–2): 133–140.
63. Gilbert, B.C., D.M. King, et al. (1984). Radical reactions of carbohydrates, Part 5. The oxidation of some polysaccharides by the hydroxyl radical: An ESR investigation. *Carbohydrate Res.* **125**(2): 217–235.
64. Angeloni, S., J.L. Ridet, et al. (2005). Glycoprofiling with micro-arrays of glycoconjugates and lectins. *Glycobiology* **15**(1): 31–41.
65. Chevolot, Y., O. Bucher, et al. (1999). Synthesis and characterization of a photoactivatable glycoaryldiazirine for surface glycoengineering. *Bioconj. Chem.* **10**(2): 169–175.
66. Chevolot, Y., J. Martins, et al. (2001). Immobilisation on polystyrene of diazirine derivatives of mono- and disaccharides: Biological activities of modified surfaces. *Bioorg. Med. Chem.* **9**(11): 2943–2953.
67. Ulman, A. (1996). Formation and structure of self-assembled monolayers. *Chem. Rev.* **96**(4): 1533–1554.
68. Kanaoka, Y. (1978). Photoreactions of cyclic imides. Examples of synthetic organic photochemistry. *Acc. Chem. Res.* **11**(11): 407–413.
69. Yoon, U.C., and P.S. Mariano (2001). The synthetic potential of phthalimide SET photochemistry. *Acc. Chem. Res.* **34**(7): 523–533.
70. Wang, D., G.T. Carroll, et al. (2006). Immunogenic sugar moieties of the Bacillus anthracis exosporium (anthrose-based compositions and related methods). United States Provisional Patent Application, Dkt 76323-pro, filed September 15, 2006.
71. Daubenspeck, J.M., H. Zeng, et al. (2004). Novel oligosaccharide side chains of the collagen-like region of BclA, the major glycoprotein of the Bacillus anthracis exosporium. *J. Biol. Chem.* **279**(30): 30945–30953.

# Chapter 10
# Expression Profiling Using Microfluidic Living Cell Arrays

**Kevin R. King, Martin L. Yarmush, and Arul Jayaraman**

## 10.1 Introduction to the Living Cell Array Concept

The cellular microenvironment is remarkably complex. In the small space near each cell, growth factors are liberated from extracellular matrix, cytokines are secreted by neighboring cells, and hormones arrive from distant endocrine organs through the circulation. These soluble cues are detected by surface or cytoplasmic receptors and integrated using complex signal transduction cascades to modulate the activity of *transcription factors* (TFs), the primary regulators of gene expression. Transcription factors serve as points of convergence between the vast number of extracellular signaling molecules and the equally vast number of target genes. For perspective, the human genome contains approximately 1500 identified transcription factors regulating more than 20,000 target genes [1].

Adding further complexity to the picture, transcription factors often cooperate, compete, and regulate each other, forming transcriptional regulatory networks with rich possibilities to control cell behavior. Under normal conditions, network regulators are activated in a defined temporal sequence and function as a transcriptional regulatory program to coordinate physiological adaptations to changes in the external cellular microenvironment. When transcriptional programs are dysregulated, they can lead to inappropriate pathological responses that result in clinical disease.

One example of such a transcriptional regulatory network is that which regulates insulin secretion by hepatocyte nuclear factors (HNFs; [2]). The expression of 3 HNFs—HNF-1alpha, HNF-4alpha, and HNF-3alpha—are positively regulated by HNF-3beta. However, HNF-3alpha also acts as a negative regulator of HNF-1alpha and HNF-4alpha, which has been attributed to competition for the

K.R. King and M.L. Yarmush
Center for Engineering in Medicine, Massachusetts General Hospital, and Harvard Medical School, Boston, MA

A. Jayaraman
Artie McFerrin Department of Chemical Engineering, Texas A&M University, College Station, TX

K. Dill et al. (eds.), *Microarrays: Preparation, Microfluidics, Detection Methods, and Biological Applications*,

HNF3 binding site. Any change in the expression of a single HNF can alter the dynamics of the entire regulatory network, and the regulation of insulin signaling as well [3, 4]. This is indeed observed in monogenetic forms of diabetes, maturity onset diabetes of the young (MODY), where mutations in specific HNF genes are thought to underlie the disease state [5]. Therefore, understanding the organization and dynamics of transcriptional networks, and their ability to synthesize dynamic cellular microenvironment inputs to coordinate adaptive responses is a key question in systems biology.

Transcriptional regulatory programs are difficult to study experimentally because it requires monitoring several molecules whose expression continuously evolve in time. Furthermore, the signaling pathways linking external cues to transcription factors are highly nonlinear (exhibiting thresholds, saturation, feedback, and cross-talk; [6]), as the transcription factor response dynamics depend on the timing of the input stimulus [7] and the initial state of the cell (e.g., phase of the cell cycle, cell shape, or degree of cell–cell contact).

Conventional biochemical methods for investigating the expression and function of transcription factors such as Western blots and DNA binding assays rely on high molecular specificity of antibodies or nucleotide sequences to identify the transcription factors activated by the different stimuli in the cellular environment. However, such methods are destructive (i.e., involve disruption of cells for the assay), require a large number of cells for the assay that results in averaging of responses, and perhaps most important, are suited only for low-throughput investigations. The advent of high-throughput microarray technologies has dramatically improved upon the number of gene expression events that can be monitored in parallel for a given experiment at a single time point; however, because most transcriptional regulatory networks are comprised of a comparatively small number of genes (~10–20), the parallel monitoring of thousands of genes can complicate the subsequent bioinformatic analysis.

In addition, both conventional assays and DNA microarrays are not well suited for investigating transcriptional network dynamics or for measuring many different input conditions. For example, because dynamic information is limited by the number of time points at which expression is profiled, one can potentially miss important transient expression events. Therefore, there is significant interest in developing methods that enable dynamic monitoring of a small set of transcriptional regulators, in a high-throughput format that mimics in vivo complexity.

We recently developed a microfluidic "living cell array" (LCA) platform to study stimulus–response dynamics of transcriptional regulatory networks in living cells [8–10]. This platform combines two enabling technologies—microfluidics and GFP reporter systems—to allow precise yet flexible control of the cellular microenvironment, while enabling simultaneous real-time monitoring of transcription factor network dynamics in living cells using time-lapse fluorescence microscopy. A schematic of the LCA platform is shown in Fig. 10.1.

Microfluidic systems offer a powerful set of tools for controlling media composition (metabolites, cytokines, hormones, and small molecule inhibitors)

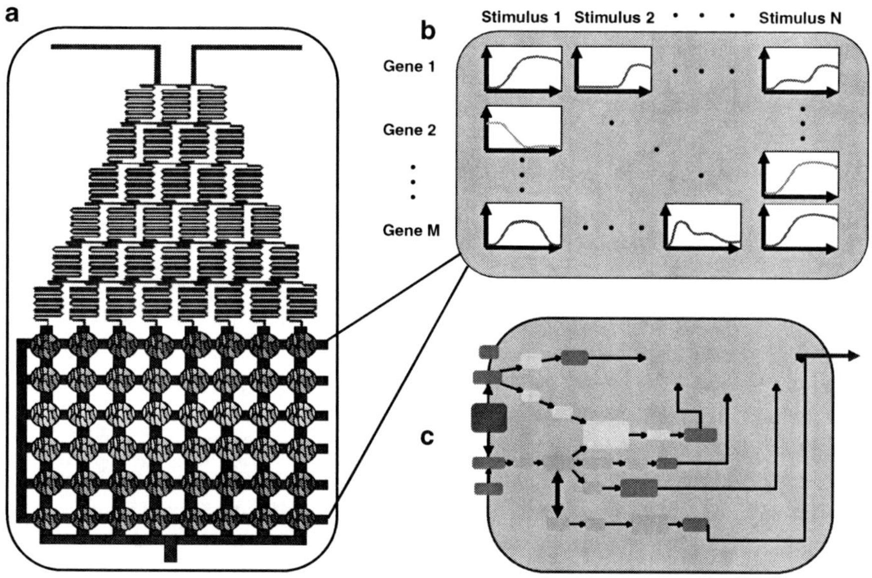


**Fig. 10.1** Microfluidic living cell array concept. (A) Upstream microfluidic circuits prepare soluble stimuli for delivery to downstream GFP reporter cell arrays. (B) Reporter responses are monitored by time lapse fluorescence microscopy and quantified by automated image analysis routines. (C) Dynamic responses can then be used to construct, revise, and validate dynamic models of transcriptional regulatory networks

and perfusion flow rate (delivery of nutrients, removal of waste, and mechanical shear stress), the "network inputs" [11–13]. Fluorescent reporters, by enabling nondestructive quantification of transcription factor activities, offer an equally important window into the "network state" and its dynamics. Ultimately, the inputs and network dynamics can be correlated with cellular responses such as cell spreading, migration, proliferation, and apoptosis, the "system outputs", to comprehensively characterize the network system properties.

Towards this goal, we have constructed a library of fluorescent reporter cell lines, each expressing a destabilized green fluorescent protein (GFP) reporter when a different transcription factor of interest is active. Monitoring the dynamics of the entire library in response to diverse inputs in a microfluidic living cell array provides a dynamic systems-level picture of transcriptional regulatory programs that complements existing targeted single-time-point techniques in an effort to understand relationships between cells and their local microenvironments during health and disease.

This chapter describes the development of the GFP reporter cell lines and the microfluidic living cell array platform, followed by several demonstrations of microfluidic reporter assays in the context of liver inflammation, and concludes with perspectives on potential applications for this high-throughput technology in systems biology.

## 10.2 Enabling Technologies for the Living Cell Array

### *10.2.1 Green Fluorescence Protein Reporter*

GFP is a widely used tool for studying cellular processes in living cells that can be used to derive information on different aspects of gene expression (e.g., transcription, protein–protein interaction, protein–DNA binding, etc.). Numerous GFP variants have been developed to overcome limitations of the original GFP and optimize it for specific studies. For example, enhanced GFP (EGFP) was developed to achieve high fluorescence while minimizing the need for damaging excitation [14]. Spectral variants such as yellow (EYFP) and cyan (ECFP) fluorescent proteins have been developed through site-directed mutagenesis so that multiple molecular events can be tracked in single cells. Destabilized variants were generated by fusing the mouse ornithine decarboxylase (MODC) degradation domain to the C-terminus of EGFP [15, 16]. Reporters based on this technology are particularly attractive for monitoring gene expression dynamics because they do not accumulate indefinitely like the more stable native GFP.

GFP reporters have provided significant information about cellular dynamics and their underlying control. They have been used to quantify noise in gene expression [17, 18] and investigate the stochastic nature of transcription and translation [19]. An advantage with GFP reporter systems is that they can be used to generate information on different levels at which gene expression can be controlled. In mammalian cells, live cell studies have focused primarily on the use of GFP fusion proteins, and several studies have revealed unanticipated dynamics such as oscillations in NF-κB [20] and p53 [21], thus providing complementary information to that obtained using conventional expression profiling methods. Similarly, incorporation of GFP reporters has led to the characterization of protein–protein interactions through fluorescence resonance energy transfer (FRET), an aspect that cannot be studied using RNA or protein profiling, but yet is an important regulator of gene expression [22, 23].

Several issues need to be considered when using fluorescent reporters for expression profiling studies that often depend on the objective of the study. GFP transcriptional reporters, which are homogeneously distributed throughout the cell, are commonly analyzed using fluorescence flow cytometry. However, this technique cannot be used to measure dynamics of individual adherent cells as cells need to be detached prior to fluorescence measurement; instead, dynamics must be inferred by measuring population distributions at several time points.

The magnification at which fluorescence is imaged is also an important consideration. High magnification provides increased spatiotemporal resolution, but minimizes the sample size (i.e., the number of cells in which fluorescence is monitored). At the other extreme, low magnification imaging methods allow measurement of large cell populations, but sacrifice signal level and spatial resolution. For transcriptional reporters based on GFP, an intermediate magnification might be most appropriate as it retains some cellular resolution while increasing the population sampling [24, 25].

## *10.2.2 Microfluidic Circuits*

### 10.2.2.1 Principles of Design: Hydrodynamics

Flow in microfluidic channels is dominated by viscous forces. As a result, the flow is purely laminar and there is no turbulence. In this regime, the normally nonlinear Navier–Stokes equation reduces to a linear equation where volumetric flow rate $Q$ in a straight channel of constant cross-section is directly proportional to the pressure difference $P$ between the inlet and outlet given in Equation 10.1.

$$P = QR, \tag{10.1}$$

where $R$ is the constant of proportionality, the fluidic resistance. In other words, microchannels behave as fluidic analogues of discrete electrical resistors, allowing complicated microchannel networks to be designed and modeled using linear circuit theory (Fig. 10.2a). The resistance of a rectangular microchannel depends on the fluid properties and the geometry of each channel and can be calculated using Equation (10.2), where $\mu$ is the fluid viscosity, $l$ is the channel length, $w$ the width, and $h$ the height.

$$R = \frac{12\mu L}{wh^3}\left(1 - \sum_{n=0}^{\infty}\frac{\tanh\left((2n+1)\frac{\pi w}{2h}\right)}{(2n+1)^5}\right)^{-1}. \tag{10.2}$$

In microfluidics, the width of the channel is often much greater than the height. In this case, the channel resistance can be approximated by the parallel plate resistance give by Equation (10.3).

$$R_{parallelplate} = \frac{12\mu L}{wh^3}. \tag{10.3}$$

When cells are seeded on the channel floor and exposed to flow (Fig. 10.2b), they experience a cell surface shear stress $\tau_{ss}$ that can be approximated by Equation (10.4).

$$\tau_{ss} = \frac{6\mu Q}{wh^2}. \tag{10.4}$$

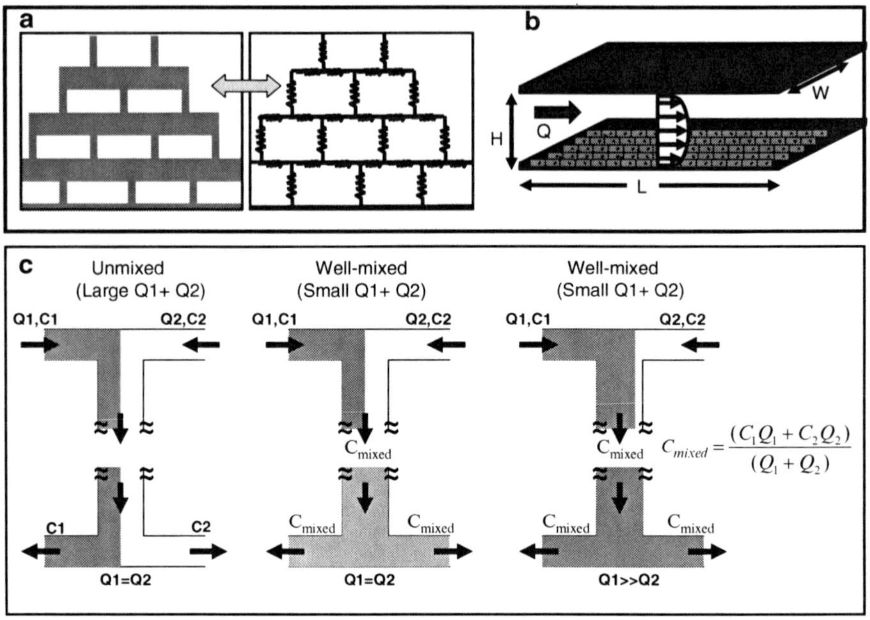


**Fig. 10.2** Microfluidic design principles. (A) Microfluidic channels can be modeled as electrical resistors because pressure and flow are linearly related. The constant of proportionality, the fluidic resistance, is determined by the channel geometries – width (W), height (H), and Length (L). Complicated networks of channels can be modeled and simulated by treating each channel as a straight branch and calculating their resistances using linear circuit theory. (B) Flow in microfluidic channels is typically laminar, velocity profiles are parabolic, and cells seeded on channel surfaces experience mechanical force due to the fluid shear stress. (C) In microfluidic channels, mixing of solutions with different concentrations (C1 and C2) occurs primarily by lateral diffusion. Three operating regimes are shown. When the total flow rate (Q1+Q2) is large, there is little time for mixing, and the two solutions exit at nearly the same concentration as they entered (left). When the total flow rate is small, diffusive mixing can take place before the solutions exit the channel, and the solutions leaving the outlet are well-mixed (middle). The well-mixed concentration is the flow-weighted average of the two solutions such that unequal flow rates can be used to generate different well-mixed concentrations at the channel outlet (right)

Studies in endothelial cells and hepatocytes suggest that shear stresses greater than ~1 dyne/cm$^2$ are associated with gene expression changes and altered cellular function [26].

#### 10.2.2.2 Principles of Design: Mass Transport

Because microfluidic flows are strictly laminar, mixing of solutions occurs almost exclusively by diffusion. Therefore, by controlling channel geometries and operating conditions, microfluidic circuits can decouple transport of fluid from mixing of dissolved solutes such as proteins and metabolites. Most microfluidic devices are designed for use in either well-mixed or unmixed regimes (Fig. 10.2c).

Consider a fluidic junction where two channels, one containing a soluble stimulus such as a cytokine or a drug, converges on a single channel and then branches downstream to feed the rest of the circuit. If the flow is fast compared to the rate of diffusion, then there will be little mixing before the two streams are separated at the branch point downstream. However, if the flow is slow compared to the rate of diffusion, then the two inlet solutions will fully mix and each of the branching channels downstream will receive the same well-mixed concentration. The relative rates of flow and diffusion can be compared by computing a dimensionless number, the ratio of the transit time (time for a particle in the flow to traverse the channel) and the mixing time (95% of complete mixing) where Q is the volumetric flow rate.

$$\frac{\tau_{transit}}{\tau_{mixing}} = \frac{lwh/Q}{w^2/D} = \frac{Dlh}{Qw}. \tag{10.5}$$

In summary, microfluidic circuits can be precisely designed to control flow rates, surface shear stress, and chemical composition in each branch of complex networks for use in live cell assays. In the following section, we describe how such microfluidic devices are fabricated.

## 10.3 Fabrication of Microfluidic Devices

### *10.3.1 Silicon Master Mold Fabrication*

Microfluidic living cell array devices are constructed using microfabrication, soft lithography, and rapid prototyping [27]. In this process (Fig. 10.3), fluidic circuit designs are first drawn using a computer-aided design program such as AutoCAD and printed on mylar film using a high-resolution printer. Microfabricated master molds are then fabricated by performing standard photolithography on polished silicon wafers, a process involving spin-coating of a photosensitive material (commonly SU-8 photoepoxy), selective polymerization by exposing the material through the high-resolution printed photomask, and removal of the soluble unexposed material, leaving permanent structures on the silicon substrate that can serve as a master mold for making polymer replicas. The lengths and widths of channels are determined by the mask drawing (resolution ~10 μm) and the fidelity of the photolithographic process that transfers the pattern to the silicon substrate. Channel heights, on the other hand, are determined by the thickness of the photoresist (1–1000 μm), which is a function of the photoepoxy viscosity as well as the speed and duration of spin coating.

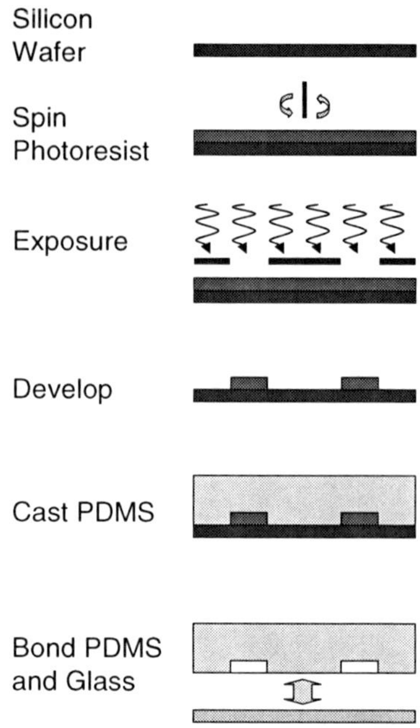


**Fig. 10.3** Fabrication of microfluidic devices. Microfabricated master molds are generated using standard photolithography techniques from a high-resolution device design. The master mold contains permanent structures corresponding to the different features of the device and can be used for making polymer replicas that has the device network embedded in it. The polymer is then irreversibly bonded to a glass slide to complete the device

## *10.3.2 Polymer Microfluidics Fabrication*

After fabricating microfabricated master molds, the channel structures can be repeatedly transferred to other polymers using a cast, cure, and peel process. One commonly used polymer is polydimethylsiloxane (PDMS), a silicone elastomer that can be polymerized simply by mixing polymer resin and curing agent 10:1, degassing, casting on the microfabricated master, curing at 65°C for several hours, and peeling the rubbery replica.

PDMS has become a favorite material for creating microfluidic replicas because it precisely conforms to the shape of the master mold prior to curing and because it does not plastically deform after curing. Furthermore, it can be bonded to commonly used culture substrates such as glass or other PDMS surfaces to form enclosed channel structures. PDMS-glass microfluidic devices are particularly attractive for live cell experiments because they are optically transparent and not directly toxic to cells. Furthermore, spin-coating PDMS can be used to create thin membranes of controlled thicknesses, allowing fabrication of deformable structures and construction of integrated microscale pumps and valves [28–31]. Finally, PDMS surfaces can be chemically modified using silane chemistry to immobilize a range of biologically specific cell–material interfaces on the channel surfaces. This microtechnology toolkit is rapidly expanding and promises to enable development of more powerful devices for performing integrated live cell assays. The devices described in the remainder of this chapter are each fabricated using this approach.

## 10.4 Microfluidic Reporter Assays

The microfluidic devices described in the previous sections can be used as massively parallel cell culture systems by sterilizing, coating surfaces with extracellular matrix components such as fibronectin, seeding with cells, and culturing or stimulating cells under continuous flow to study their dynamic responses. Macroscale perfusion systems accommodate only one condition per experiment. For transcriptional responses, which typically evolve over many hours or even days, this becomes prohibitively time-consuming, and ultimately limits the total number of conditions that can be reasonably explored. In contrast, microfluidic culture systems are highly parallel, and have the potential to explore hundreds of conditions in a single experiment, opening tremendous opportunities for systematically characterizing dynamic cellular responses. In the remainder of this section, we provide examples of three different microfluidic circuits that allow parallel control of stimulus concentration, timing, and location in the LCA, thereby demonstrating the power of the fluidic array.

### *10.4.1 Microfluidic Dose–Response Experiments*

One of the most common experiments in cell biology involves studying the cellular response to different doses of a stimulus such as a growth factor or cytokine. Therefore, we constructed a microfluidic device, inspired by a gradient generating circuit [32], that operates in the complete mixing regime by taking a concentrated stimulus solution and progressively diluting it to generate multiple stimulus concentrations which are delivered to an integrated cell culture chamber downstream [8] Fig. 10.4). The network consists of two inlets and a single outlet. Culture medium is delivered through the inlets, with one inlet containing the experimental stimulus (e.g., TNF-α) and the other without stimulus.

Fig. 10.4 illustrates the generation of eight distinct TNF-α concentrations in a single device circuit. NF-κB reporter cells were exposed to the various cytokine concentrations and the resultant GFP fluorescence was monitored using time-lapse fluorescence microscopy. We found that the timing of NF-κB activation was independent of TNF-α concentration (i.e., all concentrations resulted in NF-κB activation after the same time) whereas the magnitude of the reporter cell response increased with increasing levels of TNF-α.

### *10.4.2 Microfluidic Dynamic Stimulation Experiments*

In addition to dose responses, microfluidic circuits can also be used to control the timing of stimulation. Because biological signaling systems are nonlinear, it is

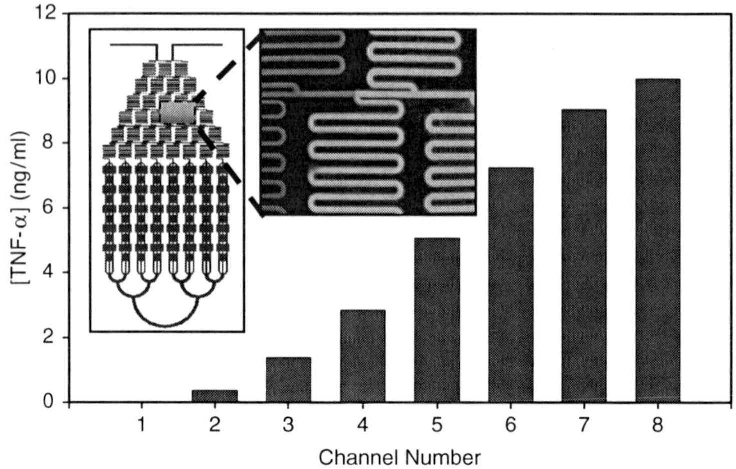


**Fig. 10.4** Generation of soluble mediator gradients. (A) Eight concentrations of a soluble mediator (TNF-α) are generated in a single device. The device consists of two inlets and a single outlet. Culture medium is delivered through the inlets, with one inlet also containing a 10 ng/mL of TNF-α. As the cytokine flows through the microfluidic network, it is gradually diluted and generates eight distinct concentrations that are used to stimulate reporter cells in the downstream microfluidic array. The inset shows mixing of a flourescent dye (of comparable MW to TNF-α) in the network. The concentration of TNF-α exiting the network and entering each cell culture channels is also shown.

nontrivial to predict responses to new stimulation patterns or construct a mathematical model of signaling using parameters extracted from step response experiments. Instead, measuring responses to more complicated stimulus patterns such as pulses of different durations or pulse sequences is required for comprehensively characterizing signaling pathways. We have developed a circuit capable of delivering diverse temporal sequences controlled by a single pressure input [33] (Fig. 10.5).

The circuits, which we call "flow-encoded switching networks," use pressure to control the ratio of two input flow rates which ultimately determine which cells are exposed to stimulus. By varying this input pressure in time, we change the state of the network dynamically to deliver different stimulus regimens to each channel in the array. By controlling the input pressure appropriately, we have demonstrated systematic variation of pulse duration, pulse frequency, and pulse train length. To demonstrate the utility of this network, we studied the activation of NF-κB in response to several durations of TNF-α.

Our data indicate that the response to transient TNF-α exposures less than 45 minutes increases with increasing duration. However, for stimulus durations longer than 45 minutes, we found that the response magnitude was independent of the stimulus duration. These results are consistent with those obtained using conventional but substantially more time-consuming techniques such as electromobility shift assays (EMSA), and demonstrate the power of microfluidic living cell arrays in investigating transcriptional activation in response to transient changes in metabolites as well as periodic stimuli such as hormone oscillations.

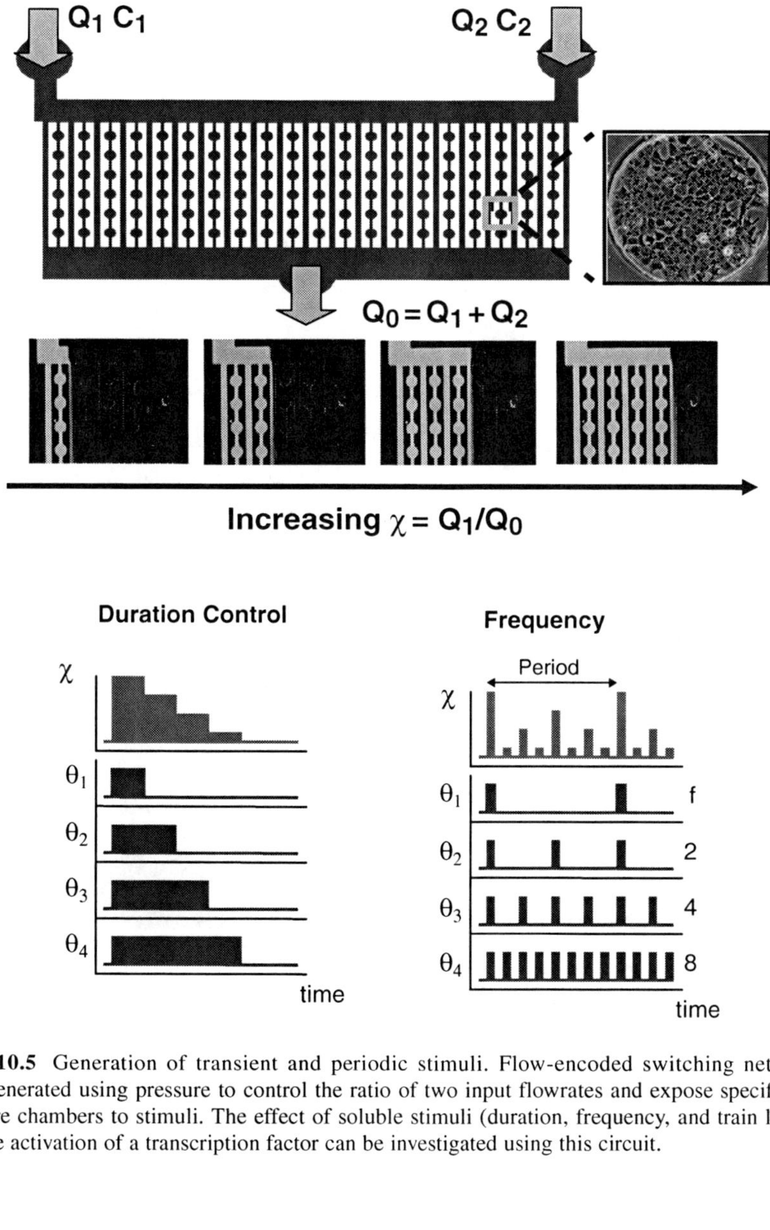


**Fig. 10.5** Generation of transient and periodic stimuli. Flow-encoded switching networks are generated using pressure to control the ratio of two input flowrates and expose specific cell culture chambers to stimuli. The effect of soluble stimuli (duration, frequency, and train length) on the activation of a transcription factor can be investigated using this circuit.

## *10.4.3 Multireporter Microfluidic Living Cell Array*

The previous sections illustrated how microfluidic circuits can be used to control the stimulus or the transcriptional network inputs. In this section, we describe the reporter array itself and illustrate how it can be used to study multiple nodes in a signaling network and characterize their dynamics under different experimental conditions. In order to study an entire transcriptional network under different stimu-

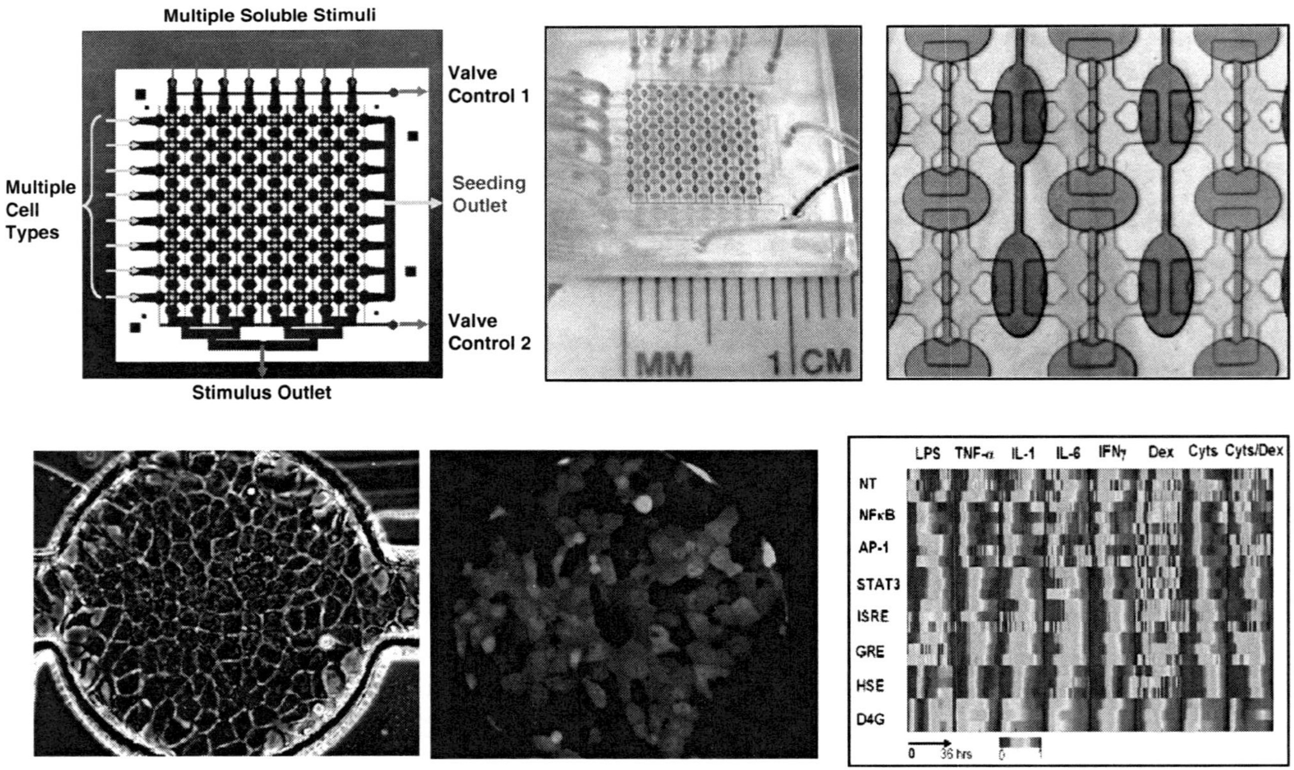


**Fig. 10.6** Multi-reporter living cell array. (A) & (B) The device consists of a fluidically addressable array of nano-liter scale cell culture chambers that can be alternately separated into rows or columns using integrated microvalves. (C) Valves (green and orange) can be operated using pressure to allow cell culture chambers (yellow) to communicate with one another. (D) Reporter cell growth in the array. (E) GFP fluorescence from reporter cells. (F) Heat map indicating the extent of response of multiple reporters to multiple stimuli from a single experiment.

lus conditions in a single experiment, we designed a fluidically addressable array of cell culture chambers (nanoliter volume) that could be alternately separated into rows or columns using two independently activated systems of integrated microvalves ([10]; Figs. 10. 6a,b).

The valves are constructed in two layers, a top network of valve-control channels and a bottom network of cell culture chambers (Fig. 10.6c). Separating the two layers is a thin flexible membrane, such that when the pressure in the layer 2 control line is changed, the interposed membrane is deformed, either opening or closing the valve. The valves are designed such that they are closed without applied pressure, but when negative pressure is applied to either control line, the underlying valves are lifted and chambers on either side can communicate. When the array is in "seeding configuration," the rows of culture chambers allow simultaneous seeding of each reporter cell line. After cell attachment, the array can then be placed in "stimulation configuration," and molecular stimuli, prepared by upstream microfluidic circuits can be delivered to the reporters in the orthogonal direction. Each reporter cell line (reporting on a single transcription factor) is seeded in a different row of the fibronectin-coated array.

After the cells attach and spread to reach confluence (Fig. 10.6d), the array is converted from rows to columns, allowing each reporter to be exposed to each stimulus, generating a matrix of stimulus–response relationships. Each experiment can be performed in quadruplicate to characterize errors due to differences in cell numbers, arrangement, and image analysis.

To demonstrate the platform, we seeded eight different cell lines in separate rows and measured their responses to eight different stimuli, including cytokines, hormones, endotoxin, and combinations of stimuli. In addition to the expected responses of the reporters to their classical cytokine and hormonal inducers (Fig. 10.6e), we also observed several unexpected gene expression responses, which demonstrates the power of the LCA approach (Fig. 10.6f). This includes induction of heat shock element-mediated transcriptional activity in response to inflammatory cytokine TNF-$\alpha$ and IL-1$\beta$. It is important to note that the induction of HSE and NF-$\kappa$B reporters to TNF-$\alpha$ occurred at different times, and might have been overlooked if single-time-point measurement techniques were used. The dynamic reporters in the living cell array enable unbiased characterization with respect to inducers and time points and by revealing different activation kinetics, provide insights into the differences in pathway activation mechanisms in transcription factor networks.

## 10.5 Summary and Applications

In this chapter, we have described the motivation underlying the living cell array platform and described its development using fluorescent reporter systems and microfluidic networks as enabling technologies. We briefly described the principles underlying microfluidic networks and demonstrate the functioning of the living cell array platform for dose–response and pulse stimulation studies, which are fundamental to most signal transduction investigations. Lastly, we also demonstrated multidimensional expression profiling by using multiple inputs (stimuli) and outputs (reporter cell lines for different transcription factors) in an addressable array. The power and potential of the living cell array as a discovery tool is evident from both the comprehensive characterization of known stimulus–transcription factor interactions as well as the identification of hitherto unidentified interactions between the different inputs and outputs.

Although we describe the living cell array platform in the context of the hepatocyte inflammatory response, this tool is equally applicable to other problems in biology and medicine where time-dependent behavior of a small number of regulatory molecules is of interest. For example, this approach can be applied in toxicogenomic investigations where the interactions of different toxic compounds in the activation of different signaling pathways (e.g., estrogen and arylhydrocarbon signaling) and their effect on downstream target gene expression (e.g., cytochrome P450) needs to be studied in a dose-dependent manner. In this case, the microfluidic network mixing module can be used to generate different combinations of pollutants and their effect on target gene expression investigated in a single experiment.

Another potential application of the living cell array platform is in cell–cell communication that underlies the function of most tissue and organ systems. In this case, the ability to address different chambers in the living cell array can be leveraged to culture different cell types in the cell chambers and investigate the soluble mediator-based communication between them based on the sequential activation of target genes in the different cell types. In summary, the living cell array platform has several potential applications in biology and medicine for investigating gene expression dynamics of a small number of target molecules and obtaining systems-level data that are complementary to high-throughput methods such as DNA microarrays.

**Acknowledgments** This work was supported in part by a Texas Engineering Experiment Station Award to AJ and grants from the NIH (BRP AI063795 and P41 EB002503) to MLY.

## References

1. Pennisi, E. (2002). Genomics. Sequence tells mouse, human genome secrets. *Science* 298(5600), 1863–1865.
2. Duncan, S., M. Navas, D. Dufort, J. Rossant, and M. Stoffel. (1998). Regulation of a transcription factor network required for differentiation and metabolism. *Science* 281(5377), 692–695.
3. Odom, D.T., N. Zizlsperger, D.B. Gordon, G.W. Bell, N.J. Rinaldi, H.L. Murray, et al. (2004). Control of pancreas and liver gene expression by HNF transcription factors. *Science* 303(5662), 1378–1381.
4. Odom, D.T., R.D. Dowell, E.S. Jacobsen, L. Nekludova, P.A. Rolfe, T.W. Danford, et al. (2006). Core transcriptional regulatory circuitry in human hepatocytes. *Molecular Systems Biology* 2, 2006 0017.

5. Yamagata, K. (2003). Regulation of pancreatic beta-cell function by the HNF transcription network: Lessons from maturity-onset diabetes of the young (MODY). *Endocrinology Journal* 50(5), 491–499.
6. Levchenko, A. (2003). Dynamical and integrative cell signaling: Challenges for the new biology. *Biotechnology & Bioengineering* 84, 773–782.
7. Hoffmann, A., A. Levchenko, M. Scott, and D. Baltimore. (2002). The IkappaB-NF-kappaB signaling module: Temporal control and selective gene activation. *Science* 298, 1241–1245.
8. Thompson, D.M., K.R. King, K.J. Wieder, M. Toner, M.L. Yarmush, and A. Jayaraman. (2004). Dynamic gene expression profiling using a microfabricated living cell array. *Analytical Chemistry* 76, 4098–4103.
9. Wieder, K.J., K.R. King, D.M. Thompson, C. Zia, M.L. Yarmush, and A. Jayaraman. (2005). Optimization of reporter cells for expression profiling in a microfluidic device. *Biomedical Microdevices* 7, 213–222.
10. King, K.L., S. Wang, D. Irimia, A. Jayaraman, M. Toner, and M.R. Yarmush. (2007). A high-throughput microfluidic real-time gene expression living cell array. *Lab-on-Chip* 7, 77–85.
11. Gu, W., X. Zhu, N. Futai, B. Cho, and S. Takayama. (2004). Computerized microfluidic cell culture using elastomeric channels and Braille displays. *Proceedings of the National Academy of Sciences USA* 101, 15861–15866.
12. Kim, L., M. Vahey, H. Lee, and J. Voldman. (2006). Microfluidic arrays for logarithmically perfused embryonic stem cell culture. *Lab-on-Chip* 6, 394–406.
13. Lee, P., P. Hung, V. Rao, and L. Lee. (2006). Nanoliter scale microbioreactor array for quantitative cell biology. *Biotechnology & Bioengineering* 94, 5–14.
14. Cubitt, A., R. Heim, S. Adams, A. Boyd, L. Gross, and R. Tsien. (1995). Understanding, improving and using green fluorescence proteins. *Trends in Biochemical Sciences* 20, 448–455.
15. Li, X., X. Zhao, Y. Fang, X. Jiang, T. Duong, C. Fan, et al. (1998). Generation of destabilized green fluorescent protein as a transcription reporter. *Journal of Biological Chemistry* 273, 34970–34975.
16. Zhao, X., T. Duong, C. Huang, S. Kain, and X. Li. (1999). Comparison of enhanced green fluorescent protein and its destabilized form as transcription reporters. *Methods in Enzymology* 302, 32–38.
17. Ozbudak, E., M. Thattai, I. Kurtser, A. Grossman, and A. van Oudenaarden. (2002). Regulation of noise in the expression of a single gene. *Nature Genetics* 31, 69–73.
18. Pedraza, J., and A. van Oudenaarden. (2005). Noise propagation in gene networks. *Science* 307, 1965–1969.
19. Swain, P., M. Elowitz, and E. Siggia. (2002). Intrinsic and extrinsic contributions to stochasticity in gene expression. *Proceedings of the National Academy of Sciences USA* 99, 12795–12800.
20. Nelson, D., A. Ihekwaba, M. Elliott, J. Johnson, C. Gibney, B. Foreman, et al. (2004). Oscillations in NF-kappaB signaling control the dynamics of gene expression. *Science* 306, 704–708.
21. Lahav, G., N. Rosenfeld, A. Sigal, N. Geva-Zatorsky, A. Levine, M. Elowitz, et al. (2004). Dynamics of the p53-Mdm2 feedback loop in individual cells. *Nature Genetics* 36, 147–150.
22. Zhang, J., R. Campbell, A. Ting, and R. Tsien. (2002). Creating new fluorescent probes for cell biology. *Nature Reviews Molecular and Cell Biology* 3, 906–918.
23. Giepmans, B., S. Adams, M. Ellisman, and R. Tsien. (2006). The fluorescent toolbox for assessing protein location and function. *Science* 312, 217–224.
24. Niswender, K., S. Blackman, L. Rohde, M. Magnuson, and D. Piston. (1998). Quantitative imaging of green fluorescent protein in cultured cells: comparison of microscopic techniques, use in fusion proteins and detection limits. *Journal of Microscopy* 180, 109–116.
25. Furtado, A., and R. Henry. (2002). Measurement of green fluorescent protein concentration in single cells by image analysis. *Analytical Biochemistry* 310, 84–92.

26. Tilles, A., H. Baskaran, P. Roy, M. Yarmush, and M. Toner. (2001). Effects of oxygenation and flow on the viability and function of rat hepatocytes cocultured in a microchannel flat-plate bioreactor. *Biotechnology & Bioengineering* 73, 379–389.
27. McDonald, J., D. Duffy, J. Anderson, D. Chiu, H. Wu, O. Schueller, et al. (2000). Fabrication of microfluidic systems in poly(dimethylsiloxane). *Electrophoresis* 21, 27–40.
28. Unger, M., H. Chou, T. Thorsen, A. Scherer, and S. Quake. (2000). Monolithic microfabricated valves and pumps by multilayer soft lithography. *Science* 288, 113–116.
29. Thorsen, T., R. Roberts, F. Arnold, and S. Quake. (2001). Dynamic pattern formation in a vesicle-generating microfluidic device. *Physical Review Letters* 86, 4163–4166.
30. Li, N., C. Hsu, and A. Folch. (2005). Parallel mixing of photolithographically defined nanoliter volumes using elastomeric microvalve arrays. *Electrophoresis* 26, 3758–3764.
31. Irimia, D., and M. Toner. (2006). Cell handling using microstructured membranes. *Lab-on-Chip* 6, 345–352.
32. Jiang, X., Q. Xu, S. Dertinger, A. Stroock, T. Fu, and G. Whitesides. (2005). A general method for patterning gradients of biomolecules on surfaces using microfluidic networks. *Analytical Chemistry* 77, 2338–2347.
33. King, K.R., S. Wang, A. Jayaraman, M.L. Yarmush, and M. Toner. (2008). Microfluidic flow-encoded switching for parallel control of dynamic cellular environments, *Lab-on-chip* 8, 107–116.

# Chapter 11
# New Approaches to the Synthesis of Addressable Microarray Molecular Libraries

**Karl Maurer and Kevin D. Moeller**

**Abstract** Approaches for the synthesis of molecular libraries on addressable arrays of microelectrodes are presented. In each case, substrates are fixed to a polymer coating of the array in the regions proximal to the microelectrodes. Selected microelectrodes are then used to synthesize chemical reagents that initiate reactions involving the substrates. The reagents generated at the electrodes are confined to the region of the array next to the selected electrodes by placing a second substrate that consumes the reagent in the reaction solution above the array. In addition, a strategy for characterizing the molecules synthesized in this manner, and a strategy for probing their biological activity are described.

## 11.1 Introduction

The development of spatially addressable libraries of small molecules has the potential to dramatically accelerate the pace at which the thousands of gene products typically produced by a cell can be isolated, identified, and probed for the factors that govern their binding to prospective ligands. This occurs because the selective binding of a gene product to a particular set of molecules within the library separates it from other gene products in the mixture that either bind to different members of the library or fail to bind at all. Variations in the concentration and structure of the molecules within the library can then lead to quantitative data concerning the nature of the interaction between the isolated gene product and the ligand or family of ligands to which it binds.

This screening method is optimized when numerous gene products can be evaluated simultaneously on a scale consistent with the tiny amounts of material generated by biological systems. To this end, chip-based microarrays of molecules

K. Maurer
CombiMatrix Corporation, 6500 Harbour Heights Pkwy, Suite 301, Mukilteo, WA 98275

K.D. Moeller
Department of Chemistry, Washington University in St. Louis, St. Louis, MO, 63130

K. Dill et al. (eds.), *Microarrays: Preparation, Microfluidics, Detection Methods, and Biological Applications*,

have proven to be particularly useful.[1] This technology enables the assembly of large libraries of potential ligands within a tiny area, and hence allows for the development of systematic, global strategies for evaluating complex mixtures of proteins. To date, much of this work has focused on the synthesis and evaluation of DNA[1a] and peptide[1b] based libraries.

It was against this backdrop that efforts began to develop molecular libraries on chips having an array of microelectrodes.[2–4] The idea was to utilize the electrodes on the chip to synthesize the molecules in the library so that each unique set of molecules in the library (those having identical structures) wound up located proximal to a unique, individually addressable electrode. In this way, the electrodes in the array could be used to monitor the behavior of the molecules in the library toward various biological receptors.

To date, three main approaches have been taken toward accomplishing this goal. In an approach forwarded by Southern and coworkers, the electrodes in the array were used to generate reagents that then acted upon substrates fixed to a glass slide above the array. The reagents reached the substrates by diffusing through a solution placed between the plates.[4a,b] In a second approach forwarded by Heller and coworkers,[4c] free-field electrophoresis was used to transport reagents to selected locations on a chip where they were effectively concentrated and allowed to react with the molecules being built. Finally, scientists at CombiMatrix initiated an effort in which substrates were fixed to the surface of the array using a porous polymer and then the electrodes in the array used to generate chemical reagents for conducting site-selective reactions on the polymer bound substrates.[2,5,6] Due space limitations, this chapter focuses on the development of this third approach.

The overall strategy used in the CombiMatrix approach begins by coating the microelectrode array with a porous, polyhydroxylated membrane. The microelectrodes are then used to both attach monomers to the membrane and then develop the monomers into larger molecules (Fig. 11.1).[5,6] In both steps, the role of the microelectrodes is to generate chemical reagents that initiate synthetic transformations. The reagents are confined to the region of the array surrounding individual

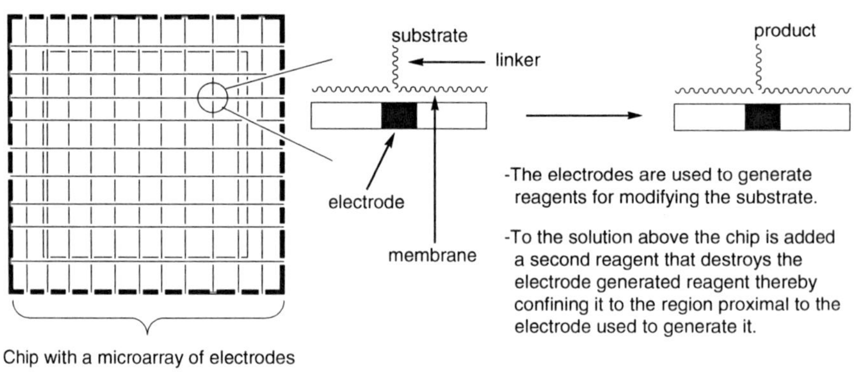


**Fig. 11.1** The strategy for doing site-selective chemical reactions on a microelectrode array

microelectrodes by covering the surface of the chip with a solution containing a second substrate.

The solution phase substrate destroys the electrochemically generated reagent before it can migrate to neighboring electrodes. In the neighborhood of a microelectrode utilized for generating the reagent, the concentration of the reagent is great enough to overwhelm the solution phase substrate. The excess reagent then reacts with the substrate bound to the polymer. However, as the distance from the selected electrode increases, the concentration of the reagent being generated decreases. In these regions, the concentration of the solution phase substrate is high enough to consume the electrochemically generated reagent, a situation that keeps the reagent from reacting with the substrate on the polymer at sites remote from the selected microelectrode.

## 11.2 Electrogenerated Acid: The Synthesis of DNA Arrays and the Deprotection of t-Boc Groups

Because of the biological relevance and overall utility of DNA arrays, much of the early work concerning synthesis on microelectrode arrays focused on the use of electrogenerated acids to catalyze the cleavage of DNA-protecting groups. The synthesis of DNA arrays was a perfect choice for this early work because the solid-phase synthetic chemistry needed was already highly developed, and the repetitive linear nature of DNA made it possible to spatially direct and contain the creation of DNA oligomers with varying compositions by developing only one electrochemically generated reagent.

As an initial strategy for deprotecting DMT groups in connection with solid-phase DNA synthesis, the use of a hydroquinone oxidation was employed for generating the necessary acid (Scheme 11.1). 2,6-Lutidine was used as the confining agent for preventing the acid generated from migrating to the neighboring microelectrodes. With a site-selective deprotection strategy in place, the standard solid-phase DNA synthesis sequence was quickly adapted to the microelectrode

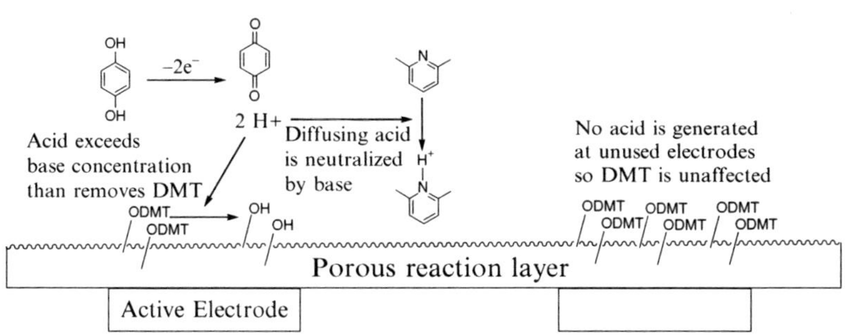


**Scheme 1**

array chip (Scheme 11.2). By iteratively repeating the four steps in this sequence, the length of the DNA oligomer could be systematically extended. Different DNA oligomers were constructed at each preselected site on the chip by varying the electrodes that were used for the DMT deprotection step in the sequence.

When the synthesis of the library was complete, the cyanoethyl protecting groups on the phosphodiester bonds were removed by treating the entire chip with ethylene diamine in ethanol. To date, libraries of DNA oligomers up to 50 bases in length have been synthesized using this protocol. The utility of this method is illustrated in Fig. 11.2.[5] The chip illustrated has on its surface a library of DNA oligomers 50 units in length that was synthesized using the procedure described above. The library was then treated with amplified gene fragments labeled with a fluorescent indicator.

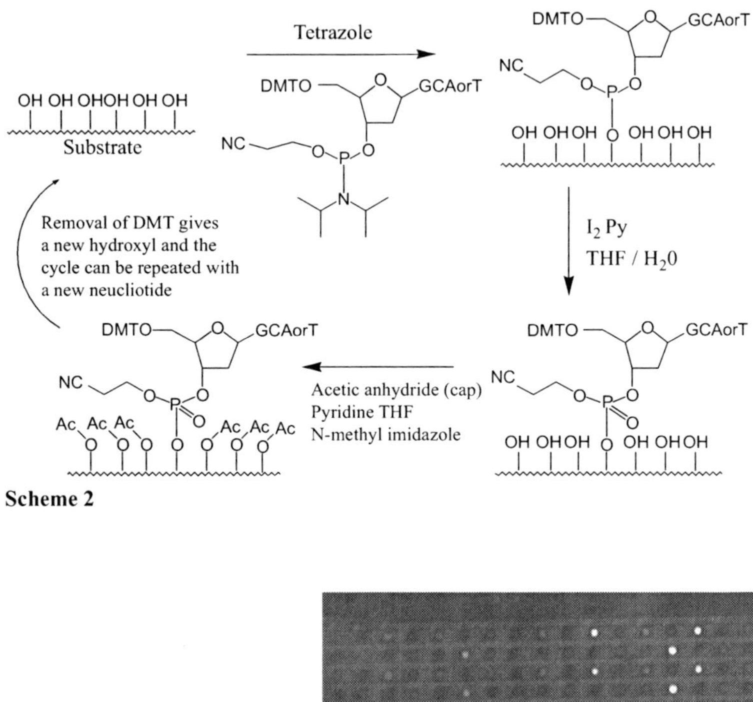


**Scheme 2**

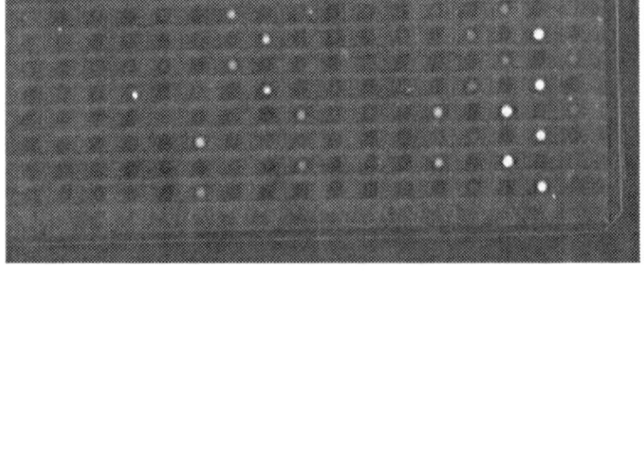

**Fig. 11.2** Evidence for the site-selective synthesis of DNA oligomers on a microelectrode array having 1028 microelectrodes cm$^{-2}$

Hybridization to DNA segments on the chip having a complementary sequence to the fragments then bound the indicator to the chip allowing for an analysis of the fidelity of the chip. As can be seen from the figure, the fluorescent indicator is localized to specific spots on the chip proximal to an individually addressable electrode. (The lighter regions on the chip that appear as a grid in the figure are the wires used to connect the electrodes to the power source for the experiment.) There was no apparent leakage of the DNA oligomers built at any given electrode to the surrounding electrodes even though the chip contains 1028 electrodes in a 1 $cm^2$ area.

Although the same approach for electrochemically generating acid can also be used for the removal of t-Boc groups from peptides,[6] an alternative strategy for electrochemically generating acid proved particularly useful for this transformation (Scheme 11.3).[7] In this case, diphenyl hydrazine was used as the precursor for acid generation. Oxidation of the hydrazine led to formation of a diazo compound plus two equivalents of acid.

Excess hydrazine provided the necessary confining agent for the electrogenerated acid. The reaction used dichloromethane as the solvent in order to minimize complexation of the acid and accelerate the t-Boc deprotection reaction. With a site-selective t-Boc deprotection strategy in place, the free amines could be treated using standard peptide coupling strategies. Iteration of the deprotection and coupling steps then allowed for the synthesis of a peptide on the microelectrode array. As with the earlier DNA synthesis, use of the microelectrodes to effect the deprotection reaction allows for different peptides to be built at various sites on the array.

The utility of this approach is illustrated in Fig. 11.3. In this experiment, five different peptides were synthesized on the chip using the t-Boc protecting group approach outlined in Scheme 11.3. Peptides with a YGGFL sequence were then imaged with a fluorescently tagged anti-YGGFL antibody. The YGGFL sequence was seen only at the electrodes selected for its synthesis. The degree of coverage on an electrode can be seen in the expanded view illustrated in the figure.

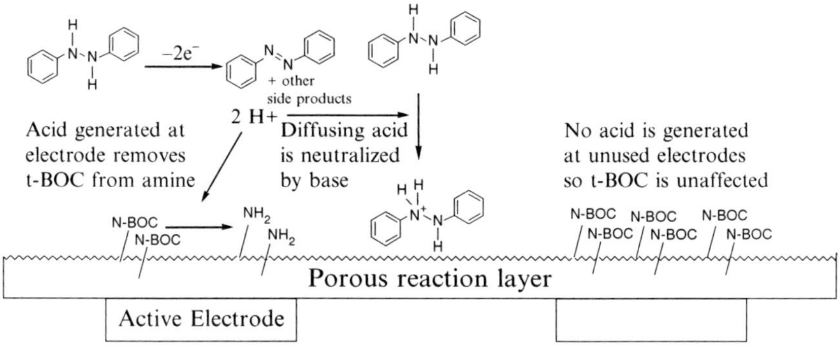


**Scheme 3**

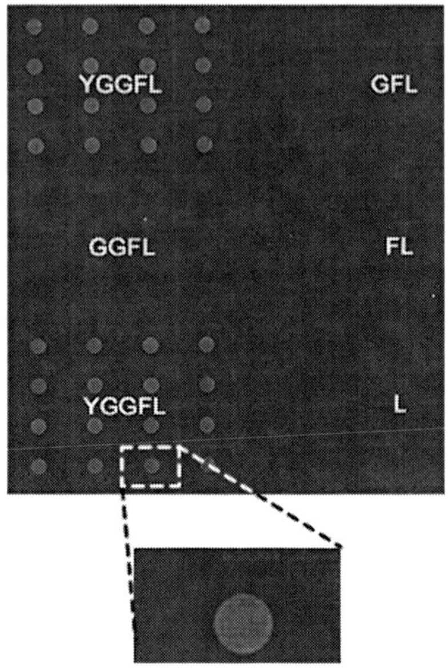


**Fig. 11.3** Evidence for the site-selective synthesis of short peptide segments on a microelectrode array having 1028 microelectrodes $cm^{-2}$

## 11.3 Electrogenerated Base: Coupling Strategies and the Deprotection of Fmoc Groups

Although oxidation reactions can be used to generate acid, reduction reactions are used to generate bases. Electrogenerated bases can also serve as powerful reagents for constructing molecular libraries on microelectrode arrays. Two examples are particularly useful for making this point. In the first, it is important to note that the monomers needed to start an array-based synthesis are frequently attached to the polyhydroxylated polymer coating the surface of the array with the use of a base-catalyzed esterification reaction between the alcohols on the polymer and an *N*-hydroxysuccinimide ester (Scheme 11.4).[8] The base catalyst is generated by the reduction of Vitamin $B_{12}$ and confined with the use of a protic solvent. Because the highest concentration of base in solution is by the electrodes, the esterification happens preferentially at those sites.

As illustrated in Fig. 11.4, the reaction illustrated in Scheme 11.4 can be conducted with a high level of confinement. To create this image, a checkerboard pattern of electrodes was used as cathodes and then using a fluorescence microscope, the pyrene was observed.

The second example of using an electrogenerated base for microarray-based synthesis involved the development of an Fmoc-based peptide synthesis strategy on the microelectrode array.[6] In this example, the electrogenerated base resulted from the reduction of azobenzene (Scheme 11.5). This reaction was performed after placing the substrate onto the polymer on the surface of the array using the

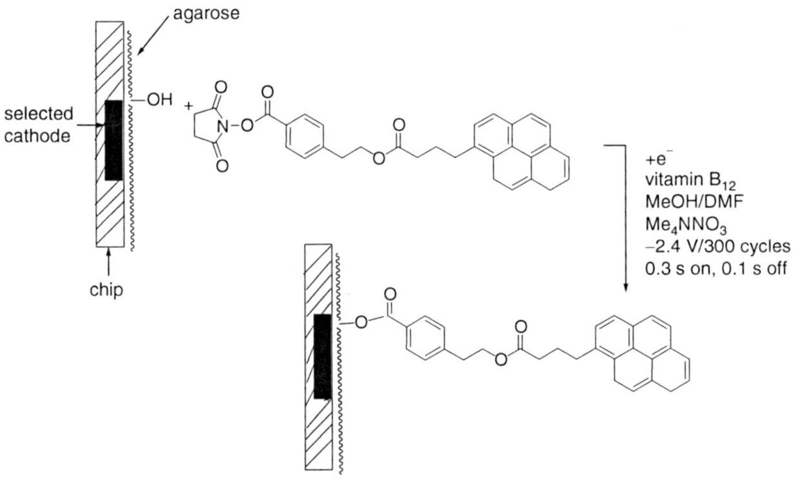


**Scheme 4**

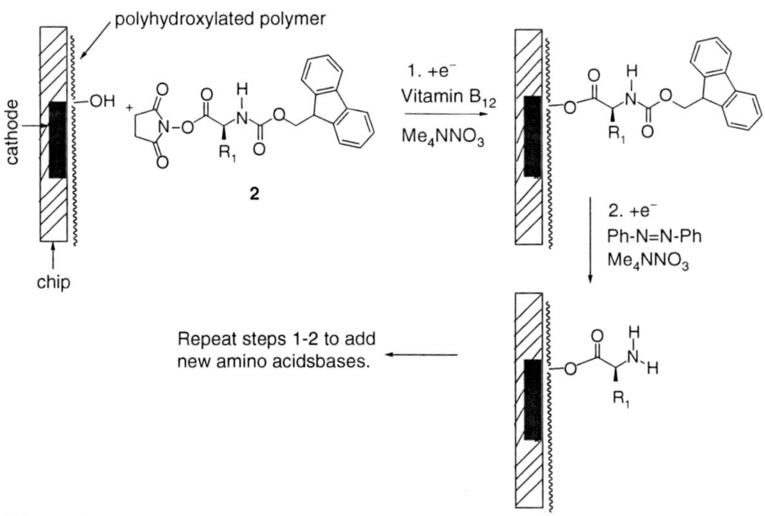


**Scheme 5**

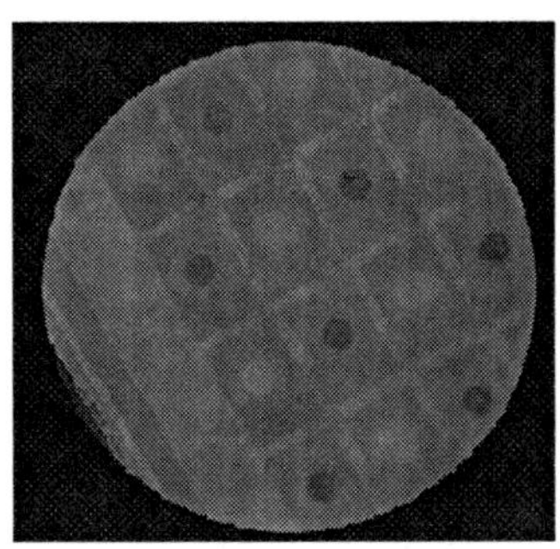

**Fig. 11.4** Evidence for the site-selective placement of substrates on a microelectrode array

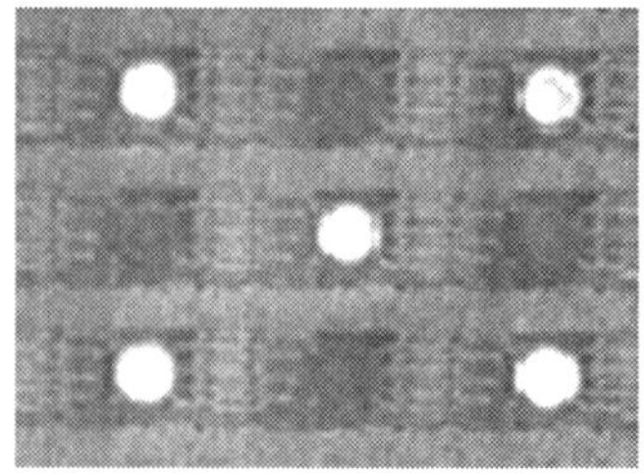

**Fig. 11.5** Evidence for the site-selective cleavage of an Fmoc protecting group

protocol discussed above. Cleavage of the protecting group then utilized a second reduction reaction.

A few items concerning this synthesis require further comment. First, cleavage of the Fmoc group requires higher concentrations of base than does the esterification reaction. Hence, the Fmoc group can survive the first base catalyzed coupling reaction (however, we have subsequently found that using a Heck reaction to place Fmoc-derived substrates on the surface of the chip to be a superior approach[9]). The need for the higher concentration of base also allowed the synthesis to proceed site-selectively in the absence of a confining agent.

In this case, the dropoff in concentration of the base as the distance from the electrode increased stopped reactions from occurring at remote sites on the array. Finally, as in the earlier processes the use of electrochemistry for the deprotection step in the sequence allows for the synthesis of a different peptide sequence at each electrode. In order to elongate the peptide the two steps outlined in Scheme 11.5 are simply repeated at the selected electrodes.

The effectiveness of this approach for peptide synthesis was again illustrated by using it to build a pentapeptide with a YGGFL sequence. In this case, an array with a checkerboard pattern of peptide was synthesized. The anti-YGGFL antibody was then used to image the chip (Fig. 11.5).

## 11.4 Transition Metal-Based Reactions

The acid- and base-derived chemistry used for DNA and peptide synthesis represents only a small fraction of the reactions that can in principle be site-selectively accomplished using a microelectrode array. This is important because small molecule libraries, particularly ones containing conformationally constrained molecules, are powerful tools for probing biological systems.[10] The synthesis of small molecule libraries makes use of much of modern organic synthesis, especially the transition metal-catalyzed reactions that have proven extremely useful for constructing new ring skeletons. For this reason, an investigation aimed at determining the utility of transition metal-based synthetic methodology for constructing molecules site-selectively on a microelectrode array has been undertaken.

This effort began by showing that well-known mediated electrolyses can be directly employed for microelectrode array-based reactions.[11] Initially, the electrochemically mediated Wacker oxidation was selected for study.[12] The Wacker oxidation involves the Pd(II) oxidation of an olefin to a ketone in water. In the electrochemically mediated version of the reaction an anodic oxidation is employed to circumvent the use of stoichiometric palladium in the process. This is accomplished by generating an amine radical cation at an anode that in turn oxidizes the Pd(0) generated during the reaction back to Pd(II).

The plan to move this process to preselected sites on an addressable array of microelectrodes was straightforward. Because during an electrolysis reaction the substrate at the electrode surface does not "know" what size the electrode is, the idea was to simply use the reaction conditions developed for the preparative process at selected microelectrodes in the array (Scheme 11.6).[8] With this in mind, the olefin substrate was deposited onto an agarose polymer coating the surface of the microelectrode array using the chemistry described above. Ethyl vinyl ether was added as a confining agent to the reaction solution above the chip. Ethyl vinyl ether undergoes a rapid Wacker oxidation in order to make ethyl acetate, a reaction that would prevent any Pd(II) oxidant generated at a selected electrode from migrating to a neighboring electrode. Every other aspect of the reaction was kept identical to the preparative scale reactions.

As illustrated in Fig. 11.6, the chip-based reaction worked beautifully. For the chip in the figure, a checkerboard pattern of electrodes was used to effect the oxidation. The ketones made at the electrodes were imaged by treating the chip with 2,4-dinitrophenyl hydrazine and then incubating the chip with an anti-2,4-DNP antibody tied to a fluorescent tag. The chip was then placed in a fluorescence

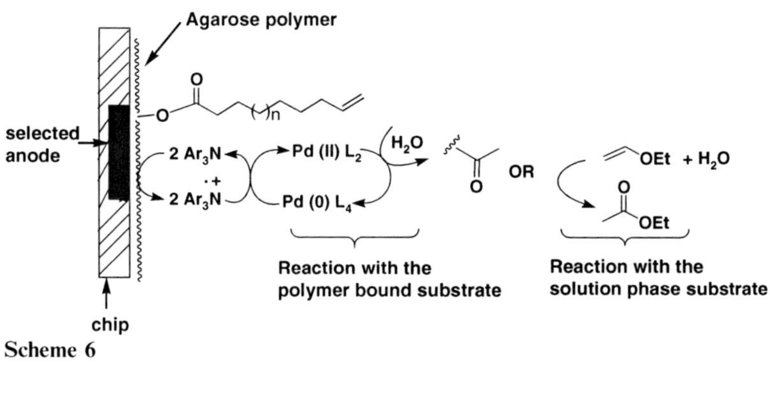


**Scheme 6**

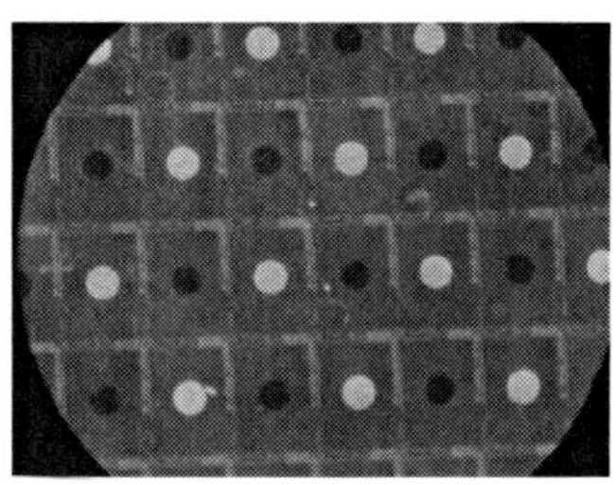

**Fig. 11.6** A site-selective Wacker oxidation on a microelectrode array containing 1028 microelectrodes $cm^{-2}$

microscope. The bright spots on the chip represent electrodes that were turned on and the dark spots are electrodes that were not used (the chip itself is fluorescent and is blocked by the Pt electrodes on the chip). Clearly the level of confinement for the Pd(II) reagent used was very high.

In order to test the generality of using Pd(II) site-selectively on the microelectrode array, the reaction conditions developed for the Wacker oxidation were employed for a site-selective reductive amination reaction (Scheme 11.7).[13] In this approach, the Pd(II) reagent generated at the selected electrodes was allowed to react with the agarose polymer covering the surface of the microelectrode array. The result was an oxidation of the alcohols on the polymer.[14,15] Ethyl vinyl ether was again used as the confining agent. A reductive amination was then used in order to convert the carbonyl generated by the oxidation into a fluorescently labeled amine. In the first step, a checkerboard pattern of electrodes was used for the oxidation and then the reductive amination used to add a red fluorescent indicator to the surface of the array. In a second step, the electrodes not used for the first oxidation were employed for the alcohol oxidation and then a reductive amination used to place a green fluorescent indicator on the surface of the chip.

The results of the experiment are shown in Fig. 11.7. Once again, the confinement of the Pd(II) oxidant on the chip was outstanding. In fact, the level of confinement in this reaction was particularly impressive. For the Wacker oxidation, the olefin substrate was placed above the electrodes. Hence, the confinement observed was for one

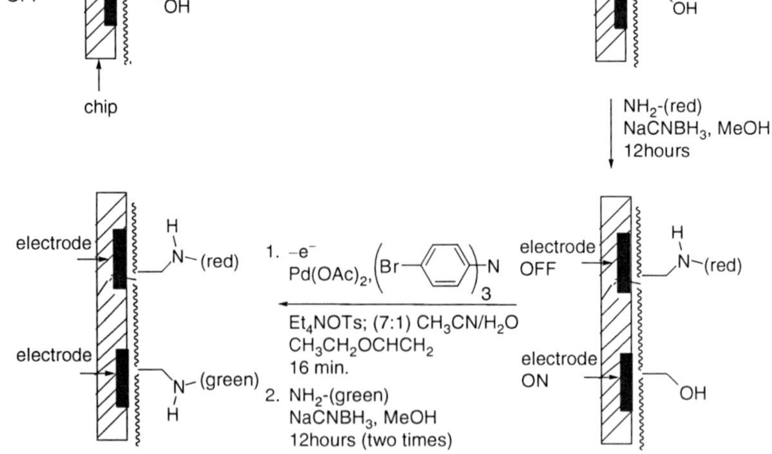


**Scheme 7**

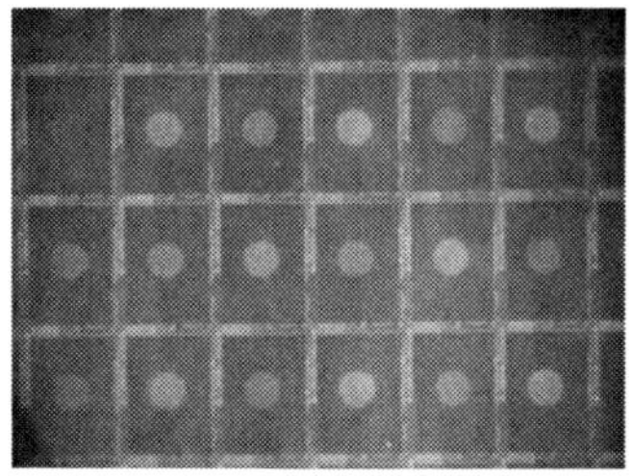

**Fig. 11.7** A site-selective reductive amination reaction

electrode over the neighboring electrodes. In contrast, for the reductive amination strategy the oxidation reaction used the polymer, coating the whole chip including the regions in between the electrodes. The confinement observed was not simply for one electrode in preference to the neighboring electrodes, but rather for an electrode in preference to the region of the chip immediately surrounding the electrode! The Pd(II) oxidations were confined to the region directly above the electrodes, an observation that indicates the confinement strategy will be compatible with an even higher density of electrodes on the chip.

The overall strategy for using transition metals on the microelectrode array is not restricted to oxidation reactions. For example, a cathodic reduction can be used to trigger site-selective Pd(0) chemistry. Along with the Pd(II) catalyzed synthesis of coumarins (see below), this chemistry represented a new challenge for electrochemical synthesis because many of the most synthetically powerful Pd(0) reactions are catalytic in palladium. Hence, rather than the traditional use of electrochemistry to convert a stoichiometric process into a catalytic one (the typical role for electrochemistry in all mediated organometallic reactions), in this case a catalytic process needed to be converted into a stoichiometric one so that it could be isolated to preselected sites on the microelectrode array.

As a starting point, a site-selective Heck reaction[16] was selected for development (Scheme 11.8). For this reaction, an aryl iodide substrate was coupled to the agarose polymer above the microelectrodes. The chip was treated with a DMF/ acetonitrile/ water solution containing $Pd(OAc)_2$, pyrenemethyl acrylate, and allyl methyl carbonate along with triphenylphosphine, tetrabutylammonium bromide, and triethylamine. The $Pd(OAc)_2$ was then reduced at selected electrodes in the array in order to generate a Pd(0) species that in turn underwent an oxidative addition with the localized aryliodide. Subsequent addition of the aryl–Pd bond across the acrylate double bond followed by β-hydride elimination and reductive elimination completed the Heck reaction and regenerated the Pd(0) catalyst. The allyl methyl carbonate in solution then served as a confining agent by converting the Pd(0) catalyst into a π-allyl Pd(II) species. Noyori and coworkers have shown that π-allyl Pd(II) species can be recycled at electrode surfaces,[17] although with the large excess of $Pd(OAc)_2$ used for the chip-based electrolysis it is unclear if such a recycling process occurs in this case.

The result of the reaction is illustrated in Fig. 11.8. Because the Heck reaction placed pyrene on the surface of the chip, the outcome of the reaction could be

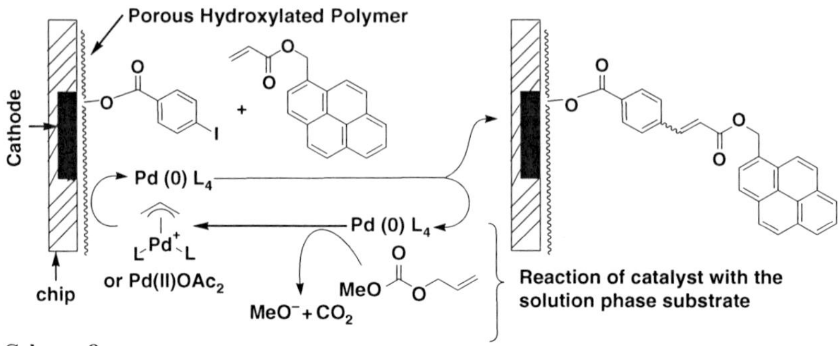


Scheme 8

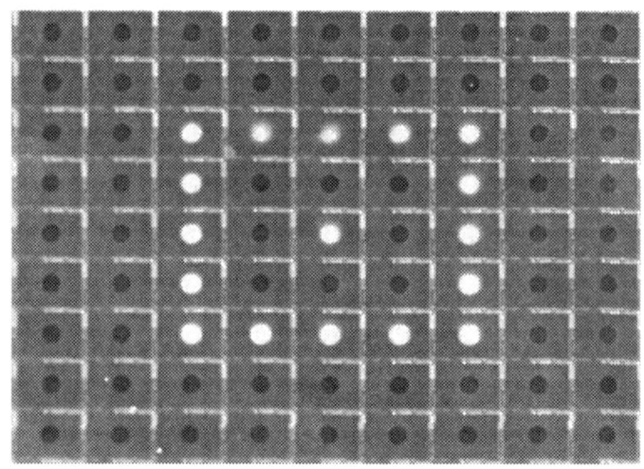

**Fig. 11.8** The site-selective Heck reaction.

directly probed using a fluorescence microscope. Confinement of the Heck reaction to selected electrodes was superb.

Although to this point all of the site-selective transition metal chemistry developed has utilized palladium, in principle the behavior of any transition metal reagent can be manipulated by control of its oxidation state. Hence, one can imagine a wide variety of transition metal catalyzed and mediated reactions being performed on a microelectrode array. For example, in related work a Cu(I) catalyzed "click reaction"[18] has been performed on a microelectrode surface in order to localize a biomolecule near the electrode.[19]

In this experiment, an azide substrate was attached to the surface of a pair of Au interdigitated array band electrodes using an alkylthiol linker. The two electrodes were separated by 10 microns. With the azide in place, the two electrodes were treated with a solution of acetylene, electrolyte, and copper(II) bis(bathopehananthroline)disulfonic acid.[20] One of the electrodes was then used as a cathode to reduce the Cu(II) reagent to a Cu(I) catalyst and trigger the click-reaction. The second electrode was used as an anode in order to oxidize any Cu(I) catalyst that migrated to its location from the first electrode thereby confining the click-reaction to the first electrode. The extent of confinement was excellent. The extension of this chemistry to a microelectrode array environment would appear to be straightforward.

## 11.5 Cleavable Linkers

With efforts to develop new site-selective reactions proceeding nicely, two other aspects of the work became a concern. First, how can the molecules synthesized on a microelectrode array be characterized, and second, how can the microelectrode array be used to rapidly screen a small molecule library for binding to a biological receptor of interest? Because these two issues are of vital importance for determining the overall utility of microelectrode arrays, two proof-of-principle experiments are of interest here.

To understand the significance of the first, it is important to note that the microelectrode array reactions described above were optimized for the level of confinement obtained. This was an excellent place to start because the feasibility of the approach had not been established. However, the fluorescence methods used for determining confinement did not afford any information as to either the quality of the products generated or how well the reactions work in terms of yield or percent conversion of starting material. These issues are critical because one of the main concerns about any small molecule library used for analyzing biological systems is quality control. How does one know that the molecules in a library that interact with a biological receptor are really the molecules that they are believed to be? With this question in mind, work began to explore methods for analyzing the molecules associated with any electrode.

One promising approach is capitalizing on TOF-SIMS techniques for monitoring microelectrode array-based reactions.[21] The TOF-SIMS experiment utilizes a charged particle beam (either gold or bismuth) to ionize molecules on a surface with a resolution of less than 100 microns. The ions generated are then analyzed by mass spectrometry. Inasmuch as an array containing 1028 electrodes/cm$^2$ contains microelectrodes having a diameter of 95 microns, the TOF-SIMS technique can be used to examine the product generated at individual electrodes in the array. Initial experiments along these lines were not successful in that the only ions observed were generated from the agarose polymer coating the chip. What was needed was a linker that would fragment in the TOF-SIMS experiment faster than the polymer covering the surface of the chip.[22]

To this end, a styrene-based linker was synthesized and analyzed for its ability to undergo McLafferty fragmentations during the experiments (cleavage site a in Fig. 11.9). The initial substrate attached to the chip's surface with the linker was the one used for the earlier Wacker oxidation experiments. Using this linker, the TOF-SIMS experiment led to the observation of the 183 parent peak associated with the olefin substrate along with a clearly delineated peak with a mass of 341 (cleavage site b in Fig. 11.5). In a similar manner, the ketone product from the Wacker

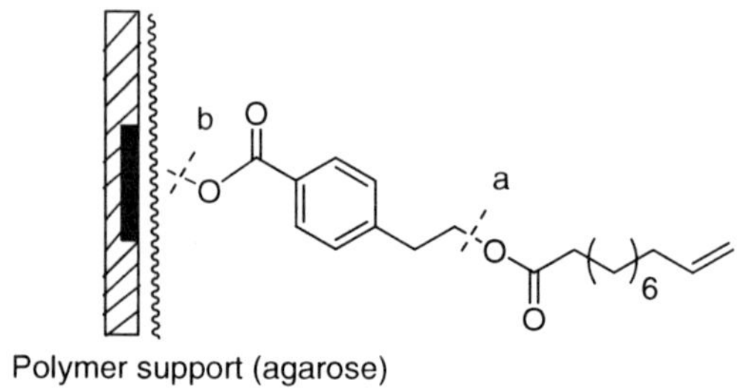


**Fig. 11.9** A mass spectrometry-cleavable linker

oxidation could be observed on the microelectrode array, this time leading to peaks with masses of 199 and 357. After developing a calibration curve, the yield for the Wacker oxidation using the initial confinement conditions was determined.

Interestingly, it was found that little if any product was made on the electrodes in these early experiments. The use of fluorescence for imaging the product was simply very sensitive and showed bright spots even though very little product was formed. Of course, the technique provided no means for determining the amount of starting material that was left. Fortunately, by simply adjusting the amounts of palladium catalyst and confining agent used for the experiment, the percent conversion of the reaction could be optimized. It was found that increasing the amount of palladium from .32 mg to 32 mg in the solution over the chip (total volume 1.5 mL) while decreasing the amount of confining agent from 50 μL to 0.5 μL led to complete conversion of the starting material to the desired product.

Clearly, the use of TOF-SIMS techniques for monitoring the percent conversion of the reactions provided a nice complement to the fluorescence techniques used for establishing the level of confinement on the microelectrode array. In the future, the combination of the two methods will allow for optimization of the reactions in a manner that makes them truly useful for site-selectively building addressable molecular libraries.

## 11.6 Chip Signaling

With rapid progress being made on the synthetic and analytical aspects of addressable library preparation, the second proof-of-principle experiment focused on the utility of the microelectrode array for rapidly screening the molecules in a library for binding to a biological receptor. Our initial plan for this process borrowed an idea from the development of electrochemical sensors.[23] In this approach, a biological receptor is placed on the surface of a gold electrode. A redox couple is then cycled between the gold electrode and a remote auxiliary electrode creating a current that can be monitored. When a molecule binds the receptor bound to the gold electrode it alters the electroconductive properties of the material on the electrode and interferes with the redox cycle. A drop in the current is measured. In this way, the receptor–ligand interaction can be monitored without the need for the subsequent washing step normally associated with an immunological assay, a step that can skew the initial result.

Several issues about a microelectrode array chip-based version of this approach were of immediate concern. First, rather than attach a large biological receptor to the electrode and then screen a series of solution phase ligands one at a time for their interactions with the receptor, the microelectrode array approach builds a series of spatially isolated small molecule ligands over the electrodes and then seeks to probe their interactions with a solution phase biomolecule or mixture of biomolecules in a parallel fashion. This should result in more rapid screening and the ability to probe the gene products from a cell in a global manner. But how many microelectrodes in an array are needed for seeing a signal? How large a molecular library can be probed?

Second, the microelectrode array-based approach does not attach the molecules to the electrode surface. Instead, it builds the molecule on a porous polymer covering the entire surface of the chip. Will the extra distance between the electrode and the interaction being monitored result in a situation where the binding event no longer interferes with the current associated with a secondary redox couple? If the answer is yes, then how can the overall approach be changed so that the molecules can be directly built on the microelectrodes? If the answer is no, then what distance between the electrode surface and the molecule being monitored can be tolerated before a loss of signal occurs?

To begin addressing these questions, coumarins were placed proximal to the microelectrodes of an addressable array. Two main advantages were associated with this choice. First, a one-step synthesis of coumarins using a Pd(II) catalyzed cycloaddition reaction between a phenol and an acetylene was known.[24] To use this reaction on a microelectrode array, a phenol substrate was coupled to an amine functionalized porous polymer on the surface of the chip (Scheme 11.9).[25] The Pd(II)-catalyzed cycloaddition reaction with the acetylene was then performed using selected electrodes in a manner identical to the Pd(II)-mediated Wacker oxidation and alcohol oxidation reactions described earlier. Once again, both the method for Pd(II) generation at the selected electrodes and the method for confinement proved to be excellent.

Second, the ability of the chip to signal a binding event could be explored by treating the coumarin functionalized chip with commercially available, coumarin-specific antibodies.[26] For this work, a chip using 12,544 electrodes/cm$^2$ was utilized.[2] Of the approximately 12,000 microelectrodes, three blocks of 121 electrodes were employed as anodes for the coumarin synthesis. On the first block, the coumarin was linked to the porous polymer through a single thymidine unit having an aminoethoxyethyl substituent (T in Scheme 11.10, $n = 1$). On the second block, the coumarin was linked to the polymer using 5 thymidines with the terminal one having the aminoethoxyethyl substituent. On the third block, the coumarin was linked to the chip using 15 thymidines with the terminal one having the aminoethoxyethyl substituent. The chip was then submerged in a reaction solution containing a ferrocene acetic acid/ferrocinium acetic acid cation redox couple and the current associated with recycling the ferrocene cation measured. This was accomplished

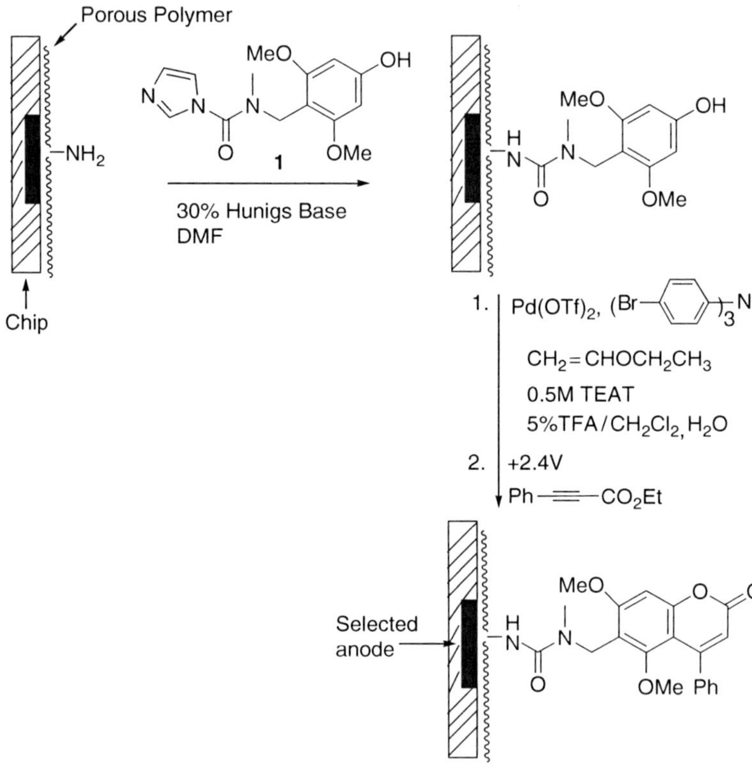


**Scheme 9**

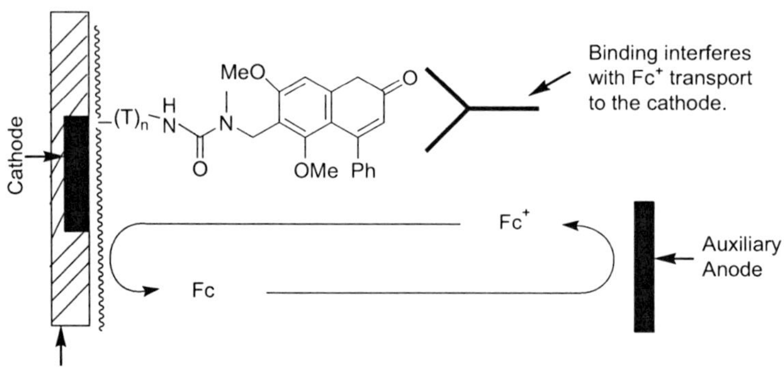


**Scheme 10**

by using the electrodes on the chip as cathodes and a remote auxiliary electrode (indium tin oxide) as the anode (Scheme 11.10).

The result of the experiment using the block of electrodes having the coumarin attached through a single thymidine unit is illustrated in Fig. 11.10. The initial

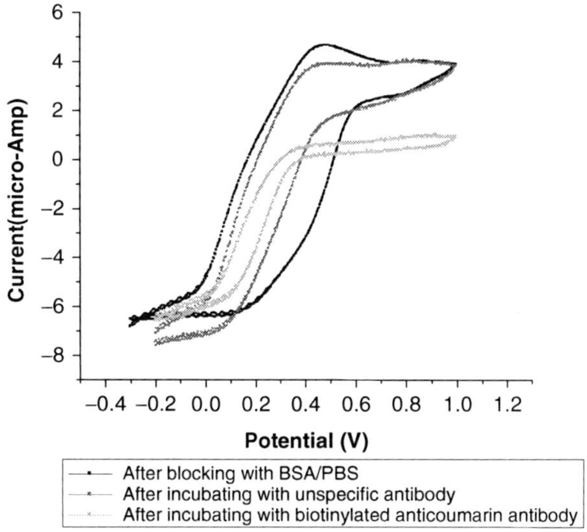


**Fig. 11.10** Dectecting the bindibg of a coumarin substrate with an anti-coumarin antibody

current measurement was made in the absence of any antibody and is represented by the black curve. The chip was then treated with an anti-2,4-DNP antibody and the current measured giving rise to the red curve illustrated in Fig. 11.10. No significant drop in current was measured indicating that there was minimal nonspecific interaction between the antibody and the chip's surface. Following this control experiment, the chip was treated with the anticoumarin antibody. The current measured gave rise to the green curve in Fig. 11.10. Clearly, a drop in current was observed indicating that the overall strategy is capable of providing a "real-time" signal in response to a small molecule–antibody binding event. The level of sensitivity obtained in this initial experiment (121 electrodes on the 12 K chip) suggests that libraries of approximately 100 molecules can already be monitored.

Only a tiny drop in current was observed when the block of electrodes using the 5-thymidine linker was used for the experiment. No drop in current was observed for the block of electrodes using the 15-thymidine linker. Hence, the signal did depend on the distance between the small molecule and the electrode surface, an observation that needs to be accounted for as work to develop the signaling strategy continues.

## 11.7 Conclusions

At the present time, the use of microelectrode arrays for building and analyzing addressable libraries appears ideal. Reactions that can be used for site-selectively synthesizing molecules by individual microelectrodes include not only the acid and base reactions typically used to make DNA and peptide libraries, but also both stoichiometric and catalytic transition metal-based methods. The range of reactions that can be performed is still limited, however, the possibilities seem endless. The effectiveness of electrochemistry for manipulating the oxidation state and hence reactivity of myriad metal- and nonmetal-based reagents means that a large percentage of the successful reactions used in solution-phase synthesis can be moved to a microelectrode array environment.

In addition, the use of microelectrode arrays is amenable to surface techniques that allow for characterization of the molecules being synthesized, and allow for the real-time detection of small molecule–biological receptor binding events. In the end, the combination of site-selective synthetic reactions, molecular characterization of products, and real-time detection strategies afforded by addressable microelectrode arrays makes them a platform for biological analysis that deserves considerable future attention.

**Acknowledgments** We thank the National Science Foundation (CHE-0314057; CHE-0518063) and CombiMatrix for their generous support of this work. We also gratefully acknowledge the Washington University High Resolution NMR facility, partially supported by NIH grants RR02004, RR05018, and RR07155, and the Washington University Mass Spectrometry Resource Center, partially supported by NIHRR00954, for their assistance. Professor Amy Walker of Washington University is gratefully acknowledged for her collaboration on the TOF-SIMS work, as is Kris Roth for his collaboration in connection with the signaling portion of this work. Finally, we wish to acknowledge the students (Dr. Eden Tesfu, Dr. Jun Tian, Ms. Ceng Chen, and Ms. Gabriella Nagy) without whom none of this work would be accomplished.

## Notes

1. (a) "High density synthetic oligonucleotide arrays." Lipshutz, R.J.; Fodor, S.P.A.; Gingeras, T.R.; Lockhart, D.J. *Nature Genetics* **1999**, *21*, 20. (b) "Spatially addressable combinatorial libraries." Pirrung, M.C. *Chem. Rev.* **1997**, *97*, 473. For more recent examples see: (c) "Gene networks in glucocorticoid-evoked apoptosis of leukemic cells." Webb, M.S..; Miller, A.L.; Johnson, B.H.; Fofanov, Y.; Li, T.; Wood, T.G.; Thompson, E.B. *J. Steroid Biochem. Molec. Biol.* **2003**, *85*, 183. (d) Shih, S.-R.; Wang, Y.-W.; Chen, G.-W.; Chang, L.-Y.; Lin, T.-Y.; Tseng, M.-C.; Chiang, C.; Tsao, K.-C.; Huang, C.G.; Shio, M.-R.; Tai, J.-H.; Wang, S.-H.; Kuo, T.-L.; Liu, W.-T. "Serotype-specific detection of enterovirus 71 in clinical specimens by DNA microchip array." *J. Virol. Meth.* **2003**, *111*, 55.
2. For a description of the chips used here see: "Antigen detection using microelectrode array microchips." Dill, K.; Montgomery, D. D.; Wang, W.; Tsai, J.C. *Anal. Chim. Acta* **2001**, *444*, 69.
3. For alternative addressable electrode arrays see: (a) "Automated electrochemical analysis with combinatorial electrode arrays." Sullivan, M.G.; Utomo, H.; Fagan, P.J.; Ward, M.D. *Anal. Chem.* **1999**, *71*, 4369. (b) "Development of a generic microelectrode array biosensing system." Zhang, S.; Zhao, H.; John, R. *Anal. Chim. Acta* **2000**, *421*, 175. (c) "Multiplexing of microelectrode arrays in voltammetric measurements." Hintsche, R.; Albers, J.; Bernt, H.; Eder, A. *Electroanal.* **2000**, *12*, 660.
4. For alternate strategies for electrode array-based synthesis see: (a) "Electrochemically directed synthesis of oligonucleotides for DNA microarray fabrication." Egeland, R.D.; Southern, E.M. *Nucleic Acids Res.* **2005**, *33*, e125. (b) "Electrochemical treatment of surfaces for stepwise synthesis of oligonucleotides or other oligomers." Southern, E. PCT Int. Appl.

(1993), 27 pp. CODEN: PIXXD2 WO 9322480 A1 19931111. (c) "Self-addressable self-assembling microelectronic systems and devices for molecular biological analysis and diagnostics." Heller, M.J.; Tu, E. PCT Int. Appl. (1995), 86 pp. CODEN: PIXXD2 WO 9512808 A1 19950511.

5. For the lead patent on DNA-related work see: "Electrochemical solid phase synthesis of biopolymers." Montgomery, D.D. PCT Int. Appl. (1998), 91 pp. CODEN: PIXXD2 WO 9801221 A1 19980115.
6. For peptide-based libraries see: "Microarrays of peptide affinity probes for analyzing patterns of protein synthesis and their synthesis and methods for analyzing gene products." Rossi, F.M.; Montgomery, D.D. PCT Int. Appl. (2000), 52 pp. CODEN: PIXXD2 WO 0053625 A2 20000914. For recent work see: "Self-assembling protein arrays using electronic semiconductor microchips and in vitro translation." Oleinikov, A.V.; Gray, M.D.; Zhao, J.; Montgomery, D.D.; Ghindilis, A.L.; Dill, K. *J. Proteome Res.* **2003**, *2*, 313 as well as Note 2 above.
7. "The removal of the *t*-BOC Group by electrochemically generated acid and use of an addressable electrode array for peptide synthesis." Maurer, K.; McShae, A.; Strathmann, M.; Dill, K. *J. Comb. Chem.* **2005**, *7*, 637.
8. For a description of experimental details see: "Building addressable libraries: The use of electrochemistry for generating reactive Pd(II) reagents at pre-selected sites on a chip." Tesfu, E.; Maurer, K.; Ragsdale, S.R.; Moeller, K.D. *J. Am. Chem. Soc.* **2004**, *126*, 6212.
9. Unpublished results with Ms. Ceng Chen.
10. For recent reviews see: (a) "Compound collections: Acquisition, annotation and access." Richardson, R. In *Exploiting Chemical Diversity for Drug Discovery* Ed. Bartlett, P.A.; Entzeroth, M. Royal Society of Chemistry, Cambridge, UK; 2006; pp. 112–136. (b) "Toward the library generation of natural product-like polycyclic derivatives by stereocontrolled diversity-oriented synthesis." Arya, P.; Quevillon, S.; Joseph, R.; Wei, C.-Q.; Gan, Z.; Parisien, M.; Sesmilo, E.; Reddy, P.T.; Chen, Z.-X.; Durieux, P.; Laforce, D.; Campeau, L.-C.; Khadem, S; Couve-Bonnaire, S.; Kumar, R.; Sharma, U.; Leek, D.M.; Daroszewska, M.; Barnes, M.L. *Pure Appl. Chem.* **2005**, *77*, 163. (c) "Cyclic peptides, a chemical genetics tool for biologists." Horswill, A.R.; Benkovic, S.J.; *Cell Cycle* **2005**, *4*, 552.
11. For reviews concerning mediated electrolyses see: (a) "Electrogenerated reagents." Simonet, J.; Pilard, J.-F. In *Organic Electrochemistry: Fourth Edition* Lund, H.; Hammerich, O. Ed. New York: Marcel Dekker, 2001, p. 1163. (b) "Organic syntheses with electrochemically regenerable redox systems." Steckhan, E. *Topics Curr. Chem.* **1987**, *142*, 1.
12. (a) "Electrochemical Wacker type reaction with a doubly mediatory system consisting of palladium complex and Tri(4-bromophenyl)amine." Inokuchi, T.; Ping, L.; Hamaue, F.; Izawa, M.; Torii, S. *Chem. Lett.* **1994**, 121. (b) "Electrode-mediated Wacker oxidation of cyclic and internal olefins." Miller, D.G.; Wayner, D.D.M. *Can. J. Chem.* **1992**, *70*, 2485.
13. "Building addressable libraries: Spatially isolated, chip-based reductive amination reactions." Tesfu, E.; Maurer, K.; Moeller, K. D. *J. Am. Chem. Soc.* **2006**, *128*, 70.
14. For Pd(II) mediated alcohol oxidations see: "Using mechanistic and computational studies to explain ligand effects in the palladium-catalyzed aerobic oxidation of alcohols." Schultz, M.J.; Alder, R.S.; Zierkiewicz, W.; Privalov, T.; Sigman, M.S. *J. Am. Chem. Soc.* **2005**, *127*, 8499 and references therein.
15. For electrochemically mediated alcohol oxidations see: (a) "Electroorganic synthesis 66: Selective anodic oxidation of carbohydrates mediated by TEMPO." Schnatbaum, K.; Schäfer, H.J. *Synthesis* **1999**, 864. (b) "Selective oxidation of 1,2-diols by electrochemical method using organotin compound and bromide ion as mediators." Maki, T.; Fukae, K.; Harasawa, H.; Ohishi, T.; Matsumura, Y.; Onomura, O. *Tetrahedron Lett.* **1998**, *39*, 651.
16. "Building addressable libraries: The use of electrochemistry for spatially isolating a Heck reaction on a chip." Tian, T.; Maurer, K.; Tesfu, E.; Moeller, K.D. *J. Am. Chem. Soc.* **2005**, *127*, 1392.
17. "Electrochemical removal of allylic protecting groups in nucleotide synthesis." Hayakawa, Y.; Kawai, R.; Wakabayashi, S.; Uchiyama, M.; Noyori, R. *Nucleosides and Nucleotides* **1998**, *17*, 441.

18. For reviews see: (a) "Advances in 1,3-dipolar cycloaddition reaction of azides and alkynes – A prototype of "click" chemistry." Wang, Q.; Chittaboina, S.; Barnhill, H.N. *Lett. Organ. Chem.* **2005**, *2*, 293. (b) "Click chemistry: Diverse chemical function from a few good reactions." Kolb, H.C.; Finn, M.G.; Sharpless, K.B.; *Angew. Chem. Int. Ed. Eng.* **2001**, *40*, 2004.
19. "Selective functionalization of independently addressed microelectrodes by electrochemical activation and deactivation of a coupling catalyst." Devaraj, N.K.; Dinolfo, P.H.; Chidsey, C.E.D.; Collman, J.P. *J. Am. Chem. Soc.* **2006**, *128*, 1794.
20. "Discovery and characterization of catalysts for azide-alkyne cycloaddition by fluorescence quenching." Lewis, W.G.; Magallon, F.G.; Fokin V.V.; Finn, M.G. *J. Am. Chem. Soc.* **2004**, *126*, 9152.
21. Pacholski, M.L.; Winograd, N. "Imaging with mass spectrometry." *Chem. Rev.* **1999**, *99*, 2977.
22. Chen, C.; Nagy, G.; Maurer, K.; McShae, A.; Moeller, K.D. Unpublished results.
23. "An electrochemical device for the assay of the interaction between a dioxin receptor and its various ligands." Murata, M.; Gonda, H.; Yano, K.: Kuroki, S.; Suzutani, T; Katayama,Y. *Bio. Med. Chem. Lett.* **2004,** 14, 137.
24. "Direct synthesis of coumarins by Pd(II)-catalyzed reaction of alkoxyphenols and alkynoates." Oyamada, J.; Jia, C.; Fujiwara, Y.; Kitamura, T. *Chem. Lett.* **2002**, 380.
25. For a previous use of this polymer see: "The removal of the t-BOC Group by electrochemically generated acid and use of an addressable electrode array for peptide synthesis." Maurer, K.; McShea, A.; Strathmann, M.; Dill, K. *J. Combin. Chem.* **2005**, *7*, 637.
26. Available from Vector Laboratories, catalogue number BA-0606.

# Chapter 12
# eSensor®: A Microarray Technology Based on Electrochemical Detection of Nucleic Acids and Its Application to Cystic Fibrosis Carrier Screening

**Michael R. Reed and William A. Coty**

**Abstract** We have developed a test for identification of carriers for cystic fibrosis using the eSensor® DNA detection technology. Oligonucleotide probes are deposited within self-assembled monolayers on gold electrodes arrayed upon printed circuit boards. These probes allow sequence-specific capture of amplicons containing a panel of mutation sites associated with cystic fibrosis. DNA targets are detected and mutations genotyped using a "sandwich" assay methodology employing electrochemical detection of ferrocene-labeled oligonucleotides for discrimination of carrier and non-carrier alleles. Performance of the cystic fibrosis application demonstrates sufficient accuracy and reliability for clinical diagnostic use, and the procedure can be performed by trained medical technologists available in the hospital laboratory.

## 12.1 Introduction

Electrochemical test methods have been widely used in clinical chemistry, most notably in the area of blood glucose monitoring for diabetes care. Electrochemical detection has significant advantages over optical detection methods, in particular due to the low cost and simplicity of instrumentation. Despite these advantages, electrochemical detection has not, until recently, been used for molecular diagnostic applications in the clinical laboratory. Although numerous publications have described molecular diagnostic technologies based on electrochemical detection (for a review, see [1]), none have achieved practical implementation. We describe in this chapter the first such practical system, and the theoretical advantages that may lead to a broader implementation of the technology.

M.R. Reed
Osmetech Molecular Diagnostics
e-mail: michael.reed@osmetech.com

W.A. Coty
Osmetech Molecular Diagnostics
e-mail: bill.coty@osmetech.com

K. Dill et al. (eds.), *Microarrays: Preparation, Microfluidics, Detection Methods, and Biological Applications*,

## 12.2 eSensor® DNA Detection Technology

Initial efforts to develop eSensor DNA detection technology were based on discoveries made by Meade and Kayyem [2] on electrical conductivity of double-stranded DNA. Early embodiments of the technology [3, 4] show the evolution of this concept to include detection on gold electrodes, incorporation of self-assembled monolayers with embedded DNA probes, and use of single and then multiple redox labels. The current eSensor system has been refined and optimized to provide highly reliable multiplex nucleic acid target detection and genotyping, and to incorporate a simple procedure within the capability of the hospital diagnostic laboratory. We have developed the Cystic Fibrosis Carrier Detection (CFCD) Assay as the first application of this technology, and have demonstrated its practicality through successful clinical trials, FDA clearance, and commercial implementation in clinical laboratories.

### *12.2.1 Microarray Substrate*

The support for the eSensor microarray is a conventional printed circuit board, which defines the physical locations of the individual capture probes and provides the required electrodes and interconnects to measure the signals generated by the individual assays. Each printed circuit board is approximately one inch square and contains an array of 36 gold working electrodes, a gold auxiliary electrode, and a silver/silver chloride reference electrode (Fig. 12.1). Traces connect each electrode to gold "fingers" along one edge of the board; these fingers allow the interface of the electrode array to standard edge connectors on the eSensor 4800 instrument. Fingers are present on both the front and back sides of the board to provide individual connections for all 38 electrodes.

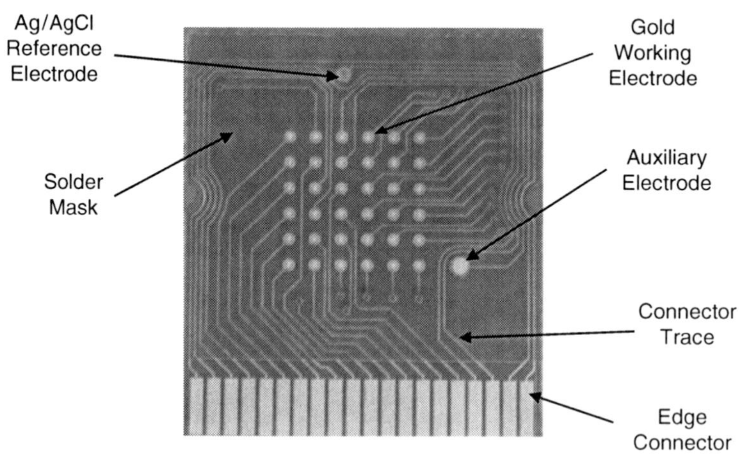


**Fig. 12.1** Printed circuit board microarray

The entire surface of each microarray is coated with an insulating solder mask, leaving only the center of the working electrodes, the auxiliary and reference electrodes, and the connector fingers exposed. Typically, a 0.010 inch diameter area of each working electrode is exposed, although this area can be varied as needed. The solder mask is applied and imaged using a photolithographic method to provide accurate registry of the mask features with the underlying electrode layout. The layout of the array can also be changed easily; new layouts can be designed and fabricated in less than one week.

### *12.2.2 Microarray Manufacture*

Each working electrode is chemically modified with an oligonucleotide capture probe to provide sequence-specific capture of single-stranded nucleic acid targets. The capture probes are applied by spotting with a small volume (50 to 100 nL) of a buffer solution containing an insulator molecule and the capture probe to form an insulating monolayer similar to that previously described [3]. Formation of the monolayer occurs by the self-assembled monolayer principle [5, 6]. Insulator and capture probe molecules form covalent gold–sulfur bonds with the electrode surface, and the molecules are oriented and aligned to form an inner hydrophobic layer and an outer hydrophilic layer similar to the structure of a cell or organelle membrane. This monolayer has a number of advantages, including minimizing non-specific binding of assay constitutents and preventing interference from sample constituents.

Capture probe oligonucleotides are synthesized by standard, phosphoramidite-based solid-phase methods [7]. Typically, capture probes are synthesized to contain the linker arm at the 3' end of the sequence by using a controlled-pore glass synthetic support containing the linker arm monomer. However, probes containing a 5'-linker arm can be synthesized using a protected phosphoramidite form of the linker. Capture probes range in size from 20 to 30 bases, and are purified by HPLC and analyzed by capillary electrophoresis (CE) to verify purity before use in preparing the array-spotting solutions.

Microarrays are spotted using a multiprobe dispenser (BioDot, Irvine, CA) in panels of 50 arrays. After a brief incubation in a humidity chamber to allow monolayer formation, the panels are rinsed in water to remove excess reagents, dried, and assembled into individual test cartridges by attachment of a plastic cover using a pressure-sensitive adhesive. The combination of printed circuit board and cover forms a hybridization chamber and inlet port that can hold approximately 100 μL.

### *12.2.3 Electrochemical Detection*

Electrochemically active labels are synthesized by attachment of ferrocene derivatives to the 2′-hydroxyl group of adenosine via a short, aliphatic linker arm [4]. The ferrocene labels are converted to protected phosphoramidites suitable for use in solid-phase DNA synthesis, and are typically incorporated into the 5′-end of signal probes, although signal probes with the reverse orientation can be made. After synthesis and deprotection, signal probes are purified by HPLC and analyzed by electrospray-ionization mass spectrometry and CE to verify purity before use in preparing genotyping reagents.

A key feature of the ferrocene labels is the ability to modify their redox potentials by derivatizing the ferrocene structure [8]. Attachment of one or more electron-donating or -withdrawing substituents to the cyclopentadiene rings decreases or increases, respectively, the redox potential of the label. In this manner, multiple labels with distinguishable redox potentials can be generated and used to detect the simultaneous binding of probes with differing sequences to a single array location.

Nucleic acid detection is based on a "sandwich" assay principle (Fig. 12.2). Signal and capture probes are designed with sequences complementary to immediately adjacent regions on the corresponding target. A three-member complex is formed among capture probe, target, and signal probe based on sequence-specific hybridization, which brings the 5′-end of the signal probe containing the ferrocene into close proximity with the electrode surface. The ferrocene labels can only be detected as described below when the signal probe is captured in this sequence-specific manner and location. In the absence of target, no specific signal can be detected. As a result, there is no need for a separation step to remove the unbound signal probe after hybridization and prior to detection, even when a relatively large amount of signal probes representing multiple target sequences is present.

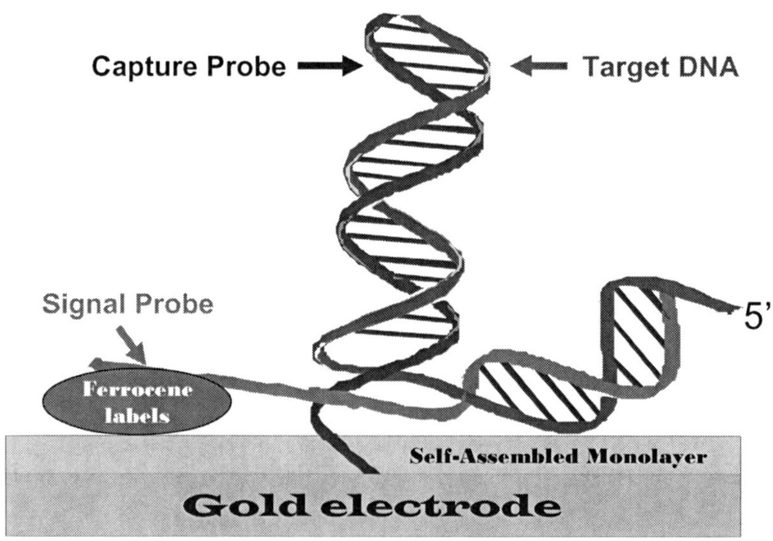


**Fig. 12.2** Electrochemical detection of DNA. Target (blue) is bound to the electrode via sequence-specific hybridization to the capture probe. Signal probe (red) hybridizes to the target sequence adjacent to the base of the capture probe, and the associated ferrocene labels are detected at the electrode surface by ACV

The ferrocene labels are detected at each electrode by alternating current voltammetry (ACV; [9]). Basically, a linear voltage ramp from −50 to +550 mV (relative to the Ag/AgCl reference electrode) over a fixed time period (typically 1 s) is applied to a working electrode together with an AC voltage variation (typically about 100 mV center-to-peak amplitude at 280 Hz). As the DC component of the voltage reaches the redox potential of a ferrocene label, the label undergoes cyclic oxidation and reduction due to the AC voltage variation, until the minimum voltage exceeds the redox potential. This process generates a combination of capacitive and faradaic currents at the working electrode, which are then detected and analyzed by electronic circuitry in the eSensor 4800 instrument.

In analyzing the resulting signals, the capacitive component of the current is independent of the bound label, whereas the faradaic component exhibits specific peaks at the redox potentials of the ferrocenes. These two components can be further differentiated by analysis of the higher-order signal harmonics generated by the AC voltage waveform. The capacitive current remains at a constant phase shift versus the excitation voltage, and so its harmonic component is minimal. Analysis of current versus voltage at a multiple of the AC frequency (e.g., the fourth harmonic is measured at 1120 Hz) is used to extract the faradaic current signals from the ferrocene labels at their characteristic redox potentials [4].

This combination of electrochemical detection with ACV, harmonic signal analysis, and use of self-assembled monolayers is highly resistant to interference from sample constituents. Constituents of blood that would normally interfere with fluorescence detection, such as hemoglobin or bilirubin, have no effect on the eSensor technology. The surface-selective nature of this electrochemical detection system allows addition of electrochemically active sample constituents such as ascorbate or acetaminophen to hybridization buffer at levels based on their concentrations in blood with no contribution to background signal. Even unprocessed sample matrices such as whole blood, urine, or soil extract can be added to the hybridization solution without generating background signal or interfering with specific signal detection.

## 12.3 Genotyping Assay Principle

Genotyping of mutations or polymorphisms is performed using the sandwich assay principle described above with allele-specific signal probes. Each genotyping test employs a first signal probe matching the wild-type sequence and containing a ferrocene label of one electrochemical potential, and a second signal probe matching the mutant sequence and containing a second, distinguishable ferrocene label (Fig. 12.3). Both the wild-type and mutant targets bind to the capture probe. Sequence-specific binding of either the wild-type or mutant signal probe to the respective target adjacent to the capture probe brings the sequence-defining ferrocene label sufficiently close to the electrode surface to allow detection.

The genotype is determined by the ratio of signals at the two potentials, expressed as a genotyping score [$\log_2$(first ferrocene signal/second ferrocene signal); Fig. 12.4];

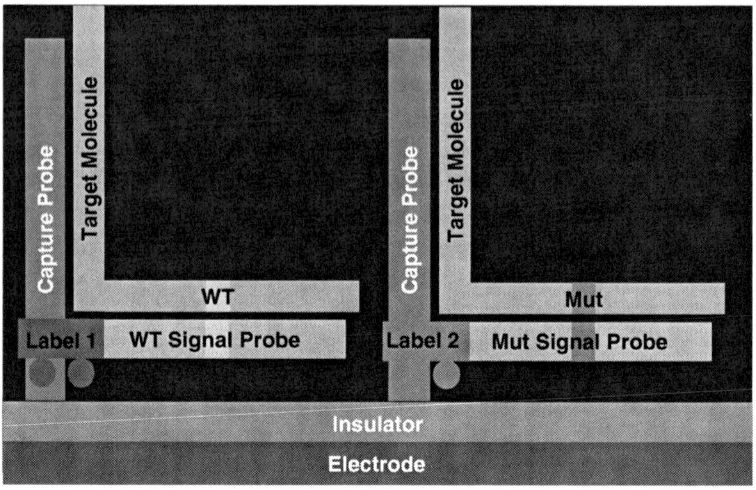


**Fig. 12.3** Genotyping assay principle

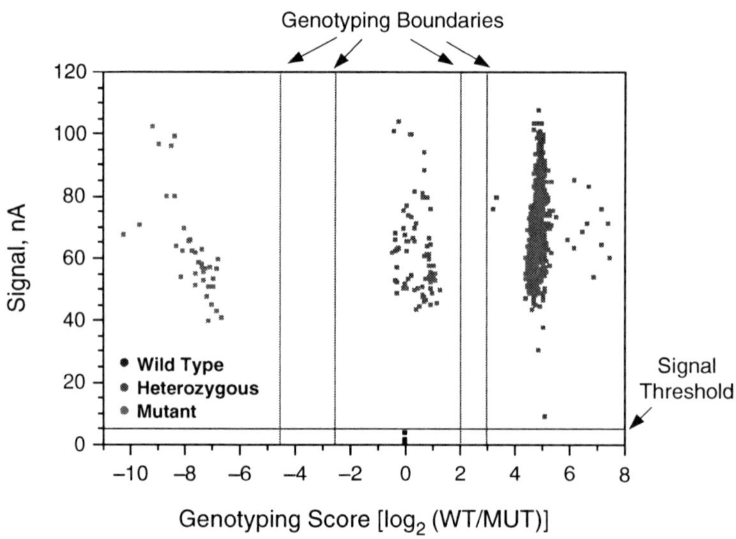


**Fig. 12.4** Scatterplot analysis of genotyping assay data

extreme scores denote a homozygous wild-type or mutant, whereas a score close to zero denotes a heterozygote. Genotyping boundaries are established based on statistical analysis of data from a large number of samples, cartridge lots, and reagent lots, and then verified and validated through internal performance testing and clinical studies.

This approach can be used to discriminate single- or multiple-base changes, insertions, and deletions. A mutation site with multiple alleles can be genotyped using additional ferrocene labels; four distinguishable ferrocene derivatives have been developed for this purpose. Also, two immediately adjacent mutations (e.g., ΔI507 and ΔF508 for the CFTR gene) can be genotyped by treating them as two alleles of the same mutation, using three signal probes for the wild-type and the two mutant sequences. Design of signal and capture probes is relatively straightforward, and is based on matching of probe melting temperatures across an array to allow hybridization and mismatch discrimination for all mutations at a single temperature. Typically, at least 30-fold ratios of binding between perfect-match and single-base mismatch probes can be achieved with this method.

## 12.4 Instrumentation and Software

The eSensor 4800 instrument performs the functions of temperature control, ACV scanning, data acquisition, and analysis of the resulting current output for up to 48 cartridges. The system is modular in design, with eight modules containing six cartridge slots. Each slot consists of a Peltier heating and cooling device, a copper block to maintain constant temperature throughout the cartridge, a 40-pin edge connector, temperature sensors, and a safety fuse. Each module contains two circuit boards which perform ACV scanning and data collection and thermal control, respectively. The instrument also contains power supplies and an interface board that routes information between external computer cable connections and the individual instrument modules.

The instrument implements the standard three-electrode electrochemical cell procedure in a multiplex mode. Digital circuit elements generate the linear voltage ramp with associated AC waveform for each electrode of a single cartridge in sequence, and then these signals are converted to analog form and transmitted to the cartridge via the edge connector. The resulting analog current signals are processed to extract and quantify the voltage-dependent faradaic signals; these data are converted to digital form for further analysis by the system software. The instrument can scan up to eight cartridges simultaneously (one from each module), after which scanning proceeds to the next cartridge in each module.

The instrument is controlled by software running on an attached personal computer (PC). The software contains the assay-specific ACV scanning protocol and uploads the protocol to the instrument. Cartridge scan data are downloaded from the instrument to the PC and analyzed to generate the final genotyping calls. The software also provides a user interface to allow sample identification and tracking, storage, and analysis of data from one or more samples, and reporting of data to the user.

## 12.5 Cystic Fibrosis Carrier Detection System

The Cystic Fibrosis Carrier Detection (CFCD) System was developed as the first clinical application to use the eSensor DNA detection technology. The system consists of the eSensor 4800 instrument, a PC with user interface and operating system software, a bar code scanner, and the reagents and cartridges for identification of carriers of cystic fibrosis. The system is designed for use in a CLIA-certified clinical diagnostic laboratory by trained medical laboratory technologists.

### *12.5.1 Cystic Fibrosis*

Cystic fibrosis (CF) is the most common lethal, autosomal recessive inherited disorder among Caucasians; in the United States, one in 3900 children are born with the disease. The median life expectancy for individuals with CF has increased with improved treatment methods to about 32 years; however, patients with CF suffer a range of debilitating symptoms. The primary consequence of CF is the secretion of an unusually thick and sticky mucus, which clogs the lungs and digestive system. This leads to recurrent lung infections, and obstructs secretion of digestive enzymes by the pancreas, leading to malnutrition and intestinal blockage [10].

CF is caused by mutations in the cystic fibrosis transmembrane regulator (CFTR) gene, which codes for an ATP-gated and phosphorylation-regulated chloride channel protein localized in the plasma membrane of cells in epithelial tissues. Since its discovery and cloning in 1989, over 1000 mutations and polymorphisms in the CFTR gene have been identified. However, 23 mutations account for 88% of detectable carriers among non-Hispanic Caucasians in the United States, and testing for additional mutations results in only a marginal gain in carrier detection in this population [11]. Therefore, the American College of Obstetricians and Gynecologists (ACOG) and the American College of Medical Genetics (ACMG; [12]) have recommended that tests for screening for carriers of cystic fibrosis be recommended to couples contemplating pregnancy, using a panel consisting of 23 mutations and one polymorphism in the CFTR gene. This panel is slightly reduced from the original recommendation of 25 mutations and one polymorphism [13], and recent policy statements have emphasized the limited value of testing for additional mutations [11].

### *12.5.2 Assay Design*

The CFCD assay kit consists of a PCR primer cocktail, thermostable DNA polymerase, exonuclease, genotyping reagents containing signal probe cocktails, cartridges, and ancillary buffer ingredients. The PCR primer cocktail contains 16 pairs of primers to amplify 16 fragments of the CFTR gene containing the 23 mutation sites and one polymorphic site recommended for genotyping by the ACMG. Primers in the set are designed with similar lengths, base compositions, and TMs for hybridization to their binding site. Each primer pair is tested in uniplex to assure specific amplification of the desired target with no side products. One primer of each pair contains a 5′-phosphate group that targets the resulting amplicon strand for later exonuclease digestion. The primer cocktail, PCR buffer, and PCR procedure were optimized by a multifactorial "design of experiments" method [14] to provide reproducible yields of all 16 amplicons.

Two cartridges are provided for each sample, each with electrodes containing capture probes for roughly half of the genotyping assays plus positive and negative controls. Two electrodes are used for each genotyping assay or control. Two genotyping buffers, each specific for one chip, are also provided; these contain a pair of signal probes for each capture probe on the corresponding cartridge, as well as a positive control synthetic target.

Signal and capture probes are designed for specificity of hybridization to the sequences of their respective targets, to eliminate direct cross-hybridization between any signal and capture probe or any pair of signal probes, and to avoid stem-loop formation within a signal or capture probe. In the case of sequence homology between or within probes, a single base change is introduced within the sequence to prevent hybridization, and the length of the altered probe is increased to compensate for the reduced TM of hybridization with the target caused by the mismatch. As a result, no detectable signal is observed on capture probes when the signal probe cocktail is hybridized in the absence of target.

### *12.5.3 Assay Procedure*

The procedure for the CFCD System begins with genomic DNA, which can be isolated from whole blood by any laboratory-validated method. The sample is amplified in a single tube by a 37-cycle multiplex PCR process, and then is treated with bacteriophage $\lambda$ 5′-exonuclease to digest the 5′-phosphorylated, nontarget strand [15] and generate the single-stranded target required for genotyping. The PCR process takes approximately 2 h and the exonuclease digestion 20 min. There are no procedures to purify the target after amplification or exonuclease digestion.

The digested target is then added directly to the genotyping buffer components and pipetted into two eSensor cartridges, which are then inserted into the instrument, incubated at 40 °C for 2 h and scanned to detect and measure the bound signal probe(s). The entire test can be performed, starting with input genomic DNA, in approximately 7 h.

### *12.5.4 Data Analysis*

Data are analyzed in a hierarchy, beginning with results from the positive and negative control electrodes. If both positive control electrodes from a cartridge fail to give a signal above a predetermined threshold and with the proper electrochemical potential, the results from that cartridge are considered invalid. Similarly, if both negative control electrodes give a signal above the threshold, the results are also invalid. If valid results are obtained for the positive and negative controls for both cartridges, then the genotyping data for that sample are analyzed.

Each mutation is genotyped using two electrodes. If one or both of these electrodes give signals above the threshold and genotyping scores within established boundaries for a carrier or noncarrier call, and the two results do not give contradictory calls, then a valid genotype is determined. If all mutations give valid calls, then the results for each mutation (either noncarrier or carrier) are provided in the final report, along with an overall call of noncarrier or carrier for that sample.

In addition to the 23 mutation results reported for each sample, the presence of the 5T allele of the intron 8 5T/7T/9T polymorphism is determined, but the results are masked by system software. In the case of a carrier call for the mutation R117H, the results of the 5T/7T/9T polymorphism test are unmasked and reported as recommended by ACMG, including further instructions for follow-up testing as needed [16].

### *12.5.5 Assay Performance*

Performance of the CFCD system was established by internal testing and clinical trials at three external sites, using multiple lots of CFCD system kits in each study. Performance parameters included reproducibility, accuracy (as judged by agreement with DNA sequencing as a reference method), and evaluation of the effects of interfering substances.

Minimum input sample was established by demonstration of reproducible genotyping calls (99% per sample call rate) after amplification of 10 ng DNA per reaction. No effect of high input DNA sample amounts was observed; addition of up to 2 μg DNA per PCR gave valid genotyping calls in agreement with DNA sequencing.

Reproducibility was determined by repeat testing of a panel of 21 cell line DNA samples which collectively gave carrier calls for all 23 mutations and one polymorphism in the panel. Testing was performed internally and at two external sites over 15 days with three lots of CFCD system kits containing independent components and raw materials; a total of 24,840 mutations were analyzed. The agreement between the eSensor CFCD system and DNA sequencing was 100% (24,810/24,812), after excluding one replicate which gave a no-call result.

Accuracy was determined by testing a total of 486 samples at the three external sites, including 127 banked gDNA samples known to contain CF panel mutations, and 329 samples collected prospectively as whole blood under institutional review board approval. Genomic DNA was extracted from the prospective samples at the three clinical sites and internally using multiple methods, and tested by the CFCD system and bidirectonal DNA sequencing without further analysis or selection. Sequencing results were obtained for all but 1 mutation in 1 sample, so results were compared for 11,663 mutation calls. After exclusion of 5 samples which gave repeated no-call results by the eSensor method, agreement with DNA sequencing was 99.97% (11,540/11,543 calls). Of the three calls in disagreement with DNA sequencing, one corresponded to a 2184delA call for a sample carrying the mutation 2183AA>G nonpanel mutation.

During internal verification and clinical trials, samples with nonpanel mutations in the CF gene were identified by DNA sequence analysis and their effect on genotyping in the CFCD System was determined (Table 12.1). Single-base polymorphisms (I506V, I507V and F508C) previously found to give false-positive results in other test systems [13] had no effect on genotyping score and gave noncarrier calls. Three mutations that occurred at the exact site as a corresponding panel mutation had varying effects on the genotyping score. R560K gave a minor shift in genotyping score, still allowing a noncarrier call, R117L gave a greater shift leading to consistent indeterminate genotyping scores, and 2183AA>G carriers gave a genotyping score identical to that of 2184delA carrier samples. It should be noted that 2183AA>G is considered a cystic fibrosis-causing mutation [17], and its

**Table 12.1** Effect of Nonpanel Mutations on CFCD System Genotype

| Number of Samples | Nonpanel Mutation | Nearby Panel Mutation | CFCD System Genotype | Phenotype | Reference |
|---|---|---|---|---|---|
| 1 | I506V | ΔI507, ΔF508 | Noncarrier | Normal | [13] |
| 1 | I507V | ΔI507, ΔF508 | Noncarrier | Normal | [13] |
| 2 | F508C | ΔI507, ΔF508 | Noncarrier | Normal | [13] |
| 3 | 2183AA>G | 2184delA | Carrier | Disease-causing | [17] |
| 1 | R560K | R560T | Noncarrier* | Variable | [18] |
| 1 | R117L | R117H | Indeterminate Score | Found in an affected individual† | [19] |
| 1 | Q552X | G551D,R553X | Noncarrier | Some CF symptoms observed | [20] |
| 2 | 711+5G>A | 711+1G>T | Noncarrier | Found in an affected individual† | [21] |
| 1 | S589N | 1898+1G>A | Noncarrier | Disease-causing | http://www.genet.sickkids.on.ca/cftr/ |
| 1 | 394delTT | G85E | Noncarrier | Found in an affected individual† | [22], [23] |

*Replicate testing yielded indeterminate score and noncarrier results.
†Presumed to be disease-causing.
Shading. Mutations shaded in grey affected genotyping calls.

frequency in the Caucasian population is just below the threshold for inclusion in the ACMG panel [12]. In contrast to these results, nonpanel mutations occurring adjacent to, but not exactly at the locus of a panel mutation had no effect on the genotyping score or call.

Studies also were performed to determine the effects of interfering substances found in blood on the results of the CFCD system. A variety of potentially interfering substances were identified, primarily on the basis of the National Committee for Clinical Laboratory Standards (NCCLS) guideline on Interference Testing in Clinical Chemistry [25], but also including substances known to affect electrochemical measurements or to inhibit PCR [26, 27]. These substances were added to whole blood at concentrations considered pathological or toxic, and then testing performed after isolation of genomic DNA. The following substances had no effect on yield of extracted DNA, yield of CFTR amplicons after multiplex PCR, or genotyping of mutations in the CF carrier screening panel: triglycerides (3000 mg/dL), high-density lipoprotein (70 mg/dL), cholesterol (250 mg/dL), bile salts (a mixture of cholate and deoxycholate; 6.4 μg/mL), human albumin (3 g/dL), human immunoglobulin G (3 g/dL), bilirubin (15 mg/dL), acetaminophen (30 μg/mL), ascorbic acid (30 μg/mL), diphenylhydantoin (phenytoin; 20 μg/mL), gentamicin (12 μg/mL), *N*-acetylsalicylic acid (200 μg/mL), nicotine (100 μg/mL), theophylline (20 μg/mL), valproic acid (100 μg/mL), vancomycin (100 μg/mL), NaCl (150 mM), KCl (5 mM), $CaCl_2$ (1.08 mM), or $FeCl_3$ (9.25 μM).

## 12.6 Future Developments

The Cystic Fibrosis Carrier Screening Assay represents the first genotyping application on the eSensor DNA Detection System. Further assays are in development, including tests for genotyping of cytochrome P450 polymorphisms important in drug metabolism. In addition, a new pharmacogenetic test for Warfarin dosing sensitivity has recently been launched as an IVD on the new esensor XT-8 system. This new platform has increased array desnsity, dramatically shortened assay time and enhanced customer convenience [29]. Additional planned applications and improvements for which feasibility has been demonstrated include detection of pathogen nucleic acid targets and quantitative measurement of target nucleic acids. Furthermore, the eSensor technology is ideally suited for inclusion in an integrated sample-to-answer system such as that described by Liu et al. [28]. Efforts are now in progress to bring these new applications and improvements to practical implementation in the clinical diagnostic laboratory.

**Acknowledgments** A large number of people have contributed to the development of the eSensor® DNA Detection Technology, beginning with the founders of Clinical Micro Sensors, Thomas Meade and Jon Faiz Kayyem, and including past employees of CMS and Motorola Life Sciences as well as current employees of Osmetech Molecular Diagnostics too numerous to mention by name, but whose creativity, enthusiasm, and hard work are gratefully acknowledged.

## References


1. Kerman, K., Kobayashi, M., and Tamiya, E. (2004). Recent trends in electrochemical DNA biosensor technology. *Meas. Sci. Technol.* 15: R1–R11.
2. Meade, T.J. and Kayyem, J.F. (1997) Nucleic acid mediated electron transfer. United States Patent 5,591,578.
3. Umek, R.M., Lin, S.W., Vielmetter, J., Terbrueggen, R.H., Irvine, B., Yu, C.J., Kayyem, J.F., Yowanto, H., Blackburn, G.F., Farkas, D.H., and Chen, Y.P. (2001) Electronic detection of nucleic acids: A versatile platform for molecular diagnostics. *J. Mol. Diag.* 3: 74–84.
4. Yu, C.J., Wan, Y., Yowanto, H., Li, J., Tao, C., James, M.J., Tan, C.L., Blackburn, G.F., and Meade, T.J. (2001) Electronic detection of single-base mismatches in DNA with ferrocene-modified probes. *J. Am. Chem. Soc.* 123: 11155–11161.
5. Finklea, H.O. (1998) Electrochemistry of organized monolayers of thiols and related molecules on electrodes. *Electroanal. Chem.: Series Adv.*, 19: 109–335.
6. Estroff, L.A., Kriebel, J.K., Nuzzo, R.G., and Whitesides, G.M. (2005) Self-assembled monolayers of thiolates on metals as a form of nanotechnology. *Chem. Rev.* 105: 1103–1169.
7. Eckstein, F. (1991). *Oligonucleotides and Analogues: A Practical Approach.* IRL Press, New York, 313 pp.
8. Gorton, J.E., Lentzner, H.L., and Watts, W.E. (1971) Bridged ferrocenes–VIII: Polarographic half-wave potentials of ferrocenophanes and related compounds. *Tetrahedron*, 27: 4353–4360.
9. Nahir, T.M. and Bowden, E.F.J. (1996) The distribution of standard rate constants for electron transfer between thiol-modified gold electrodes and adsorbed cytochrome *c*. *Electroanal. Chem.* 410: 9–13.
10. Moskowitz, S.M., Gibson, R.L., Sternen, D.L., Cheng, E., Cutting, G.R. (2005) CFTR-Related Disorders. Gene Reviews, http://www.geneclinics.org.
11. Grody, W.W., Cutting, G.R., and Watson, M.S. (2007) The cystic fibrosis mutation "arms race": When less is more. *Genet. Med.* 9: 739–744.
12. Watson, M.S., Cutting, G.R., Desnick, R.J., Driscoll, D.A., Klinger, K., Mennuti, M., Palomaki, G.E., Bradley W. Popovich, B.W., Pratt, V.M., Rohlfs, E.M., Strom, C.M., Richards, C.S., Witt, D.R., and Grody, W.W. (2004) Cystic fibrosis population carrier screening: 2004 revision of American College of Medical Genetics mutation panel. *Genet. Med.* 6: 387–391.
13. Grody, W.W., Cutting, G.R., Klinger, K.W., Richards, C.S., Watson, M.S., and Desnick, R.J. (2001) Laboratory standards and guidelines for population-based cystic fibrosis carrier screening. *Genet. Med.* 3: 149–154.
14. Box, G.E.P., Hunter, W.G., and Hunter, J.S. (1978) *Statistics for Experimenters: An Introduction to Design, Data Analysis and Model Building.* John Wiley and Sons, New York, 644 pp.
15. Higuchi, R.G. and Ochman, H. (1989) Production of single-stranded templates by exonuclease digestion following the ploymerase chain reaction. *Nucl. Acids Res.* 17: 5865.
16. Amos, J., Feldman, G.L., Grody, W.W., Monaghan, K., Palomaki, G.E., Prior, T.E., Richards, C.S., and Watson, M.S. (2005) Technical Standards and Guidelines for CFTR Mutation Testing, http://www.acmg.net/Pages/ACMG_Activities/stds-2002/cf.htm.
17. Kilinc, M.O., Ninis, V.N., Tolun, A., Estivill, X., Casals, T., Savov, A., Dagli, E., Karakoc, F., Demirkol, M., Hüner, G., Özkinay, F., Demir, E., Seculi, J.L., Pena, J., Bousono, C., Ferrer-Calvette, J., Calvo, C., Glover, G., and Kremenski, I. (2000) Genotype-phenotype correlation in three homozygotes and nine compound heterozygotes for the cystic fibrosis mutation 2183AA→G shows a severe phenotype. *J. Med. Genet.*, 37: 307–309.
18. Ferec, C., Audrezet, M.P., Mercier, B., Guillermit, H., Moullier, P., Quere, I., and Verlingue, C. (1992) Detection of over 98% cystic fibrosis mutations in a Celtic population. *Nat. Genet.* 1: 188–191.

19. Ferec, C., Novelli, G., Verlingue, C., Quere, I., Dallapiccola, B., Audrezet, M.P., and Mercier, B. (1995) Identification of six novel CFTR mutations in a sample of Italian cystic fibrosis patients. *Mol. Cell Probes* 9: 135–137.
20. Delmarco, A., Pradal, U., Cabrini, G., Bonizzato, A., and Mastella, G. (1997) Nasal potential difference in cystic fibrosis patients presenting borderline sweat test. *Eur. Respir. J.* 10: 1145–1149.
21. Bisceglia, L., Grifa, A., Zelante, L., and Gasparini, P. (1994) Development of RNA-SSCP protocols for the identification and screening of CFTR mutations: Identification of two new mutations. *Hum. Mutat.* 4: 136–140.
22. Schwartz, M., Anvret, M., Claustres, M., Eiken, H.G., Eiklid, K., Schaedel, C., Stolpe, L., and Tranebjaerg, L. (1994) 394delTT: A Nordic cystic fibrosis mutation. *Hum. Genet.* 93:157–161.
23. Claustres, M., Laussel, M., Desgeorges, M., Giansily, M., Culard, J.F., Razakatsara, G., and Demaille, J. (1993) Analysis of the 27 exons and flanking regions of the cystic fibrosis gene: 40 different mutations account for 91.2% of the mutant alleles in southern France. *Hum. Mol. Genet.* 2: 1209–1213.
24. NCCLS (2002) *Interference Testing in Clinical Chemistry; Approved Guideline*. NCCLS Document EP7-A. NCCLS, Wayne, PA, 104 pp.
25. Waleed, A.A, Jönsson, L.J., and Rådström, P. (2000) Identification and characterization of immunoglobulin G in blood as a major inhibitor of diagnostic PCR. *J. Clin. Micro.* 38: 345–350.
26. Waleed, A.A. and Rådström, P. (2001) Purification and characterization of PCR-inhibitory components in blood cells. *J. Clin. Micro.* 39: 485–493.
27. Liu, R.H, Yang, J., Lenigk, R. Bonanno, J., and Grodzinski, P. (2004) Self-contained, fully integrated biochip for sample preparation, polymerase chain reaction amplification, and DNA microarray detection. *Anal. Chem.* 76: 1824–1831.
28. Liu, R.H., Coty, W. A., Reed, M.R. and Gust, G (2008) Electrochemical detection-based DNA micro arrays. IVD Technology 7: 36–42.

# Part V
# Future Improvements in Microarray Sensing

# Chapter 13
# Use of Redox Enzymes for the Electrochemical Detection of Sequence-Specific DNA and Immunochemical Entities

**Kilian Dill and Andrey Ghindilis**

## 13.1 Introduction

There is a host of methods for the detection of immunoassays and sequence-specific DNA on a microarray chip (or biochip); see earlier chapter. Many detection technologies are optical methods, such as surface plasmon resonance, luminescence, fluorescence, and visible detection modes (absorbance or reflectance). They do require optical systems that are somewhat expensive.

As shown in previous chapters, electrochemical detection is a viable option for immunochemical and sequence-specific DNA detection. The electrochemical detection methods vary greatly, with anything from impedance measurements, to oxidation of specific nucleotdides within the duplex, to conductive interacalators, redox-intercalators, metal tags, and redox enzyme systems. Many have specific niches and amplification modes. The bottom line is that electrochemical detection is sensitive, the system footprint is small, and the system is inexpensive.

To date only a few electrochemical detection systems have been commercialized and succeeded. These include (1) the CombiMatrix ElectraSense® system, utilizing HRP as a redox enzyme (1–6); (2) the Osmeotech Esense® (Motorola Life Sciences) system, utilizing a Ferrocene tag (see Chapter 12 in this book); and (3) the Toshiba gene analyzer, utilizing a redox active dye (7). The CombiMatrix system is the only commercial system that uses a true microarray concept, up to 12,000 individual electrodes. The latter E-chem detection systems utilize a microarray that requires spotting onto gold electrodes and the number of electrodes is in the neighborhood of 15–25 per chip. In these cases, the voltage is scanned and current recorded. A substantial increase in the peak-to-peak current reflects the presence of the redox species present, and hence a duplex being formed.

In the case where redox enzymes (or products that are redox active) are utilized for the detection of specific DNA sequences and immunoassays, there is a variety of enzymes from which to choose. Many have been used before on simple electrode

K. Dill and A. Ghindilis
CombiMatrix Corporation, 6500 Harbour Heights Parkway, Suite 301 Mukilteo, WA 98275

K. Dill et al. (eds.), *Microarrays: Preparation, Microfluidics, Detection Methods, and Biological Applications*, 

systems but have not been carried over to the microarray platform. Such enzymes are glucose oxidase, β-galactosidase, glucose dehydrogenase, alkaline phosphatase, horseradish peroxidase, and laccase to name a few. In each case the product is usually detected as it is oxidized or reduced. Thus, the electrode detects the electron flux created by the enzyme near the electrode surface.

There are many considerations for choosing an enzyme system for use on a microarray. One is stability. Another is the enzyme turnover number. One must also consider soluble substrate and product. A substrate with limited solubility does not help and may cause nonlinear product output. A product that precipitates is also unwanted as it does not provide a species that migrates to the electrode and can be oxidized or reduced. Several enzymes meet the criteria provided above and we have used them successfully: laccase, horseradish peroxidase, glucose dehydrogenase, and alkaline phosphatase. Details and results are given below for each enzyme system. Again, some may not be redox enzymes, but rather the enzymic products may be redox active.

CombiMatrix has developed an electrochemical detection system whereby the electrochemical events are amplified at the electrode site, making feasible the detection of these events. The method was utilized for both specific DNA detection as well as immunoassay monitoring of specific analytes. This amplification system relies on the use of redox enzymes that convert substrates to products and then utilize the electrode to reconvert product to substrate. It is this reconversion that is detected and monitored. The enzyme used in past studies was horseradish peroxidase. However, there are many other enzymes that can be, and have been, used such as laccase, glucose oxidase, and glucose dehydrogenase, to name a few.

It would be of interest to the users of electroactive microarrays if a number of enzymes could be used on a single chip to perform multiplexed assays involving genetic analysis, such as gene expression. In this way, readings for various enzymes/substrates or specific DNA sequences can simply be accomplished by changing the potential at the electrodes and the composition and pH of the assay solution.

## 13.2 Specific Redox Enzymes

In our experiments, we used three redox enzymes that could be monitored in terms of their products formed: horseradish peroxidadse (an oxidoreductase), laccase (an oxidoreductase), and glucose dehydrogenase. The laccase and HRP are similar in action, but the secondary substrates ($H_2O_2$ and $O_2$) are different. Their chemistries are described in the following. The HRP (horseradish peroxidase) reaction scheme using ortho-phenyldiamine as the substrate is shown below.

$$\text{OPD} + H_2O_2 \text{ ----- } \rightarrow \text{ox-OPD} + H_2O$$

$$\text{Ox-OPD} + 2H^+ + 2e \text{ --- } \rightarrow \text{OPD}$$

In the first reaction, the product (ox-OPD) was detected amperometrically. For the oligomer detection application, TMB was used as the substrate. This assay was performed at −0.3 V versus Pt wire in 0.05 M Na-citrate-phosphate buffer pH 5.0 containing 0.2 M $Na_2SO_4$. OPD, and hydrogen peroxide in 1 mM concentration was used as an enzymatic substrate solution. A checkerboard pattern from the microarray showed that binding occurred at the proper locations without the presence of noise or cross-talk resulting from the bleeding over of the signal from neighboring electrodes.

The second enzyme assay (laccase) was performed at −0.3 V vs. Pt wire in 0.05 M Na-citrate-phosphate buffer pH 5.0 containing 0.2 M $Na_2SO_4$. Catechol in 1 mM concentration was used as an enzymatic substrate.

$$\text{Catechol} + O_2 \text{-----} \rightarrow \text{Quinone} + H_2O$$

The product (quinone) was detected amperometrically:

$$\text{Quinone} + 2H^+ + 2e \text{ ---} \rightarrow \text{Catechol.}$$

Glucose dehydrogenase (GDH) that employs pyrroloquinoline quinone (PQQ) as a cofactor has a potential to be used as an alternative enzyme label for ECD. GDH catalyses a reaction of glucose oxidation by an electron accepting substrate:

$$\text{Glucose} + \text{electron acceptor} \xrightarrow{\text{GDH}} \text{Gluconic acid} + \text{reduced electron acceptor (electron donor)}$$

A detection of the electron donor, which is formed as a result of enzymatic reaction, can be performed on the electrode:

$$\text{Electron donor} - \text{electron} \text{ ---------} \rightarrow \text{electron acceptor}$$

In contrast to HRP detection, GDH detection is performed at higher electrode potential and anodic current is detected. Because HRP and GDH detection are based on currents of different polarity, GDH and HRP generated signals do not interfere with each other. These two enzymes can be used to design a dual-enzyme detection system, similar to two-color optical detection.

For the use of alkaline phosphatase, the substrate para-aminophenyl phosphate may be used. Reaction of the enzyme with the PAPP substrate creates a product PAP, which can be further oxidized, and the oxidized product is stable. If the electrode system is switched around, the oxidized species may then be reduced. This shuttling of the oxidized/reduced species may be repeated. The current intensity is then related to the amount of product generated by the enzyme. As long as the enzyme produces more product with time, the shuttle current increases with time. Hence the shuttle current can be plotted as a function of time and the quantity of bound enzyme determined. The chemical and enzymatic reactions are given below. The system requires that the voltage be switched between the two electrodes present and the current monitored. These two electrodes (one system) must then be isolated from any other electrode system. In most cases, researchers and certain small companies have focused on "channels," where each channel is isolated from the other one and the sides or walls of these channels then become the electrodes that are switched (cycled) in voltage settings.

### 13.2.1 Alkaline Phosphatase Reaction Scheme (8–10)

| | |
|---|---|
| (1) Enzymic reaction | $AP + PAPP \dashrightarrow PAP_R$ |
| (2) Redox cycling | $PAP_R \xrightarrow{ox} PAP_O$ |
| (3) Redox cycling | $PAP_O \xrightarrow{red} PAP_R$ |

For most systems, the oxidation of released PAP is measured as a current increase. Now for a signal enhancement, the redox cycle may be introduced where the reduced species is oxidized and then reduced using the two electrodes. The signal is increased with time as the enzyme continually produces $PAP_R$.

### 13.2.2 Results

Figs. 13.1 and 13.2 show molecular binding schemes that led to signals being generated and detected at the chip surface. Fig. 13.1 shows a sandwich-based immunoassay where the capture-antibody is tagged with an oligomer that is complementary to the one synthesized near the electrode surface. The analyte is captured by the surface-bound antibody and then capped by a second antibody, labeled with either horseradish peroxidase (HRP) or laccase via two different biotin–streptavidin complexes (directly or indirectly linked); the enzyme is the signal generator of the amplified signal. Fig. 13.2 shows the binding scheme for a complementary DNA strand that is hybridized to its counterpart on the electrode surface. The target oligomer contains a biotin molecule that facilitates the capture of either HRP or glucose dehydrogenase (GDH). All in all, the binding events begin by the synthesis of unique 15-mers on the chip surface at select electrodes.

Figs. 13.3A and B show the data for the electrochemical detection of ricin (RCA) and AGP on a single chip surface using two different enzymes for the detection of these analytes. The data is from a CombiMatrix 1 K chip. The immunosandwich assay format was used as shown in Fig. 13.1. Note that with this enzyme set, the order of the detection process is crucial. The assay utilizing laccase must be undertaken first. The presence of residual $H_2O_2$ interferes with the laccase reaction (see below).

In the presence of catechol and atmospheric oxygen, the immunoassays for (AGP) $\alpha$1-acid glycoprotein, using a laccase tagged antibody, gave a positive signal (Fig. 13.3A). The right-hand side of Fig. 13.3A (immunoassay for ricin using an HRP tag) provided no measurable signal under the laccase assay conditions.

**Fig. 13.1** The use of the dual-enzyme system for the detection of RCA and AGP using an immunoassay format

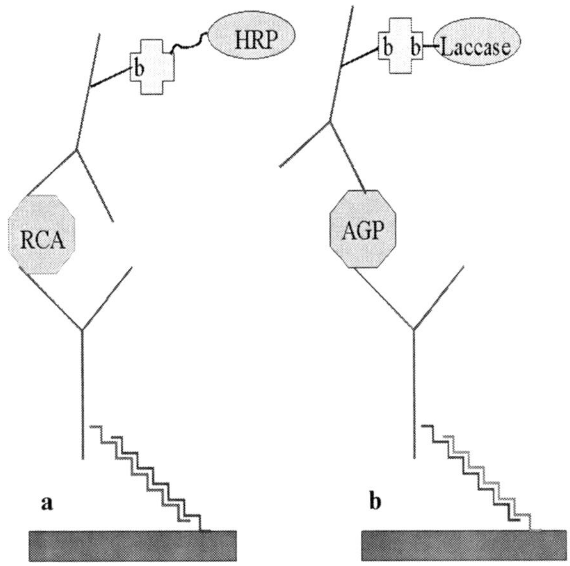


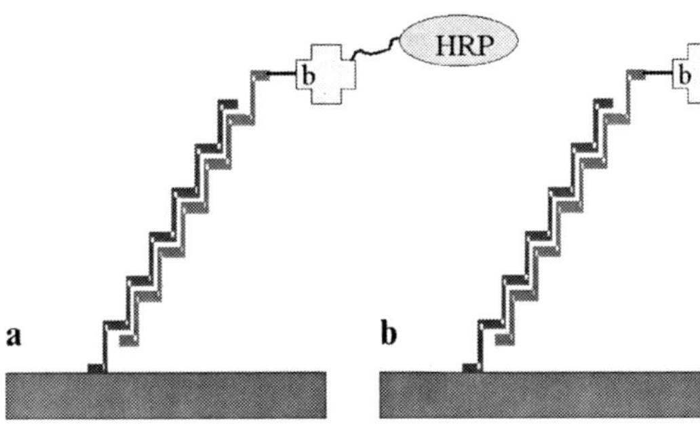


**Fig. 13.2** The dual-enzyme system used in the detection of different DNA sequences

When the chip was washed and placed into a substrate solution and buffer containing OPD and hydrogen peroxide, a positive signal was observed for ricin containing an HRP tag (right-hand side of Fig. 13.3B). However, a low-intensity response can also be observed for the laccase-conjugated antibody associated with the detection of AGP. Earlier detailed publications using HRP show that detection range is almost 5 logs and the limit of detection for certain analytes is in the fmol range (2,3).

Figs. 13.4A and B show the results for a two-enzyme system, as depicted in Fig. 13. 2. In this case biotin containing oligomers P1 and P2 were hybridized to their complementary strands on a 12 K chip in either rows 2B and 16A or 3A and 17B. When substrate, solution, and voltage settings were provided that induced

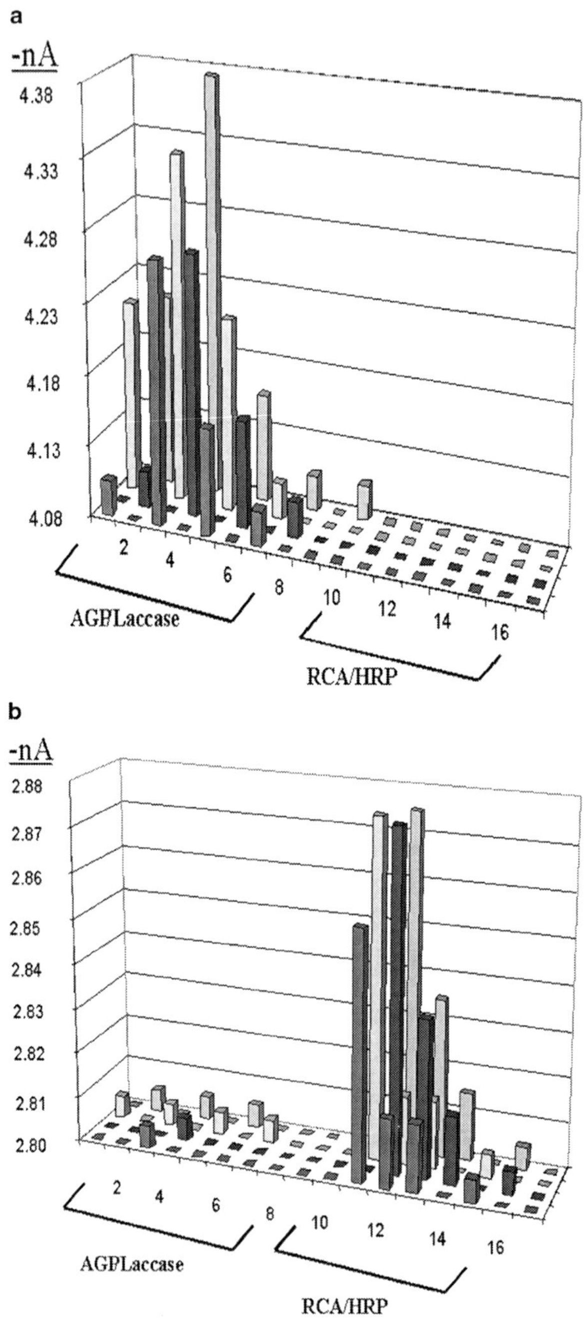


**Fig. 13.3** (A) (B) Results for the electrochemical detection of RCA and AGP using the enzymes HRP and laccase

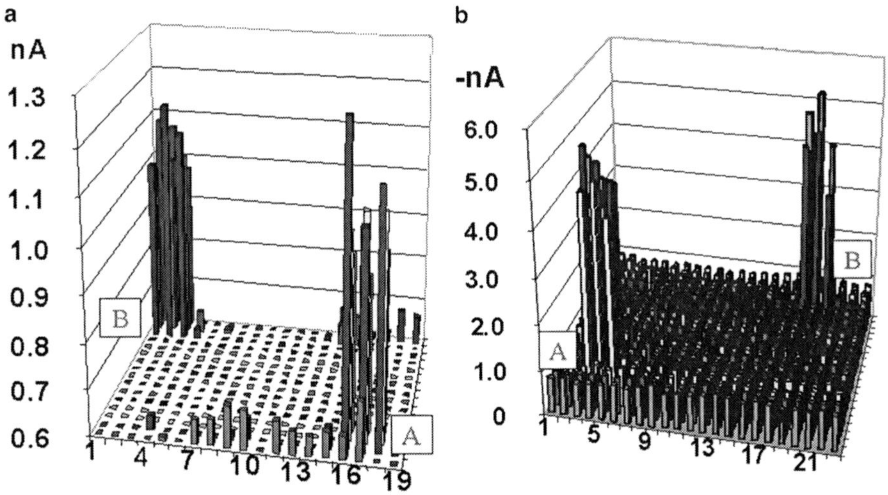


**Fig. 13.4** (A) (B) Results for the electrochemical detection of two different oligomers (P1 and P2) hybridized to complementary strands synthesized at the surface. The enzymes HRP and GDH were used as signal generators. Note that the sign in amps for GDH is positive

GDH to produce product, signals were observed in Fig. 13. 4A that were indicative of GHD being present. Note that there are no signals present in areas 3A or 17B that would indicate the presence of HRP. When the chip was washed and placed into buffer and substrate conditions favorable for HRP activation, the signals for HRP are shown in Fig. 13.4b. Although these data are somewhat noisier, no signal can be observed where one would expect GDH to be present.

There are many examples in the literature of alkaline phosphatase being used. Many of these provide for three electrode systems, which also include a reference electrode. The walls of the electrodes are coated with the immunocomplex or capture DNA. The substrate and buffers are added to the finished electrode product, with a reference electrode. As the enzyme turns over the substrate, the product is cycled to provide the current.

What makes the system so complex is the fact that so much of the electronics is required to change polarity on the electrodes. Each row (electrode) must be monitored with time to get a periodic readout and plot. Hence, the number of electrodes that can be monitored with time is low.

We have shown that for a two-enzyme system, two different DNA strands or two analytes (for immunoassays) can be observed on one chip. In the case of immunoassays, numerous enzyme systems can be used to detect various analytes, as long as each antianalyte antibody is tagged with a unique enzyme. In the case of oligomers, one would have to utilize (incorporate) unique probes into the amplified DNA strand (such as biotin, or digoxin, fluorescein, etc.). In this way recognition antibodies for these tags can be used, which contain the enzyme used in the amplification process.

**Acknowledgment** Partial funding was provided by Phase II SBIR grant (DAAD13-00-C-0033).

## References

1. A. Kumar and K. Dill, *IVD Technology*, 2005, **11**, 35–40.
2. K. Dill, D.D. Montgomery, A.L. Ghindilis, K.R. Schwartzkopf, and S. Ragsdale, *Biosens. Bioelectron.*, 2004, **20**, 736–742.
3. K. Dill, D.D. Montgomery, A.L. Ghindilis, and K.R. Schwarzkopf, *J. Biochem. Biophys. Methods*, 2004, **58**, 181–187.
4. K. Dill, Enzyme-Amplified Redox Microarray Detection Process, European Patent Application WO03019147, published March 6, 2003.
5. K. Dill and A. McShea, *Drug Discovery Today: Technol.*, 2005, **2**, 261–266.
6. K. Dill, K.R. Schwarzkopf, and A. Ghindilis, Multiplexed analyte and oligonucleotide detection on microarrays using several enzymes in conjunction with electrochemical detection, *Lab Chip*, 2006, **6**, 1052–1055.
7. K. Hashimoto and Y. Ishimori, Preliminary evaluation of electrochemical pna array for detection of single base mismatch mutations, *Lab Chip*, 2001, **1**, 61–63.
8. W.R. Vandaveer, D.J. Woodward, and I. Fritsch, *Electrochimica Acta*, 2003, **48**, 3341–3348.
9. Z.P. Aguilar, W.R. Vandaveer, I Fritsch, *Anal. Chem.*, 2002, **74**, 3321–3329.
10. Z.P. Aguilar, I. Fritsch, *Anal. Chem.*, 2003, **75**, 3890–3897.

# Chapter 14
# Biochip Platforms for DNA Diagnostics

**Anil K. Deisingh, Adilah Guiseppi-Wilson and Anthony Guiseppi-Elie**

**Abstract** This chapter looks at the use of microlithographically fabricated biochip platforms for DNA diagnostics and prognostics, although protein and RNA biochips are also briefly considered. Biodetection methods such as ion-selective electrodes (ISEs), microelectromechanical systems (MEMS) devices such as microcantilevers, optical, piezoelectric-based acoustic wave, and mass spectrometry are briefly discussed. Emphasis is given to label-free electrochemical (impedimetric, voltammetric, and amperometric) detection. The production of DNA biochips is highlighted as are the operation and design of the experiments to reveal gene expression and SNP data. Applications discussed include the monitoring of microbes, cancer classification studies, and patient stratification in drug development. Finally, challenges and issues facing the development of diagnostic and prognostic biochips are discussed in detail.

## 14.1 Introduction

Biochips, as the contraction implies, incorporate biologically derived recognition entities with the additive and subtractive techniques of microlithography in the production of useful analytical devices (1). Biochips are similar to but may be distinguished over biosensors by the density of functional recognition sites, generally $10^2 - 10^4$. Biochips are highly versatile devices that may be used for clinical

---

A.K. Deisingh and A. Guiseppi-Elie (✉)
Center for Bioelectronics, Biosensors and Biochips (C3B),
Department of Chemical and Biomolecular Engineering,
Department of Bioengineering, Clemson University, 100,
Technology Drive, Anderson, South Carolina 29625, USA.
e-mail: guiseppi@clemson.edu

A. Guiseppi-Wilson
ABTECH Scientific, Inc., 800 East Leigh Street,
Richmond, Virginia 23219, USA.

K. Dill et al. (eds.), *Microarrays: Preparation, Microfluidics, Detection Methods, and Biological Applications*,

diagnostics and prognostics, patient stratification in drug development, disease management, and forensic applications, amongst others. These biochips automate repetitive laboratory tasks by replacing standard equipment with miniaturized microfluidic assays thereby providing very sensitive detection methods (2).

In recent times, three main types of biochips have become very important for diagnostic and prognostic purposes. These are nucleic acid (DNA, RNA, and PNA-based types) and protein biochips. DNA biochips may be classified according to the two types of nucleic acid probes affixed to the chip's surface and by the density of such probes presented to a sample. Probes may be cDNA, generally derived from bacterial clone libraries, or oligonucleotides, prepared by solid-phase synthesis. Oligonucleotides may be *in situ* synthesized or presynthesized and then affixed to the chip substrate. One format for the DNA biochip uses a piece of glass, typically the size of a microscope slide, containing thousands of cDNA probes affixed to the glass platform within tiny (ca. 250 $\mu m^2$) polyacrylamide gel pads (MAGIChip™, MicroArrays of Gel-Immobilized Compounds on a Chip (3)). When fully integrated with sample preparation, microfluidic sample management, and analyte detection, these systems are commonly termed 'lab-on-a-chip', which may allow for field-portable DNA analyses. DNA biochips are widely used for the detection of mutations in specific genes and to detect the differences in gene expression levels in cells (2).

RNA-based biochips were first reported by Breaker's group at Yale (4). They used RNA-based molecular switches on a gold-coated silicon surface and arranged them in clusters. Each switch was able to bind to one specific target molecule and the researchers were able to identify different strains of *E. coli*. Protein biochips are being increasingly developed and several formats are available including (5):

(i) Antibodies arrayed to detect antigens from body fluids
(ii) Tissue extracts or purified antigens arrayed to detect serum antibodies or known proteins
(iii) 'Bait' proteins immobilized to detect interacting proteins
(iv) Large-scale proteome arraying of entire expression libraries encoded with fusion proteins

Generally, protein biochips are similar to DNA microarrays, being instead composed of arrays of immobilized proteins but with surface preparation and immobilization methods being far more complex and intricate in order to mitigate surface-mediated denaturation and to provide for preferential orientation of the active site or complement binding domain to the analyte in solution. In addition to DNA/RNA and proteins, subcellular organelles, cells, and tissues may also be the basis for biorecognition in biochips.

This chapter, however, concentrates on biochip platforms for DNA diagnostics. Considered topics include biorecognition moieties, biodetection methods, microarrays, opportunities for biochip diagnostics and prognostics, and their uses in patient stratification and drug development. The chapter concludes with a discussion of the key issues related to the production and use of diagnostic and prognostic biochips.

## 14.2 Molecular Biorecognition

Several biomolecules can be used as recognition moieties in the development of DNA biochips. These include cDNA, oligomeric DNA, and PNA as well as RNA. Integrated with solid-state devices, these biomolecular recognition entities need to be immobilized, stabilized, and presented to targets so as to maximize hybridization kinetics and hybridization fidelity as well as maximize the subjacent device sensitivity.

There are five main classes of general immobilization techniques:

(a) Retention by an inert membrane.
(b) Physical adsorption at a solid surface.
(c) Cross-linking with bifunctional agents such as glutaraldehyde, hexamethylene diisocyanate, and others.
(d) Physical entrapment within polymer matrices such as polyacrylamide and cellulosics.
(e) Covalent coupling to a functionalized solid-state support such as polystyrene, silicon, glass, or metal: this method is the most irreversible of the techniques.
(f) Covalent coupling to and entrapment within polymer matrices such as within hydrogels.

The major advantages of irreversible immobilization are:

(a) It may allow for reuse of the device or system.
(b) The immobilized molecule may be more stable than the solution species.

Regardless of the approach taken, the goal is to harness the specific recognition properties of biomolecules. This can be done either discretely or as part of a more complex system so that the recognition reaction is linked to a physicochemical transduction device and made functionally integral to an analytical system.

## 14.3 Biochip Transduction Methods

The wide range of possible detection methods gives the biochip its diversity. These include electronic devices (such as field effect transistors); microelectromechanical (MEMS) devices such as micro and nanocantilevers; simple metallic and semiconductor electrodes for electrochemical (amperometric, voltammetric, and impedimetric) detection; optical devices including fibers and fiber bundles (for absorption, fluorescence, luminescence, and chemiluminescence); quartz-crystal oscillating devices; and mass spectrometry. Amongst these methods are those described as direct and others that are indirect. Direct methods derive their signals from the hybridization reaction directly, for example, quartz crystal oscillators, impedimetric biochips, micro- and nanocantilevers, or interferometers. These detection methods exemplify responses that require no label and are therefore also referred to as label-free methods. Although label-free, their signals may be enhanced, limits of detection lowered, sensitivity increased, and dynamic ranges attenuated by the use of nanoparticle or chemically responsive labels. Indirect methods derive their signals from the measurement of a label, for example, a radioactive isotope or a fluorescence label. Each of these methods is briefly described in the succeeding paragraphs.

Ion-selective electrodes (ISEs) have allowed for greater specificities to be achieved, especially for single-use applications. In the 1970s, improvements in semiconductor technology led to the development of the field-effect transistor (FET) that usurped the ISE platform. The FET is a very high impedance transistor and most sensitive measurements of small potentials requiring very low currents are made using this technology. However, the FET proved expensive to produce and as such has yet to achieve its full commercial potential; in no small part because of difficulties with regard to its reliability in operation. Development and commercialization of pH-FETs and the emergence of organic thin film field effect transistors (TFTs) suggests that DNA FETs may become a technological and commercial possibility.

MEMS devices are proving popular as a modern method for the detection of biomolecules. They may combine mechanical parts, several unit operations, sensors, actuators, and electronics on a common silicon substrate through the utilization of microfabrication technology (6). This technology allows the integration of silicon-based microelectronics with micromachining approaches and sophisticated detection systems to allow for the development of complete systems-on-a-chip (6). Some of the enabling technologies allowed by MEMS include polymerase chain reaction (PCR), microsystems for DNA amplification and identification, biochips for the detection of chemical and biological agents, and microsystems for high-throughput drug screening and selection (6).

Electrochemical methods are generally very accurate and sensitive instrumental techniques. One of the most widely used methods is voltammetry, which makes use of a microelectrode such as a platinum wire. The potential of the electrode is varied, generally in a linear sweep, and the resulting current is recorded as a function of the applied potential. The microelectrode restricts the current to a few nanoamperes with the result that the concentration of the test substance in solution remains essentially unchanged after the experiment (7). Amperometry is the application of voltammetric measurements at a fixed potential to detect changes in the current as a function of the chemical potential of an electroactive species. Amperometric electrodes are made on a microscale ($\leq 50\,\mu m$ diameter) which allows for enhanced mass transport independent of flow, an increased signal-to-noise ratio, and measurements in high-resistance media (8). As described later, electrochemical methods allow for miniaturization of instruments for use in DNA hybridization (9).

One of the reasons for the rapid progress of the biochip field is the advances in the optics arena. The evanescent wave phenomenon is one such example. When light is reflected at an optical interface where there is a change in the refractive index, there is a decay of energy away from the point of reflection into the surrounding medium (10). When the evanescent wave is used to excite fluorescent molecules bound to the surface of the waveguide, the phenomenon is termed total internal reflection fluorescence (TIRF). If, however, there is an excitation of the electron plasma of a thin metal layer covering the surface of the waveguide, the process is called surface plasmon resonance (SPR). Both these phenomena, because of their considerable surface sensitivity, have received considerable attention in the development of DNA biochips.

Another major development has been the introduction of optical fibers and optical fiber bundles. Optical fibers have an outer, black polymeric sheath that provides protection, avoids coupling of ambient light through the walls, and gives mechanical support and strain relief to the inner core. In fiber optics, the detection circuitry is completely electrically isolated from the point of measurement, which has important safety implications for clinical diagnostic devices. In fiber optic devices, it is possible to use small polymer microspheres coated with an absorbing or fluorescent dye. These spheres can be placed within a hollow tube fixed onto the end of the fiber bundle or attached directly to the surface of the fiber optic (11,12). By using these microspheres, there is increased surface area available for immobilization of biorecognition molecules and interaction with the sample.

The application of piezoelectric-based acoustic wave devices to DNA detection has also been an important development. These are sensitive to changes in mass, density, viscosity, and acoustic coupling phenomena and, therefore, series resonance frequency can be used as a sensitive transduction parameter (13). These piezoelectric crystal devices generate and transmit acoustic waves that are dependent on frequency. Quartz is the most frequently used piezoelectric crystal because it can act as a mass-to-frequency transducer. AT-cut crystals (+35° 15′ orientation of the plate with respect to the crystal plane) are favored because of the excellent temperature coefficients in the range 10–50°C. One of the first sensors to be introduced from piezoelectric materials was the thickness-shear mode (TSM) sensor, which, if the substrate is quartz, may commonly be termed the quartz crystal microbalance (QCM) or bulk acoustic wave (BAW) sensor.

Mass spectrometric methods are becoming very important with these including standard approaches such as electrospray ionization and matrix-assisted laser desorption ionization (MALDI). Recently, newer techniques have started to make an impact and these include:

(a) Ion mobility spectrometry (IMS) which has the ability to separate ionic species at atmospheric pressure. Research is currently underway to develop low-pressure IMS systems.
(b) Atomic pressure ionization (API) and proton transfer reaction (PTR) techniques. Both are rapid, sensitive, and specific and allow measurements in real-time. In addition, they do not suffer drift or calibration problems.

A key feature of each detection modality is its ability to produce a measurable response, whether directly or indirectly, that is above noise and is consequent to the DNA hybridization reaction.

## 14.4 DNA Microarrays

Microarray technology "promises not only to dramatically speed up the experimental work of molecular biologists but also to make possible a whole new experimental approach in molecular biology" (14). DNA microarrays exploit an ordered, two-dimensional presentation of biorecognition probe entities, fluorescence tagging of targets, and scanning confocal imaging of the recognition–target complex or hybridization product. The most attractive features of microarrays are throughput and consequently, speed of analysis. The potential for miniaturization with its attendant reduction in sample volume does exist, but is still to be fully realized. This allows DNA microchip technology to have great potential for rapid multiplex analysis of nucleic acid samples. Examples of these include the diagnosis of genetic diseases, detection of infectious agents, measurements of differential gene expression, patient stratification in drug development, drug screening, and forensic analysis (15).

These chips are often fabricated from glass, silicon, or plastic supports and are usually composed of thousands of reaction zones (10–250 μm in diameter) onto which individual cDNA and/or oligonucleotides have been deposited and tethered. This results in densities up to $10^6$ sites/$cm^2$ in a typical 1–2 $cm^2$ chip (16). These high-density DNA arrays will usually require the use of physical delivery such as microjet deposition technology, which involves the dispensing of picoliter volumes onto discrete locations on the chip (15). In addition, contact pin arraying (spotting of 75–150 μm diameter spots) and microsolenoid dispensing (dispensing of 150–250 μm diameter spots) are also widely used. However, high-density DNA arrays generally benefit from the activation of the surface for a covalent attachment of the cDNA or oligonucleotide probes.

The successful development of DNA chip technology requires a multidisciplinary approach with various technical requirements to be satisfied. These include the development of algorithms and informatics for defining oligonucleotide probe sequences and lengths, methods for fabricating the probe arrays, detecting the target hybridization, algorithms for analyzing the data, and reconstructing the target sequence (15). As may be discerned, these are not simple procedures and careful experimental work is a necessity. With the advent of automated gene chip systems, there has been a decrease in the time required for analysis and there has been a reduction of human error.

In general terms, a microarray for gene expression analysis works by exploiting the ability of a given mRNA molecule to be reverse transcribed and have the resulting cDNA RT product successfully hybridized with high fidelity to the complementary strand of the DNA template from which it originated. If the microarray contains many DNA probes (typically $1–3 \times 10^4$), it is possible to determine the expression levels of thousands of genes within a cell, performed in a single experiment, by measuring the amount of cDNA bound to each site on the array (17).

To achieve meaningful results with microarrays, it is necessary to invest in proper design of the experiment. Usually, the following steps are involved (17).

(i) Prepare the DNA chip using the probe DNA molecules (cDNA or oligonuleotides).
(ii) Isolate the mRNA from the properly defined cells or tissue and evaluate this for extent of degradation if any.
(iii) Perform a reverse transcription (RT) and fluorescence labeling of the RT cDNA product. This may be achieved via *in situ* labeling using d-NTPs or posttranscriptional labeling using amino allyls.
(iv) Generate a suitable hybridization buffer solution containing a mixture of the fluorescently labeled cDNAs.
(v) Incubate the hybridization mixture containing the fluorescently labeled cDNAs with the DNA chip.
(vi) Using confocal scanning laser technology, the bound cDNA is detected and the acquired data stored on a computer.
(vii) Analyze the data using informatics tools and approaches established by the design.

The fluorescent tags are excited by the laser and, by using a confocal microscope and camera in tandem, a digital image of the array is created.

To date, microarray technology, a largely semiquantitative analytical technique, has been most valued in the basic research arena (18). Studies using microarrays have served to advance understanding of disease processes, and as the technology evolves it will become a tool for clinical medicine, providing a rich source of information on disease susceptibility, diagnosis, and prognosis. As a research tool, DNA microarrays have already been used in the study of heart, blood vessel, and lung disease; cystic fibrosis; human immunodeficiency virus (19); cancer (20); and single nucleotide polymorphisms (21). They have been used more broadly as well, to study arabidopsis, rat, yeast, and *E. coli* genomes and mouse models. Additionally, DNA microarrays are being increasingly used to monitor gene expression in humans (15). Researchers have used RNA expression in biochips to identify differential gene expression relevant to different biological states. Clontech (Palo Alto, CA) have produced the Atlas™ microarrays that provide sensitive detection of gene expression by using fluorescent dyes and glass or nylon substrates (15).

The evolution of microarrays is to use the information gleaned from genomic microarrays in the development of pathway-specific, or diagnostic/prognostic microarrays that employ smaller suites of genes in highly focused assessments (22). This evolution towards so-called "theme arrays" has begun. SuperArray Bioscience Corporation has developed the Human Th1-Th2-Th3 Gene Array which contains 96 genes relevant to understanding helper T cell biology. These genes include the cytokines specifically expressed by both Th1 and Th2 subtypes. The array also contains genes encoding transcriptional factors that regulate the expression of these cytokines as well as other markers of CD4+ T lymphocytes. Simple side-by-side hybridization allows relative expression of these genes in experimental RNA. Related products include: Cancer Drug Resistant and Metabolic; Common Cytokine; Inflammatory Cytokines and Receptors; Chemokines and Receptors; Interleukins and Receptors. IntelliGene™ DNA microarrays are medium-density cDNAs arrayed on standard 2.5 cm × 7.5 cm glass slides for standard dual color analysis using high-resolution fluorescent detection. Arrays target human cancer, human cytokines, endocrine disruption, as well as cyanobacterial ORF, *Arabidopsis*, and mouse and *E. coli* gene analysis. This movement towards confocally imaged, targeted microarrays is converging with DNA biochips that use detection technologies other than confocal fluorescence imaging.

One of the areas where the use of DNA microarrays is proving to be of immense importance is in the monitoring of microbes. These can serve as detection and identification tools for clinical applications. Commercially available biochips include (23): *Helicobacter pylori* arrays from MWG Biotech; *Mycobacterium tuberculosis*, *Plasmodium falciparum*, and *Candida albicans* chips from OPERON; *E. coli* arrays from Pan Vera; *E. coli*, *Pseudomonas aeruginosa* and severe acute respiratory syndrome (SARS) biochips from Affymetrix; and *E. coli* and *M. tuberculosis* arrays from Sigma-Genosys. Other companies such as Siebersdorf Research (Germany) and Agilent will create custom-made chips for the organism of interest (23). In a related area, Nanogen Inc. (San Diego, CA) has developed an electronically controlled sample preparation process for the dielectrophoretic separation of *E. coli* from blood cells. The bacteria are then lysed by high-voltage pulses. The company has fabricated several microelectronic chips including those containing 25, 100, 400, 1600, and 10,000 addressable test sites. The latter is being developed for drug discovery methodologies.

In a major academic development, researchers at St. George's Hospital Medical School in London have formed a group called BμG@S (Bacterial Microarray Group at St. George's). This has been funded by the Wellcome Trust and brings bacterial genome researchers around the world into an organization to develop whole-genome arrays for 12 pathogens including *Campylobacter jejuni*, *Haemophilus influenzae*, *Yersinia pestis,* and *Salmonella typhi* (23). This collaborative effort is expected to have immense benefits as the collective research will generate more results on a faster basis. These can then be used to lead to more efficient identification procedures that will have a positive impact on health care systems.

Bekal and co-workers (24) have described a method for the rapid identification of *E. coli* pathotypes by virulence gene detection. An *E. coli* virulence factor DNA microarray composed of 105 DNA PCR amplicons printed on glass slides and arranged in eight subarrays corresponding to different *E. coli* pathotypes was developed. Fluorescently labeled genomic DNA sequences were hybridized to the virulence gene microarrays for optimization and validation. It was reported that this type of microarray is a powerful tool for both gene quantitation and subtyping.

Finally, *Listeria* species have also been identified by a microarray-based assay. Six species of *Listeria* were amplified by a multiplex PCR and subsequently hybridized to individual oligonucleotide probes specific for each *Listeria* species (25). It was demonstrated that the method allowed unambiguous detection of all six species based on sequence differences in the iap gene. The authors concluded that microarrays are valuable for the identification and characterization of bacterial pathogens. It must be borne in mind, however, that the huge amount of data generated from these microarrays will require experts to interpret the results.

DNA microarrays have also been used for the identification of viruses, although this area is still in its infancy. Recent applications include the detection of the human immunodeficiency virus (HIV) responsible for AIDS (26) and the influenza virus (27). In the former, two methods of nucleotide sequencing were compared for the detection of drug resistance mutations in HIV type 1 reverse transcriptase (RT) in viruses isolated from highly RT inhibitor-experienced individuals (26). It was found that of the 11,677 amino acids deduced from population PCR products by both cycle sequencing and sequencing by hybridization to high-density arrays of oligonucleotide probes, 97.4 % were concordant by both methods with discordances mainly due to genetic mixtures within or adjacent to discordant codons. For isolates evaluated by additional sequencing of molecular clones of PCR products by both methods, the discordance between methods was less frequent.

In the latter example, a model DNA microarray was shown to facilitate typing and subtyping of human influenza A and B viruses (27). RT-PCR was used to prepare cDNAs encoding about 500 bp influenza virus gene fragments that were cloned, sequenced, reamplified, and spotted onto a glass support. The target DNAs included multiple fragments of the hemagglutinin, neuraminidase, and matrix protein genes. Cy3- or Cy5-labeled fluorescent probes were hybridized to the target DNAs and the arrays were scanned to locate the probe binding site(s). The researchers indicated that the hybridization pattern agreed well with the known grid location of each target and the signal-to-background ratio varied from 5 to 30. No cross-hybridization could be detected. With further testing, this method may prove to be clinically useful. Further differentiation of amplified molecules in complex mixtures can be obtained by hybridizing combinations of Cy3- and Cy5-labeled DNAs. A particularly attractive feature of this research is the identification of several sets of multiplex primers that collectively target influenza A and B virus strains. This allows DNA microarray technology to act as a supplement to the information obtained from PCR-based diagnostic methods.

Additionally, researchers at the Center for Biologics Evaluation and Research and at the National Institute of Allergy and Infectious Diseases have developed a system to identify isolates of rotaviruses. These organisms are responsible for causing diarrhea in infants in the developing world (28). Five clinically relevant G genotypes (G1 to G4 and G9) were studied. The genotype-specific oligonucleotides on the glass slides were allowed to bind to multiple target regions within the VP 7 gene which are highly conserved among individual rotavirus genotypes. Nested PCR was performed and the identification of rotavirus genotype was based on hybridization with several individual genotype-specific oligonucleotides. An advantage of this microarray approach over PCR is that there is unambiguous identification of all rotavirus serotypes (28). This approach combines the high sensitivity of PCR with the selectivity of DNA–DNA hybridizations. Furthermore, the presence of random mutations allows each individual virus isolate to produce a unique hybridization pattern. This allows the differentiation of different isolates of the same genotype.

De Risi's lab (29) has reported on an approach for highly parallel viral screening. A long (70 mer) oligonucleotide DNA microarray capable of simultaneously detecting hundreds of viruses was developed. The different viruses were detected by using virally infected cell cultures and related viral serotypes were distinguished by the unique hybridization pattern of each virus. This offers an advantage over existing techniques for screening a broad spectrum of viruses as it allows for the comprehensive and unbiased analysis of viral prevalence in a biological setting. In addition, individual viruses which were not explicitly represented on the microarray were detected indicating that this method may find use in virus discovery. An important aspect of this research was the ability to detect multiple viruses in human respiratory specimens without the use of sequence-specific or degenerate primers. This method allows for discrimination among viral subtypes.

DNA arrays have also found use in cancer classification studies. With over 200 different types of cancer, it is difficult to differentiate some cancer subtypes; for example, cells of acute myeloid leukaemia (AML) and acute lymphoblastoid leukaemia (ALL) look very similar. It is essential to achieve a correct diagnosis as the treatment regimens may be different. It has been reported that, by analyzing the expression patterns of 50 genes on an array representing about 7000 genes, accurate classification was possible for both AML and ALL. Also, the array allows a more robust diagnosis than tests based on a single protein or the activity of one enzyme. In another application, DNA microarrays were used in the discovery of new cancer subtypes; for example, analysis revealed two previously unrecognized subtypes of non-Hodgkins lymphoma.

Gene expression patterns in renal cell carcinoma were assessed by complementary DNA microarray (30). Renal cell carcinoma comprises several histological types each with different clinical behavior and thus accurate pathological characterization is essential. The authors describe gene expression profiles in 41 renal tumors determined by DNA microarrays containing 22,648 unique cDNAs representing 17,083 different UniGene Clusters, including 7230 characterized human genes. Differences in the patterns of gene expression among the different tumor types were readily observed (30).

The genomic classification of brain tumors and brain tumor subtypes is also being pursued at the Center for Bioelectronics, Biosensors and Biochips and the Neuroanatomy Project. Using a 10,000 gene spotted oligonucleotides (50 mers) microarray, the expression profiles of various tumor types are being defined. Such an effort involves very close working among bioengineers, pathologists, biostatisticians, and neurosurgeons. As a result of the semiquantitative nature of today's microarrays, the likely diagnostic arrays of the future must be brought into a more quantitative and clinically relevant format (31). Also, the diagnostic community will need to accommodate decisions and interventions based on quantitative risk-based assessments in a fashion similar to that of the environmental community.

To achieve these goals, the design and fabrication of the microarrays themselves need to be brought into a more quantitative framework with greater attention paid to well-defined and highly reproducible surface chemistries, DNA probe attachment protocols and procedures, and more rigorously defined hybridization protocols (32). Furthermore, present biochip readers are slow and expensive. Current confocal scanning methods address signal quality issues optically. However, electrical signal processing methods can also be used to reduce interference. These include cross-talk cancellation and multiplexing schemes. Plextek (Essex, UK) claims that further improvement can be obtained by the design of a linearized detector sensitivity control. The current emphasis on statistical manipulation of data to address engineering limitations and interoperator variability that contributes noise, poor precision, and poor accuracy needs to be addressed by improved design, fabrication, and implementation protocols of DNA microarrays.

Current normalization techniques implemented in most software for microarrays assume that fluorescent background within spots is essentially the same as that found throughout the microarray slide and can be measured by fluorescence surrounding the spots (33). This assumption, however, is not valid if the background fluorescence is spot-localized and inaccurate estimates of background fluorescence under the spot create a source of error, especially for genes of low transcript copies (33). Such nonspecific adsorption of target cDNA to oligo or cDNA probe spots may be addressed by including *Arabidopsis* genes on human microarrays and subtracting the intensity of the nonspecifically adsorbed signal from the probe signal. Finally, DNA detection schemes using metal nanoparticles or quantum dots as labels, rather than fluorescent organic dyes that are subject to photobleaching, have been developed. These are based on the physical properties of metal nanoparticles: large extinction and scattering coefficients, catalytic ability, surface electronics, and efficient Brownian motion in solution. These properties have resulted in nanoparticles overcoming some of the limitations of fluorescent labels such as cost, ease of use, selectivity, and sensitivity.

## 14.5 Opportunities for DNA Biochips

Opportunities for applications of DNA biochips are most attractive in the area of human health and specifically in the area of personalized medicine. The R&D investments, marketing, and sales and distribution costs, given today's models for development of technologically driven companies, do not support similar opportunities in the environmental or industrial biotechnology sectors. The human health sector presents opportunities for near-patient (bedside), physician's office, clinical laboratory, and molecular diagnostics laboratory (hospital). In the short term, diagnostic biochips will likely have more impact in the area of drug development than in clinical molecular diagnostics. However, clinical molecular diagnostics represents a looming opportunity separated only by the regulatory challenges and broader societal concerns.

### *14.5.1 Diagnostic Biochips and Links to Drug Development*

A primary goal of drug research is to determine, on as small a sample as possible, and for as many different molecule pairs as possible, the formation of a biologically active complex. Pharmaceutical companies must perform biochemical assays during all phases of the drug development cycle; from the initial screening of libraries of available compounds, through the focused evaluation of promising new drug candidates, to the clinical testing of drugs being readied for market. These assays are poised to become more important as the industry moves towards the delivery of drugs that target specific genetic profiles. A critical element in this process is the comprehensive evaluation of patient response based on molecular indicators. It is unlikely that populations will be prospectively screened, stratified, and selected prior to joining a study, which limits market size and may even be socially objectionable. It is more likely that subpopulations identified by the study will be screened retrospectively to establish a logical genetic basis for exclusion of nonresponders and identification of overresponders within the target population (34). This limits the cost associated with development and is less disruptive of the well-established drug development process. A key feature will be the procurement and banking of tissue and/or fluid specimens from patient participants in the study. Diagnostic biochips will feature prominently as a means for providing the critical molecular data for drug discovery (21), and improving the drug discovery process by enabling proper patient stratification during clinical trials.

One of the most important developments in recent times has been the use of single nucleotide polymorphisms (SNPs pronounced "snips"). A SNP is the most common type of variation in the human genome and it arises as a result of a single base difference in the coded protein between individuals. These variations occur about once every 1000 base pairs in the genome, making up the majority of the three million variations in the genome. Furthermore, the frequency of a particular polymorphism remains stable in the population. Usually, the effect of a single SNP on a gene is not large but small effects can influence susceptibility to common diseases such as Alzheimer's and the risk of heart disease. Oligonucleotide (DNA) chips can be used to discriminate between alternative bases at the site of a SNP. These DNA chips allow many SNPs to be arranged in parallel, which is necessary for large-scale association or pharmacogenomic studies. A DNA sequence containing a SNP is hybridized to a DNA chip and discrimination of alternative bases (termed "typing") is carried out at the polymorphic site. The signal that corresponds to the specific identified base is detected.

Two typing methods are widely used. The first uses allele-specific hybridization where short DNA sequences on the chip represent all possible variations at a polymorphic site. A labeled DNA will only stick if there is an exact match. The fluorescent signal indicates where the base is located. Secondly, the oligonucleotide on the chip may stop one base before the variable site and typing depends on allele-specific primer extension. A DNA sample attached to the chip is used as a template for DNA synthesis with the primer being the immobilized nucleotide. The four nucleotides, to which are added fluorescent labels and DNA polymerase, are also introduced. The incorporated base is identified by the fluorescent signal, although mass spectrometry may also be used. These methods are suitable for high-throughput SNP typing, usually for large-scale studies of populations. Two applications are: (a) association studies, which try to correlate SNP profiles with susceptibility to disease, and (b) pharmacogenomic studies, which link SNP profiles with drug response patterns. A disadvantage of these chip-based assays is that new SNPs cannot be readily incorporated onto the chip. However, by using bead arrays, this problem is being solved.

In 1999, a SNP consortium was set up by the Wellcome Trust in collaboration with ten big pharmaceutical companies and several genome research institutions. The goal was to produce a public database of SNPs that could be easily accessed in the fight against major disease. Each company contributed US $10 million to this endeavor and by September 2001, 1.5 million SNPs (five times the original anticipated number) were discovered (35).

Recently, a new initiative has been established to provide a haplotype map of the genome. Within the human genome, different genetic variants within a chromosomal region (haplotypes) are not found in all possible combinations; certain combinations are more common than others. Differences in haplotypes may be associated with different susceptibilities to disease and, by mapping the haplotype structure of the genome, the genetic basis of some diseases may be identified. A five-nation consortium of Canada, China, Japan, the United Kingdom, and the United States, has pledged US $100 million over three years (2002–2005) to construct a haplotype map based on 200–400 genetic samples from each of four different populations: the Han Chinese; the Yorubas in Nigeria; the Japanese; and individuals in the United States with northern or western European ancestry. It is hoped that the haplotype map will simplify the search for medically important DNA sequence variations and provide insights into human population structure and history.

Haplotypes can be used for the mapping of disease genes. This is possible because mutations responsible for a genetic disease always enter the population within an existing haplotype ("ancestral haplotype"). Over several generations, recombination events occur within the haplotype but the disease allele and the closest SNPs will still be inherited as a group. Once this haplotype can be identified in a group of patients with the disease, typing the alleles within the haplotype allows a conserved region to be identified. Because many SNPs are present, genes can be mapped accurately. The International HapMap project aims to speed the discovery of genes related to common illnesses such as asthma, cancer, diabetes, and heart disease (36).

An alternative to genotyping is phenotyping. Thus, an antidepressant drug such as fluoxetine (Prozac®) is metabolized by the cytochrome P450 family of enzymes to produce nor-fluoxetine. Both drug and metabolite molecules will appear in the urine. Using a recombinant form of the enzyme in a suitably designed biochip allows the analysis of two orders of magnitude in variation of metabolic activity found in the population for this drug. Companies such as ABTECH Scientific are currently researching this approach, and although it does not measure DNA directly, it gives an indication of the clinical manifestation of this variation in the population. The core technology is based on interdigitated microelectrodes, which are arrays of tiny electrodes derived from metal sputtered onto a substrate (37). A microlithographically fabricated chip that can detect biological relevant analytes has already been produced.

### 14.5.2 Diagnostic Biochips for Clinical Molecular Diagnostics

The era of personalized molecular medicine is here (38). Ushered in by a now universal appreciation of the molecular basis of disease, molecular analyses, both genetic or proteomic, to improve a patient's outcome in the management of his disease, is revolutionizing healthcare. With the potential to (i) optimize therapy and enable appropriate patient dosing, (ii) detect diseases when in their early stages, (iii) greatly reduce adverse drug reactions and drug–drug interactions, and (iv) potentially improve patient compliance, clinical molecular diagnostics is a reality of growing significance (39).

Several clinical molecular diagnostic (MDx) areas are rapidly emerging; (i) MDx for the optimal drug dosing of patients, that is, the selection of the patient-specific dosage for a more universal therapy; (ii) MDx for the selection of patient-specific therapy, that is, the application of therapeutic drugs to patients who by the presence or absence of a particular biomarker will have a high likelihood of favorable drug response; and (iii) MDx for the molecular grading and staging of disease, particularly cancer, that is, the development of diagnostic and prognostic tests to more accurately predict patient outcomes to multistage disorders. In many cases, differential drug response arises from subtle differences in the genes that code for the production of drug-metabolizing enzymes, drug transporters, or molecular drug targets such as receptors. Particularly noteworthy amongst drugs demonstrating differential drug response are those drugs metabolized by the liver enzyme, Cytochrome p450. This is because of the very large variation in CYP activity that is apparent in the population.

Related in principle to patient stratification in drug development, the development of diagnostic tests to measure specific protein levels or gene copy numbers has allowed the emergence of such drugs as Herceptin® and Gleevec®. Herceptin is an antibody drug that specifically inhibits the cell surface protein, human epidermal growth factor receptor 2 (HER2). This receptor has been found to be overexpressed in approximately 30% of breast cancer patients histopathologically defined classes. Molecular diagnostic tests now exist that can measure HER2 protein levels or gene copy numbers and so allow an identification of that subset of patients for whom Heceptin is indicated. The beneficial corollary is that women who are HER2-negative need not be given Herceptin as a first course of treatment.

Gleevec is a tyrosine kinase inhibitor that specifically binds to the ATP binding site and inhibits the action of the abnormally formed BCR–ABL protein, a kinase enzyme. The BCR–ABL produced kinase is responsible for an uncontrolled increase in white blood cell population; the basis for chronic myelogenous leukemia (CML) (40). The BCR–ABL protein is itself a fusion of two normal proteins that results because of a chromosomal rearrangement. A molecular diagnostic test for the gene that codes for BCR–ABL allows targeted prescription resulting in improved response rates, lower toxicity, and a high probability of complete remission (41). There are many other highly regarded examples, some already available and several yet emerging, that foretell the success of molecular medicine and personalized care. Such examples are most persuasive when the diagnostic test is linked to an available therapy. When such MDx tests are not accompanied by an established therapy they may be the source of frustration for patients, providers, and insurers alike. However, such tests, during a period of ambivalence may well be the basis for proper patient stratification that enables the future development of appropriate therapies.

Another source of complication arises for those diseases and disease states that are multigenetic. Here the challenge is first arriving at a suitable panel of genes that may be diagnostic and prognostic for the disease and its outcomes. Second, is the challenge of implementing that panel of genes along with suitable analytical controls onto a suitable geneosensor or biochip platform. The first challenge may be defined in part as the challenge of class comparison, that is, the identification of genes that are differentially expressed among the predefined classes of that disease. The first challenge may also be defined in part as the challenge of class prediction, that is, the identification of genes that are differentially expressed among patients for whom the disease outcomes are or may become known. In both cases the technical challenge is exacerbated by the likely emergence of genetic subclasses revealed by the very attempt at class comparison and by the multiple interventions that affect patient outcomes. A looming question then is whether the health care community is prepared to accept a risk based on a deterministic paradigm for health care in an era of molecular diagnostics and personalized medicine.

### 14.5.3 Biochip for Diagnostic Classification and Prognostic Stratification of Primary Brain Tumors

One important nexus for future DNA microarrays is found in clinical diagnostics and prognostics. This entails the development of diagnostic tools based on microarray technology that utilize a targeted suite of genes directed at specific diagnostic screening applications (42). At the Center for Bioelectronics, Biosensor and Biochips we are engaged in the development and deployment of a diagnostic and prognostic biochip for primary brain tumors. This biochip seeks to combine a focused panel of genes that is capable of genetically delineating the several World Health Organization (WHO) histopathological classes of primary brain tumors (astrocytomas) with microfluidics technologies that aims to support the several unit operations needed to realize a "tissue-in-data-out" bioanalytical paradigm. Two parallel efforts were established: the conduct of retrospective genetic class comparison, class delineation, and class prediction studies using differential gene expression profiling of tissues taken from the Tissue Data Bank at Virginia Commonwealth University; and the development of electroanalytical techniques that may be suited for rapid hybridization detection of transcripts in a clinical molecular diagnostics format (43).

Microarrays are now being widely applied to the study of differential gene expression profiles to improve class prediction for many different cancer types, including colon, lung, esophageal, and breast cancers. Various statistical methods have been developed to allow improved class prediction using microarray data. In parallel, there is considerable attention now being given to rendering microarray data quantitatively more rigorous by (i) optimization of oligonucleotide probe length and definition relative to the 3′ end, (ii) reproducible covalent immobilization of probes to qualified surfaces resulting in exacting surface coverage, (iii) minimizing nonspecific adsorption of targets to reference areas, (iv) the implementation of a larger number of control features, and (v) the development of advanced reagent sets that allow improved detection of low copy number transcripts and more stringent hybridizations.

The C3B 10k oligonucleotide microarray was designed using the MWG 10kA human oligonucleotide library (Cat # 2190-000000, MWG) as the base gene library. Seventeen additional "housekeeping" genes and eleven control genes that are found on the Affymetrix Hu133A chip (Table 14.1) were added to the 9984 5′-$C_6$-amine-terminated and HPLC purified 50 mer oligonucleotides in the MWG set. These additional genes serve as internal controls and allow cross-platform data analysis and harmonization, an on-going project of the C3B. The gene library was

**Table 14.1** Control Genes that were added to the MWG 10k A Pan Human Oligo Set Used to Produce the C3B 10KO Oligo Array

| **Control Genes** | Accession Number |
|---|---|
| GAPDH 5′ | M33197 |
| GAPDH 3′ | M33197 |
| Beta Actin 5′ | X00351 |
| Beta Actin 3′ | X00351 |
| ISGF 5′ | M97935 |
| ISGF 3′ | M97935 |
| Bio B | J04422 |
| Bio C | J04423 |
| Bio D | J04424 |
| Thr C | X04603 |
| Phe B | M24537 |
| Phosphofructokinase, platelet | M64784 |
| Asparagine synthetase | M27396 |
| Aldolase A, fructose-bisphosphate | M11560 |
| Phosphoglycerate mutase 1 (brain) | XM_083842 |
| Glucose-6-phosphate dehydrogenase | M12996 |
| Ribosomal protein S3 | AB061838 |
| Non-POU-domain-containing, octamer-binding | XM_088688 |
| Ribosomal protein s27a | NM_002954 |
| Lactate dehydrogenase A | NM_005566 |
| Phosphoglycerate kinase 1 G | NM_000291 |
| Mitochondrial ribosomal protein L19 | NM_014763 |
| Rho GDP dissociation inhibitor (GDI) alpha | AA453756 |
| Lactate dehydrogenase A | NM_005566 |
| Beta-2-microglobulin | NM_004048 |
| Phosphofructokinase, platelet | M64784 |
| Aldolase A, fructose-bisphosphate | M11560 |
| Ribosomal protein S27a | NM_002954 |

further supplemented with 78 custom-designed 50 mer oligos corresponding to genes previously documented in the literature as differentially expressed in brain tumors but which were not in the MWG library. The complete list of gene specific probes is publicly available (44).

The C3B oligo library was printed on γ-glycidoxy-modified (3-glycidoxypropyltrimethoxysilane; 0.1 wt% in anhydrous toluene for 30 min at 42°C then cured for 20 min at 110°C) 1.0' × 3.0' borosilicate glass microscope slides (31, 39) using a Cartesian PixSys 5500 Microarrayer. The arrays were produced by contact printing with a 4 × 2 pin arrangement using eight Parallel Synthesis spotting pins (Parallel Synthesis, Santa Clara, CA). The arrays were printed at room temperature under an air atmosphere of 50% relative humidity. The spotting buffer (pH = 5.2) was composed of 25 mM oligo in 1.5 X SSC and 0.75 M betaine [Diehl, 2001 #151]. The C3B10KO has a total of 10,584 independent genes that were spotted in duplicate creating an array with 21,168 total features divided into 4 × 12 (48) subgrids of 21 × 21 (441) spots (Fig. 14.1). The C3B10KO oligomicroarray has been the basis for cross-platform performance comparisons (31) for the development of cross-platform correlation estimates (32) and is intended to reveal patterns of gene expression to enable retrospective genetic class comparison, class delineation, and class prediction using differential gene expression profiling of IRB-approved banked and acquired tissues.

A parallel effort at the C3B has been the development of impedimetric and amperometric DNA hybridization detection platforms consisting of microlitho-

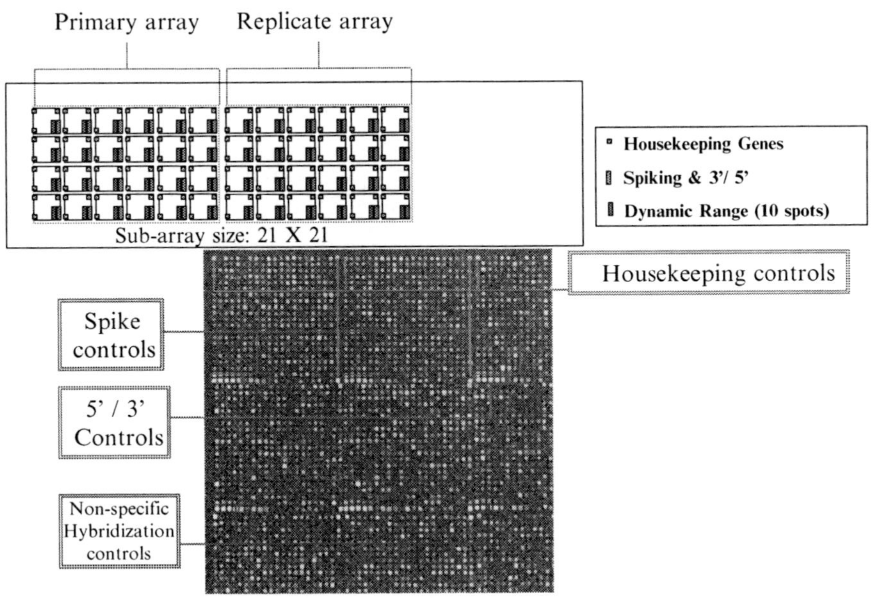


**Fig. 14.1** Schematic illustration of the layout of the C3B10KO oligo array showing the 4 × 12 (48) sub-grids, each with 21 × 21 (441) spots, including the various controls that resulted in the inclusion of 10,584 independent genes, spotted in duplicate, and creating an array with 21,168 total features

graphically fabricated interdigitated microsensor electrodes and microdisc array electrodes (45). These approaches focus on devising means to directly detect the hybridization of DNA on a multielement microelectrode device that will serve as the detector in a fully integrated molecular diagnostics system (46). The beBiochip-32™ and beBiochip-64™ consist of 32 regions of interdigitation and 64 microdisc pads that comprise 32- and 64-element arrays, respectively (Fig. 14.2). The beBiochip-32 functions by impedimetric detection of nanoparticle labeled and unlabeled DNA hybridization whereas the beBiochip-64 allows amperometric detection of redox-labeled DNA hybridization reactions. Fig. 14.2a shows an impedimetric array of 10 opposing fingers, each 2 microns wide and 3 mm long and separated by 1 micron spaces. Fig. 14.2b shows a voltammetric or amperometric sensor element comprising a hexagonal close-packed arrangement of microdiscs in the form of a microdisc electrode array, a large area (100 times the area of the microdisc working electrode) counter electrode, and a reference electrode.

The electrodes of this three-electrode electrochemical cell-on-a-chip were similarly fabricated from 100 nm magnetron sputter-deposited gold or e-gun deposited platinum on a 10 nm titanium/tungsten adhesion promoting layer. The substrate used was either oxidized silicon with a minimum 100 nm of thermally grown oxide or the highly polished, electronics-grade, Schott D262 borosilicate glass. The microdisc array working electrode comprises a fluoro-etched opening through a 0.5 mm thick silicon nitride ($Si_3N_4$) layer that was deposited onto the previously patterned noble metal beneath. Collectively these biochips allow: (1) label-free impedimetric detection of DNA hybridization, (2) enhanced impedimetric detection using colloidal gold nanoparticles as labels on reverse-transcribed mRNA, and (3) the use of an electroactive layer of poly(pyrrole-*co*-pyrrolyl butanate) to provide covalent attachment of DNA probes and enhanced redox detection sensitivity with electroactive labels such as ferrocene. Significant changes in electrochemical impedance values (both real

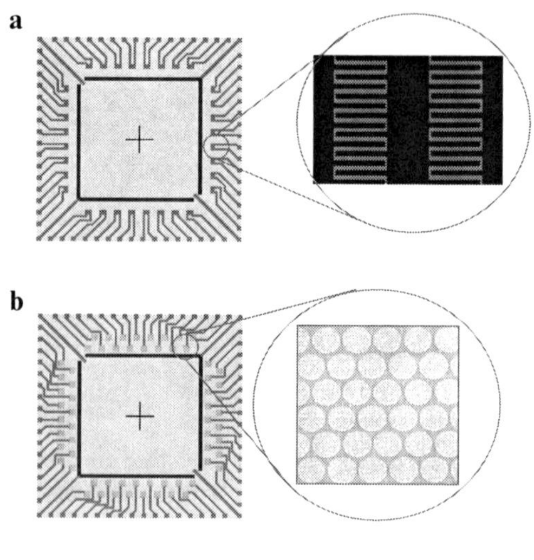


**Fig. 14.2** Schematic illustration of a microfabricated multi-element array; a) comprising 32 interdigitated microsensor electrodes, and b) comprising 64 independently addressable microdisc voltametric electrodes. Each device shows the large area counter electrode (middle) and the reference electrode as a band arond the counter electrode.

and imaginary components) (11% increase in impedance modulus at 120 Hz) have been detected after hybridization of covalently immobilized oligonucleotide probes to their complement (47).

Fig. 14.3a shows an impedimetric sensor element comprising a circumferentially arranged interdigitated microsensor electrode array, a large area (100 times the area of the interdigitated working electrodes) counterelectrode, and a reference electrode. The electrodes of this four-electrode electrochemical cell-on-a-chip were fabricated from 100 nm magnetron sputter-deposited gold or e-gun deposited platinum on a 10 nm titanium/tungsten adhesion promoting layer. The substrate used was either oxidized silicon with a minimum 100 nm of thermally grown oxide or the highly polished, electronics-grade, Schott D262 borosilicate glass. These independently addressable interdigitated electrodes could be used for two-electrode electrical impedance or in combination with the reference electrode, in three-electrode electrochemical impedance.

Finally, the two interdigitated electrodes could be shorted and used as a single working electrode within which the large area counterelectrode supports the electrochemical oxidation or reduction of appropriate electroactive species. As an example, electroconductive polymers that were grown by electropolymerization onto each region of interdigitation, was accomplished in this format. Finally, a reference electrode of silver/silver chloride accompanied each sensor element of the device and this was prepared by silverization of the gold or platinum layer of that electrode. This reference electrode provided the reference potential for the electrochemical impedance or amperometric determination of each multiplexed senor element of the array. Fig. 14.3b shows the modified sensor element following silverization of the reference electrode. Here the bright silver prior to chloriodization is shown. Fig. 14.3c shows

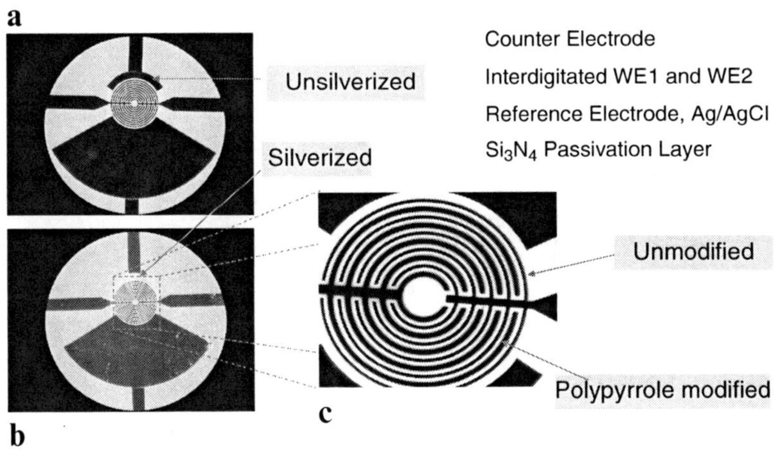


**Fig. 14.3** Optical micrographs (x 50) showing one sensor element of the microfabricated electrochemical cell-on-a-chip interdigitated microsensor electrode (ECC IME) device. a) Showing the various electrodes of a single four-electrode electrochemical cell, b) showing the silver plated reference electrode prior to chloridization, and c) with electropolymerized polypyrrole (PPy).

the modified sensor element following electropolymerization of a layer (exaggerated) of polypyrrole onto the region of interdigitation (48). Thin films of polypyrrole provide the anchorage chemistry for the covalent coupling of oligonucleotides.

Fig. 14.4 shows a voltammetric or amperometric sensor element comprising a hexagonal close-packed arrangement of microdiscs in the form of a microdisc electrode array, a large area (100 times the area of the microdisc working electrode) counterelectrode and a reference electrode. The microdiscs are not independently addressable but rather exploit ultramicroelectrode electrochemistry in improving sensitivity, dynamic range, and signal-to-noise performance for a single DNA probe. The electrode designs of Figs. 14.3 and 14.4 may also be implemented as individual sensor elements in a microfluidic system. Fig. 14.5 shows a pair

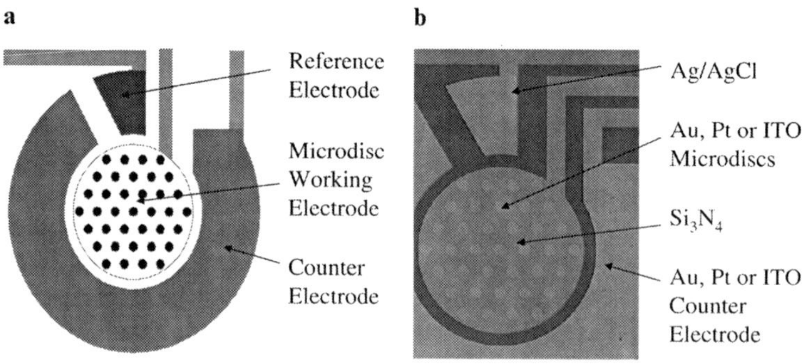


**Fig. 14.4** a) Schematic illustration of the microfabricated voltammetric or amperometric microdisc array sensor element sowing the three electrodes of the electrochemical cell-on-a-chip. b) Optical micrograph (x 50) of the microfabricated voltammetric or amperometric microdisc array sensor element showing the materials of construction

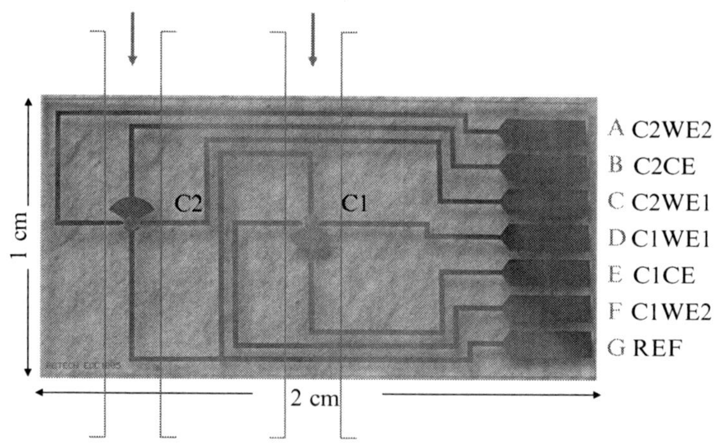


**Fig. 14.5** Optical micrograph of a microfabricaed, dual-channel, electrochemical cell-on-a-chip interdigitated microsensor electrode (ECC IME) device showing the fluid flow over the microelectrode arrays. A=C2W2, B= C2CE, C=C2W1, D=C1W1, E=C1CE, F=C1W2, G=REF. C2 is Cell 2, C1 is Cell 1, W is working electrode, CE is counter electrode and REF is reference electrode.

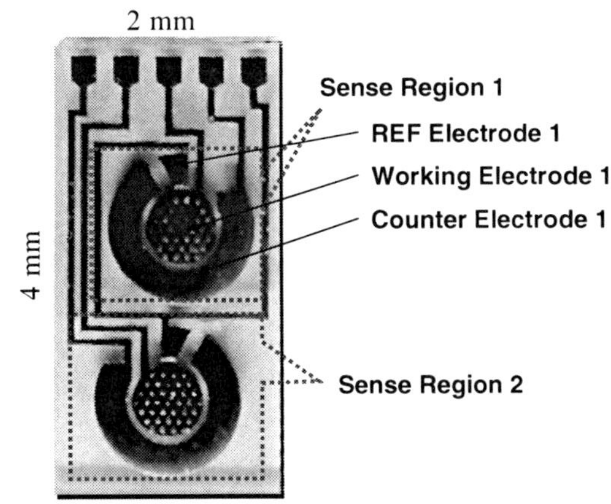


**Fig. 14.6** Optical micrograph of a microfabricated, dual-channel, electrochemical cell-on-a-chip voltammetric microsensor electrode (ECC MDEA) showing W is working electrode, CE is counter electrode and REF is reference electrode.

of the four-electrode sensory elements of Fig. 14.3 that has been fabricated onto a single 1 cm × 2 cm × 0.05 cm glass chip. Fig. 14.6 shows a pair of the three-electrode sensory elements of Fig. 14.4 that has been fabricated onto a single 4 mm × 2 mm × 0.5 mm glass chip. Both chips offer a dual channel or a ratiometric detection scheme.

## 14.6 Impact and Challenges for Biochip-Based Molecular Diagnostics

Biochips are expected to have the greatest impact in the molecular diagnostics market. Here the major challenges go well beyond the usual technological challenges of acceptable clinical detection limits, levels of sensitivity and specificity, dynamic range, repeatability and reproducibility, response time, and immunity from false positives and false negatives. Although important, these analytical parameters must be matched to the specifics of the assay that is targeted and to the decision context of the acquired data.

One additional major challenge is the selection of appropriate molecular targets that are decidedly linked to disease susceptibility or drug response. This should *prima facie* be the point of initiation of all research and development activities on diagnostic biochips. The second major challenge is to overcome the fact that a biochip diagnostic will simply be an additional format for delivering an assay that could otherwise be done using the same reagent set and assay approaches that are implemented on the chip. This is not a trivial matter and should not be overlooked. After all, identification of the molecular target as "appropriate" would have originated using lab-based assays. The biochip format must therefore offer some significant competitive advantage over its lab-based equivalent.

Areas of competitive advantage for diagnostic and prognostic biochips may include th following.

(i) *Higher throughput*. The biochip format, when automated, may permit the handling of many samples. This will reduce the overall cost per test performed. The consequence is that the platform migrates away from the point of concern and is then favored in the clinical or molecular diagnostics laboratory.
(ii) *Smaller sample volume*. The biochip format allows smaller sample volumes to be used. This is relevant when samples are available in very small quantities such as in some tumor biopsies.
(iii) *Integration of sample preparation*. Sample preparation and/or workup are a major source of error and cost in many molecular diagnostics. The ability to integrate these unit operations (such as cell sorting, lysis, protein separation, PCR) onto the same biochip platform is a major advantage. Although this does not save time, it can limit the need for operator involvement and allow the diagnostic test to migrate closer to the site of interest (near patient: physician's office or bedside).

Additional major challenges include the size of the market, access via distribution channels, and repayer issues. The last is likely to be the most important forestaller of what could potentially have the largest impact on the way health care is provided.

The impact of diagnostic biochips in molecular diagnostics must await the results of further research and targeted product development. Several companies, including Millennium Predictive Medicine, Abbott Diagnostics, Bayer, and ABTECH Scientific are actively pursuing such diagnostic products. Once these products are developed, validated, clinically tried, and approved by the FDA, they must be sold to a traditionally conservative medical community. This community has been slow to adopt new diagnostic paradigms which are due, in part, to many overhyped claims for other approaches in the past. Genomics-derived molecular diagnostics are likely to have a significant impact on the market by the end of the current decade.

## 14.7 Issues Related to Diagnostic Biochips

A burgeoning issue, unique to biochip technologies, is the plurality of technological approaches being researched, developed, and commercialized. Market forces (capital, speed to market, product positioning) rather than policy requirements will drive end-user choices among the several dozen competing biochip technologies available. Each of these requires sophisticated software programs and powerful computing capabilities in the design of chip functionality as well as in the capture, processing, and analysis of the massive amounts of data they will generate. As an example, the absence of a standard format for comparing and transferring microarray data has produced proposals to create a standard "language" to facilitate the sharing of data among scientists (49). Open structure databases such as Gene X, may emerge with the capability to accept Affymetrix GeneChip® data, cDNA data, and spotted oligonucleotide microarray data. When successfully integrated with clinical databases, these will be the forerunners of the dedicated systems to be used in clinical diagnostics.

A second major issue for the diagnostic biochip is the need to define an appropriate complement of genes and a large number of alleles (various forms of the same gene) that are appropriate for the decision-making requirements in a given clinical context. This creates pressures from two opposing sides: the need to have licensed access to a sufficiently large number of genes to achieve the diagnostic purpose and the competing need to limit the total number of genes to make the diagnostic commercially viable. To eliminate this downward pressure, patent licensing strategies may evolve to facilitate the exchange of rights between the competing entities who "own" the rights in the legion of genes utilized on the biochip platforms. "An equitable and streamlined mechanism for licensing genes and alleles would help to promote continued research, development and commercialization" (50). One proposal that has been advanced has been to "create a compulsory licensing scheme for DNA sequences used on gene chips, modeling it on the statutory licensing arrangement for music under the Copyright Act" (51).

A third and confounding issue pits genetic screening using genomic microarrays against the more targeted diagnostic biochips. The likely economics of multiple genetic tests makes it preferable to screen for many diseases and susceptibilities at once using genomic microarrays. Although screening may have benefits, it raises several questions:

(i) Are there adequate interventions or therapies revealed by the screen?
(ii) Are the privacy rights of patients adequately protected?
(iii) Are genetic counselors suitably prepared to guide the interpretation of the myriad risk-based assessments associated with a broad genomic screen?
(iv) Can informed consent be obtained given the broad range of possible outcomes of the screen?

Some of these questions may be adequately addressed with established procedures such as institutional review boards (IRBs), tiered informed consent, recontact or follow-up protocols, and anonymizing or keycoding of patient data. The potential for economies of scale in the screening of patients must be balanced by the patient's right to privacy protection. The potential for a full genomic scan to reveal not just what is currently known about an individual, but also, once entered into a database, that which may be known in the future, not just about an individual but also about their progeny, siblings, and lineage, must dampen enthusiasm for such a genomic screen. The solution is the development of specific gene chips, diagnostic and prognostic biochips that target specific disease states or genetic conditions. It is evident that the current genomic gene chips will be replaced by more dedicated Oncochips®, arthritis chips, and so on, which assure individual protection and meet the rigors required for IRB approval.

A fourth issue related to diagnostic biochip testing arises with the accumulation of detailed health information in databanks. "Such databanks are powerful and valuable tools to help understand and counteract disease processes" but raise questions of equity of access and parity of use in structuring inquiry for societal benefits (52). Of course, such issues must be balanced by the intellectual property rights of those who invest in the development of such databases.

## 14.8 Concluding Remarks

The diagnostic biochip will emerge as a highly quantitative, clinically focused molecular diagnostic tool for the modern diagnostics laboratory. However, the road to this destination is filled with regulatory requirements and societal concerns, to which the technological and practitioner communities must be cognizant and responsive. Furthermore, the technology, in its current format, is labor-intensive and quite expensive. There is need for alternative formats that focus on specific diagnostic or prognostic applications within targeted disease states. Better methods of manufacture and detection are being developed, including direct detection of hybridization by electric current, based on the fact that single-stranded DNA conducts electrons at a different rate than double-stranded DNA. In the near future, nucleic acid nanotechnology may prove very useful in DNA diagnostics. Research has led to advanced nucleic acid nanostructures and devices, semisynthetic DNA–protein conjugates, and efficient assembly of individual oligonucleotide-functionalized nanoparticles in two- and three-dimensional networks (53). Some of these assemblies have already proven useful for diagnostic purposes (54, 55) and the construction of nanometer-sized DNA-based nanowires has been achieved (56). The latter have been incorporated into a DNA-based two-dimensional network of functional scaffolds for protein arrays. It will only be a short time before similar DNA arrays are developed.

**Acknowledgments** This work was supported by the industrial consortium of the Center for Bioelectronics, Biosensors and Biochips and by the Commonwealth Technologies Research Fund (CTRF) Grant SE2002-02. This chapter is dedicated to the memory of Dr. Anil K. Deisingh who passed away on October 21st, 2005 and was at the time a member of the technical staff of the Analytical Chemistry and Microbiology Services Unit, Caribbean Industrial Research Institute, University of the West Indies Campus, St. Augustine, Republic of Trinidad & Tobago.

## References

1. Guiseppi-Elie, A. (2003) Biochip platforms for DNA diagnostics. *Business Briefing: PharmaTech*, London, World Markets Research Centre, p. 87.
2. Persidis, A. (1998) *Nature Biotechnology*, **16**, 981–983.
3. Proudnikov, D., Timofeev, E., and Mirzabekov, A. (1998) Immobilization of DNA in polyacrylamide gel for the manufacture of DNA and DNA–oligonucleotide microchips. *Analytical Biochemistry* **259**(1), 34–41. Llewellyn, B., Lebed, J., and Chik, V. (2002), www.promega.com/geneticidproc/ussymp11proc/content/llewellyn.pdf.

4. Seetharaman, S., Zivarts, M., Sundarsan, N., and Breaker, R.R. (2001) *Nature Biotechnology*, **19**, 336–341.
5. Wagner, P. and Kim, R. (May 2002) www.currentdrugdiscovery.com, pp. 23–28.
6. Gad-el-Hak, Mohamed, Ed. (2001) *The MEMS Handbook*. CRC Press, Boca Raton, FL. The MEMS Exchange (2002), http://www.mems-exchange.org/MEMS/what-is.html.
7. Christian, G.D. (1994) *Analytical Chemistry*, 5th ed., Chapter 13. Wiley, New York, 384–397.
8. Cunningham, A.J. (1998) *Introduction to Bioanalytical Sensors, Techniques in Analytical Chemistry*. John Wiley & Sons, Inc. New York.
9. A. Guiseppi-Elie and L. Lingerfelt "Impedimetric Detection of DNA Hybridization: Towards Near Patient DNA Diagnostics" In Immobilization of DNA on Chips I (2005); Christine Wittmann, Ed.; *Topics in Current Chemistry* Vol. 260, Springer Berlin, Heidelberg. pp 161–186.
10. Turner, A.P.F. and Newman, J.D. (1998) *Biosensors for Food Analysis*, A.O. Scott (Ed.). Royal Society of Chemistry, Cambridge, pp. 1–10.
11. Steemers, F.J. and Walt, D.R. (1999) Multi-analyte sensing: from site-selective deposition to randomly-ordered addressable optical sensors. *Microchimica Acta*, **131** (1–2): 99–105.
12. Brogan, K.L. and Walt, D.R. (2005) Optical fiber-based sensors: Application to chemical biology. *Current Opinion in Chem. Biology*, **9**, 494–500.
13. Hall, E.A.H. (1990) *Biosensors*. Open University Press, Milton Keynes,UK.
14. Blohm, D.H. and Guiseppi-Elie, A. (2001) *Current Opinion in Biotechnology*, **12**, 41–47.
15. Wang, J. (2000) *Nucleic Acids Research*, **28** (16), 3011–3016.
16. Schena, M. (2002) *Microarray Analysis*. Wiley-Liss, New York.
17. NCBI (2003) http://www.ncbi.nlm.nih.gov/About/primer/microarrays.html (accessed on May 5th, 2008).
18. Draghici, S. (2003) *Data Analysis Tools for DNA Microarrays*. Chapman & Hall/CRC, Boca Raton, FL.
19. Vahey, M., Nau, M.E., Barrick, S., Cooley, J.D., Sawyer, R., Sleeker, A.A., Vickerman, P., Bloor, S., Larder, B., Michael, N.L., and Wegner, S.A. (1999) *Journal of Clinical Microbiology*, **37**, 2533–2537.
20. Golub, T.R., Slonim, D.K., Tamayo,P., Huard, C., Gaasenbeek, M., Mesirov, J.P., Coller, H., Loh, M.L., Downing, J.R., Caliguiri, M.A., Bloomfield, C.D., and Lander, E.S. (1999) *Science*, **286**, 531–537.
21. Taylor, S., Smith, S., Windle, B., and Guiseppi-Elie, A. (2003) Impact of surface chemistry and blocking strategies in DNA microarrays. *Nucleic Acids Research*, **31**(16), e87.
22. Windle, B. and Guiseppi-Elie, A. (2003) Microarrays and gene expression profiling applied to drug research. In *Burger's Medicinal Chemistry*, 6th ed., D.J. Abraham (Ed.). Wiley, New York.
23. Willis, R.C. (2003), Monitoring microbes. *Modern Drug Discovery*, January, 16–21.
24. Bekal, S., Brousseau, R., Masson, L., Prefontaine, G., Fairbrother, J., and Harel, J. (2003) *Journal of Clinical Microbiology*, **41**, 2113–2125.
25. Volokhov, D., Rasooly, A., Chumakov, K., and Chizhikov, V. (2002) *Journal of Clinical Microbiology*, **40**, 4720–4728.
26. Hanna, G.J., Johnson, V.A., Kuritzkes, D.R., Richman, D.D., Martinex-Picado, J., Sutton, L., Hazelwood, J.D., and D'Aquila, R.T. (2000) *Journal of Clinical Microbiology*, **38**, 2715–2721.
27. Li, J., Chen, S., and Evans, D.H. (2001) *Journal of Clinical Microbiology*, **39**, 696–704.
28. Chizhikov, V., Wagner, M., Ivshina, A., Hoshino, Y., Kapikian, A.Z., and Chumakov, K. (2002) *Journal of Clinical Microbiology*, **40**, 2398–2407.
29. Wang, D., Coscoy, L., Zylberberg, M., Avila, P.C., Boushey, H.A., Ganem, D., and DeRisi, J.L. (2002) *Proceedings of the National Academy of Sciences of the USA*, **99**, 15687–15692.
30. Higgins, J.P.T., Shinghal, R., Gill, H., Reese, J.H., Terris, M., Cohen, R.J., Fero, M., Pollack, J.R., vandeRijn, M., and Brooks, J.D. (2003) *American Journal of Pathology*, **162**(3), 925–932.

31. Archer, K.J., Dumur, C.I., Scott Taylor, G., Chaplin, M.D., Guiseppi-Elie, A., Buck, G., Grant, G., Ferreira-Gonzalez, A., and Garrett, C. (2008) A disattenuated correlation estimate when variables are measured with error: Illustration estimating cross-platform correlations. *Statistics in Medicine*, **27**(7), 1026–1039.
32. Archer, K.J., Dumur, C.I., Taylor, G.S., Chaplin, M.D. Guiseppi-Elie, A., Grant, G., Ferreira-Gonzalez, A., and Garrett,,C. (2007) Application of a correlation correction factor in a microarray cross-platform reproducibility study. *BMC Bioinformatics*, **8**, 447.
33. Martinez, J.M., Aragon, A.D., Rodriguez, A.L., Weber, J.M., Timlin, J.A., Sinclair, M.B., Haaland, D.M., and Werner-Washburne, M. (2003) *Nucleic Acid Research*, **31** (4), e18.
34. van Brunt, J. (2003) *Signals Magazine* (online), 25th April 2003, http://www.signalsmag.com/signalsmag.nsf/0/6D4F7034D24D25988256D130011E957?Open.
35. SNP Consortium (2001) http://snp.cshl.org/about/2001_TSC_project.overview.shtml. (accessed on May 5th, 2008)
36. NIH New Advisory (October 2002) http://www.genome.gov/page.cfm?pageID=10005336. (accessed on May 5th, 2008)
37. McNeely, G. (May 5th, 2003) *Small Times On-Line IEEE* http://www.smalltimes.com/articles/article_display.cfm?Section=ARCHI&C=Profi&ARTICLE_ID=268701&p=109 (accessed on May 5th, 2008)
38. Gupta, R., Kim, J.P., Spiegel, J., and Ferguson, S.M (2004) Developing products for personalized medicine: NIH research tools policy applications. *Personalized Medicine* **1**(1): 115–124.
39. Ginsburg, G.S. and Angrist, M. (2006) The future may be closer than you think: a response from the Personalized Medicine Coalition to the Royal Society's report on personalized medicine. *Personalized Medicine*, **3**(2): 119–123.
40. Giles, F.J., Cortes, J.E., and Kantarjian, H.M. (2005) Targeting the kinase activity of the BCR-ABL fusion protein in patients with chronic myelogenous leukemia. *Current Molecular Medicine* **5**(7):615–623.
41. Druker, B.J.,Talpaz, M., Resta, D.J., Peng, B., Buchdunger, E., and Ford, J.M. (2001) *The New England Journal of Medicine*, **344**(14), 1031–1037.
42. Case-Green, S.C., Mir, K.U, Pritchard, C.E., Southernn, E.M. (1998) Analysing genetic information with DNA arrays. *Current Opinion in Biotechnology* **2**, 404–410.
43. Anthony Guiseppi-Elie, Scott Taylor, Louise Lingerfelt, Chris Nixon, Ryan Georgiana, Joy Kim, Stephanie Smith, Brad Mangrum and Nicholas Farell "Studies of the Interaction of Platinum Drugs with DNA Using Oligonucleotide Microarrays" *Macromolecular Symposia* (2006) 235(1), 115–120
44. CTRF Consortium of Virginia http://www.ctrf-cagenomics.vcu.edu/publiclyavaildata.htm. (accessed on May 5th, 2008)
45. Lingerfelt, L., Karlinsey, J., Landers, J., and Guiseppi-Elie, A. (2008) Impedimetric detection for DNA hybridization within microfluidic biochips. In *Microchip-Based Assay Systems Methods in Molecular Biology*, Pierre N. Floriano, Ed.; Royal Society of Chemistry. Humana Press, NJ. vol. 385, Chapter 8, pp 103–120.
46. Katz, E. and Willner, I. (2003) Probing biomolecular interactions at conductive and semiconductive surfaces by impedance spectroscopy: Routes to impedimetric immunosensors, DNA-Sensors, and enzyme biosensors. *Electroanalysis*, **15**, 913–947.
47. Hang Tin, C. and Guiseppi-Elie, A. (2004) Frequency dependent and surface characterization of DNA immobilization and hybridization. *Biosensors and Bioelectronics* **19**, 1537–1548.
48. Guiseppi-Elie, A., Brahim, S., and Wilson, A. (2007) Biosensors based on electrically conducting polymers. In *Handbook of Conducting Polymers: Conjugated Polymer Processing and Applications*; 3rd ed., T. Skotheim and J.R. Reynolds (Eds.). Taylor and Francis, New York. Chapter 12, pp. 12:1–12:45.
49. Davenport, R.J. (2001) Microarrays: Data standards on the horizon. *Science*, **292**, 414–415.
50. Williams, E. (2002) *Gene Chips: Science and Policy Basics,* 2002 Virginia Biotechnology Summit and Governor's Conference on Technology Transfer and University Research, McLean, Virginia, October 14–16.

51. Johns, D.J.; Brettwisch, R., and Lebovitz, R. (1991) Patenting DNA: Letter to the editor. *Science*, **254** (5036), 1276.
52. Williams, E.D. (2001) The policy and ethics of DNA chip technologies. In N. Fujiki, M. Sudo, and D.R.J. Macer (Eds.), *Bioethics and the Impact of Human Genome Research in the 21st Century: Pharmacogenomics, DNA polymorphism and Medical Genetics Services*, 104–109, Eubios Ethics Institute, Tsukuba, Japan.
53. Wengel, J. (2004) *Analyst*, **2**, 277–280.
54. Niemeyer, C.M. (2002) The developments of semisynthetic DNA-protein conjugates. *Trends Biotechnol.*, **20**, 395–401.
55. Niemeyer, C.M. (2002) Nanotechnology: Tools for the biomolecular engineer. *Science,* **297** (5578), 62–63.
56. Yan, H., Park, S.H., Finkelstein, G., Reif, J.H., and La Bean, T.H. (2003) *Science*, **301**, 1882–1884.

# Chapter 15
# MagArray Biochips for Protein and DNA Detection with Magnetic Nanotags: Design, Experiment, and Signal-to-Noise Ratio

**Sebastian J. Osterfeld and Shan X. Wang**

**Abstract** MagArray™ chips contain arrays of magnetic sensors, which can be used to detect surface binding reactions of biological molecules that have been labeled with 10 to 100 nm sized magnetic particles. Although MagArray chips are in some ways similar to fluorescence-based DNA array chips, the use of magnetic labeling tags leads to many distinct advantages, such as better background rejection, no label bleaching, inexpensive chip readers, potentially higher sensitivity, ability to measure multiple binding reactions in homogeneous assays simultaneously and in real-time, and seamless integration with magnetic separation techniques. So far, the technology of MagArray chips has been successfully used to perform quantitative analytic bioassays of both protein and nucleic acid targets. The potential of this technology, especially for point-of-care testing (POCT) and portable molecular diagnostics, appears promising, and it is likely that this technology will see significant further performance gains in the near future.

## 15.1 Principle of Operation and Technological Benefits

The majority of magnetic-label based biochips, or MagArrays, share a common principle of operation. At the heart of the technology stands an elaborate magnetoresistive (MR) multilayer thin film, the resistance of which varies with the orientation and strength of the proximate magnetic field. This MR film is sputter-deposited onto a nonconducting substrate wafer and patterned into a sensor array by lithographic means. With a reasonable degree of circuit complexity, each sensor of such an array can be independently and simultaneously read out. The MR response is very fast (on the time scale of nanoseconds or less) and is a direct function of

S.J. Osterfeld and S.X Wang
Stanford University, Department of Materials Science and Engineering
e-mail: osterfeld@stanfordalumni.org
sxwang@stanford.edu

K. Dill et al. (eds.), *Microarrays: Preparation, Microfluidics, Detection Methods, and Biological Applications*,

the proximate magnetic field; this is not the case for inductive sensors such as pick-up coils, which respond only to changing magnetic fields. The MR thin film in our MagArray chip is a spin valve (SV) structure which at room temperature can achieve a magnetoresistance of $\Delta R/R_0 = 12\%$. A simple linear stripe of this SV thin film can be used as the actual sensing element on a MagArray chip, where its resistance would be monitored for tiny fluctuations induced by nearby magnetic nanoparticle labels.

In a very simplistic thought experiment, a miniature permanent magnet could be used to label a biological molecule of interest. If this molecule then attaches to the sensor, for example, due to a specific binding reaction, a small change in resistance could be registered. To a first-order approximation, the signal induced in an MR sensor by a properly oriented magnetic label would be approximately proportional to the inverse of the sensor-to-label separation cubed; that is, it would follow the strength of the dipole field from the miniature permanent magnet label. However, it should be noted that in practice there is an optimal sensor-to-label separation, as shown in Fig. 15.1, mainly due to the stray field from the sensor and the magnetic field line curvature which on approach will ultimately reduce the effective in-plane component to which the sensor responds [1].

Because of the finite observation volume shown in Fig. 15.1 properly designed MR sensors are ideal for detecting surface-bound labels (magnetic nanotags). Unbound magnetic labels, if they are adequately stable in suspension, are unlikely to be inside the observation volume in significant concentrations, and will then contribute very little background signal. The rejection of the background signal from excess labels is therefore very high, so high that excess labels do not need to be removed and homogeneous assays that omit the washing step can be performed.

MagArray chips are expected to have several additional benefits when compared to more traditional fluorescence-based biochips. Because the surface concentration of magnetic labels is directly translated into a linearly proportional electrical

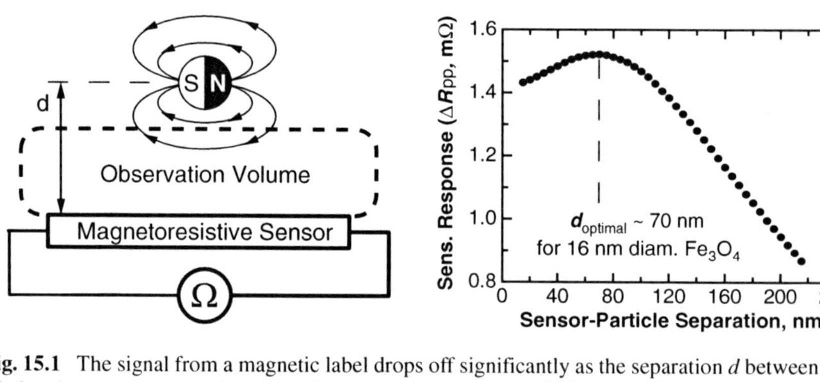


**Fig. 15.1** The signal from a magnetic label drops off significantly as the separation *d* between the label and sensor increases beyond a few hundred nanometers. This results in an observation volume that encompasses primarily surface-bound labels, whereas the background signal from unbound labels is relatively small. Shown at the right is the simulated MR signal dependence on separation for a single 16 nm Fe3O4 nanoparticle above a 200 nm wide sensor

signal [2], the required instrumentation (chip reader) is simple, inexpensive, and very suitable for miniaturization. Single magnetic label detection is possible, photobleaching is not an issue, and although biological molecules can autofluoresce, they have no intrinsic magnetic signal. Magnetically labeled molecules can also be collectively manipulated with magnetic fields, for example, to preconcentrate the analyte, especially in combination with microfluidics.

In reality, using miniature permanent magnets as labels would not be feasible, because the labels would tend to cluster and precipitate together with the molecules that they are attached to, and because their free orientation might result in a small net signal. For that reason, superparamagnetic labels are typically used which have no remnant magnetic moment. To generate a magnetic signal from a collection of superparamagnetic labels, a directed magnetic "tickling" field is applied which drives the sensor away from its equilibrium resistance. At the same time, the superparamagnetic labels become magnetized by the tickling field and generate a small dipolar magnetic field, opposite in direction to the tickling field. As a result, the magnetic labels slightly reduce the sensor's response to the tickling field. Using an electromagnet to generate an AC tickling field also opens up the possibility for modulation schemes such as narrowband lock-in detection, which make competitive signal-to-noise ratios possible.

## 15.2 Prior Work in the Field of Magnetic Biosensors

It appears that the use of magnetic nanoparticles as labels in immunoassays was first reported in 1997 by a group of German researchers who used a superconducting quantum interference device (SQUID) to detect the binding of antibodies [3]. Although their experiment was successful, it was performed in a magnetically shielded room, and the SQUID magnetometer required cooling with liquid helium.

At around the same time, giant magnetoresistive (GMR) stacks [4] and spin valves (SV), which had been introduced in hard disk drives as read head sensors in 1995 [5], were reaching sufficiently high performance levels at room temperature to become suitable for MagArray chips. Modern SV read heads are sensitive enough to detect magnetic data bits from a hard disk at temperatures up to about 100°C. Each magnetic bit typically contains a few hundred cobalt alloy magnetic nanoparticles, but the SV sensors in hard disk drives operate at very high frequencies (up to ~500 MHz) and benefit from the high signal modulation rate which is beyond the 1/f noise range of the detection process.

This advantage is absent in biological detection assays, where the magnetic fluctuations that need to be detected occur much more slowly. Slow changes permit longer sampling times and correspondingly a better resolution of the absolute signal level, but on the other hand, this also means that the requirements with respect to 1/f noise, interference, drift, and long-term measurement stability are much more stringent when GMR and SV sensors are used on biochips.

One of the earliest papers on biomagnetic detection assays using GMR sensors was published in 1998 by a research group at the U.S. Naval Research Laboratory (NRL). Their bead array counter (BARC) chip was able to detect a single 2.8 μm diameter polystyrene bead containing dispersed maghemite grains [6], albeit in a dry state. Their data showed that the signal-to-noise ratio improved significantly as the sensor width was decreased from 20 μm to 5 μm. Due to its potential for miniaturization, the BARC sensor was later proposed for use in a portable detector for biological warfare agents [7]. In this paper, the NRL group also demonstrated the application of a magnetic force to manipulate the magnetic beads and improve the assay outcome. In 2003 the same group, using a multisegment GMR sensor, measured a signal change resulting from biologically bound 2.8 μm beads in an aqueous solution. However, the binding event could not be recorded in real-time, apparently because the application of the tickling field that magnetizes the beads would also lead to clustering of the particles and hence obscure the natural binding process [8].

One of the first groups to publish real-time particle capture curves used a short single-segment SV sensor, and a magnetic gradient to concentrate the particles in the vicinity of the sensor. They also reported particle clustering problems with 400 nm high magnetic content particles, which were, however, resolved through the use of 2 μm lower magnetic content microspheres [9].

A direct performance comparison of magnetic biochips with a fluorescent detection method for DNA hybridization was carried out by researchers in Germany, who defined the relative sensitivity of each assay as the signal ratio between positive probes and negative probes, which generate only the signal from nonspecific adsorption. The conclusion of this group was that the performance of the magnetic detection method was superior to the fluorescent method, primarily because at low concentrations the fluorescent method had a higher background signal level [10], which may stem from autofluorescence of the negative probes.

Our own group at Stanford was one of the first to focus on truly nanometer-sized magnetic labels. Unlike other groups which mostly used particles that ranged from 200 nm to 3 μm, our aim had been to develop a biochip based on high-moment monodisperse 11 nm diameter Co nanoparticles [11] and 16 nm diameter $Fe_3O_4$ nanoparticles [2]. To complement our approach of using nanoparticles, we have also investigated the feasibility of using very thin passivation layers [12]. Similarly, we have focused on using SV sensors with line widths below two micrometers. In an earlier implementation, such sensors with widths of 0.2 μm have already been shown to detect a few tens of these 16 nm particles in a dry environment before and after capture experiment [13], and later in this chapter we present a real-time assay based on 40 nm composite particles and 1.5 μm wide sensors.

GMR and SV sensors remained the dominant read head technology in hard disk drives until roughly 2005, when magnetic tunnel junctions (MTJ) began replacing GMR and SV sensors in hard disk drives. However, whether MTJs will also become the primary sensors in magnetic biochips remains to be seen; after all, the different requirements in biological applications such as the need for low drift and data collection over a large binding surface may well favor GMR and SV sensors over MTJs, which are better suited for highly localized measurements. Thus far, MTJs were used in a real-time detection of 2.8 μm beads in an aqueous solution, albeit without biological binding events [14].

Although the field of magnetic biochips is still in its early stage, it has also attracted attention from several companies. The NRL lab joined forces with NVE Corporation and more recently Seahawk Biosystems Corporation to advance the development of the BARC sensor. The group who demonstrated real-time particle detection with MTJs collaborates with Micro Magnetics Inc., and Philips Research in the Netherlands is working on a magnetic biochip for its line of point-of-care diagnostic medical devices.

Some representative research groups that are actively developing magnetic biochips are listed in Table 15.1. This list is certainly far from complete, but it can serve as a starting point for a further literature search. Some of the basic parameters of each group's platform are also included, but it should be noted that each group typically evaluates several different designs at any given time; for example, our group is also investigating single-segment submicron SV sensors and highly monodisperse 16 nm $Fe_3O_4$ nanoparticles which are not yet commercially available.

The great variety of magnetic biochip designs under investigation is a reflection of the novelty of the field, in which definite design guidelines have not yet been established.

**Table 15.1** Some representative research groups developing magnetic biochips

| Institution and site | Principal investigators | Magnetic nanotags | Sensor technology | Sensor passivation |
|---|---|---|---|---|
| NRL, Washington NVE, Eden Prairie | Whitman, LJ Tondra, M | Dynal M280 2.8 μm | GMR, Multi-Segment 1.6 x 8000 μm, 42 kΩ | $Si_3N_4$ 1000 nm |
| IST, Lisbon, Portugal | Ferreira, HA Freitas, PP | Nanomag-D 250 nm | SV, Single-Segment 2.5 x 100 μm, 1 kΩ | $Al_2O_3/SiO_2$ 100/200 nm |
| University of Bielefeld, Germany | Reiss, G Brueckl, H | Bangs CM01N 350 nm | GMR, Spiral 1 x 1800 μm, 12 kΩ | $SiO_2$ 100 nm |
| Stanford University, Stanford | Wang, SX Pourmand, N | Miltenyi MACS 40 nm | Multi-Segment SV 1.5 x 2800 μm, 45 kΩ | $SiO_2/Si_3N_4/SiO_2$ 20/20/20 nm |
| Brown University, Providence | Xiao, G | Dynal M280 2.8 μm | MTJ, Elliptic Patch 2 x 6 μm, 142 Ω | $Au/SiO_2$ 200/200 nm |
| Philips Research, Netherlands | Prins, M | Ademtech 300 nm | GMR, Gradiometer 3 x 100 μm, 250 Ω est. | Unknown >1000 nm est. |

GMR = Giant Magnetoresistive Stack
SV = Spin Valve
MTJ = Magnetic Tunnel Junction

**Note:** Basic design parameters are listed that can be used to estimate the theoretical performance limit of each platform. However, in practice the performance of a magnetic biochip may depend on additional factors such as the choice of binding chemistry, modulation, and signal processing.

## 15.3 Sensors, Passivation, and Magnetic Nanotags

The three types of magnetoresistive elements commonly used in magnetic biochips are giant magnetoresistive stacks, spin valves, and magnetic tunnel junctions, all of which are examples of spintronic (magnetoelectronic) sensors, meaning that spin interactions are used to modulate the electronic properties of the structure [15].

As electrons travel through a magnetized material, they tend to align their spin with the magnetization of the material surrounding them. If such spin-polarized electrons cross an interface and enter a differently magnetized region, their spins will need to realign, which causes an increase in the apparent electrical resistance of the overall structure as illustrated in Fig. 15.2, where electrons emerge from a magnetic reference layer with a fixed (pinned) magnetization and cross into a soft magnetic layer with variable magnetization. The magnetization of the free layer closely follows the direction and magnitude of the surrounding magnetic tickling field $H$, whereas the magnetization of the reference layer is largely independent of $H$. These two layers are typically separated by a nonmagnetic metal or insulator layer. A more detailed explanation of the physical principles of spin valves and other magnetoresistive sensors is beyond the scope of this text, but due to the leading developments in the magnetic storage technology sector there are many good books available on the subject [5, 16].

Spin valves and GMR stacks use a noble metal layer to separate the magnetic layers and thus have a low resistivity, which makes the design of sensors with small electrical cross-sections and in-plane measurements possible. For this reason, spin valves and GMR stacks are preferred for narrow linear sensors that are folded back and forth to cover a large sensing area. A single GMR or SV sensor can thus easily cover an area of about 100 μm in diameter, a size that is comparable to a typical DNA array spot formed with manual or robotic spotting.

In contrast to spin valves and GMR stacks, magnetic tunnel junctions (MTJs) utilize spin-dependent tunneling across an insulating barrier and accordingly have a much higher resistivity. As a result, they need to be patterned into sensor elements

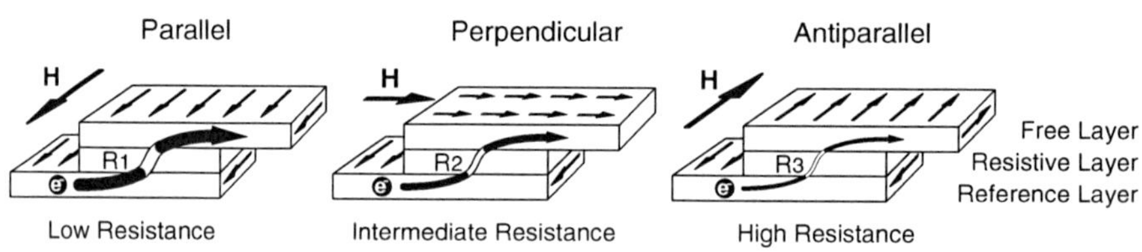


**Fig. 15.2** In general, spin valves and similar devices are based on spin-dependent electron transport, where the resistance varies in response to the degree of magnetic alignment of the magnetic layers that sandwich a nonmagnetic layer. Whereas the magnetization of the reference layer is pinned and largely independent of any applied fields, the free layer will easily rotate and align itself with the applied tickling field $H$. The actual current path depends on the electrical contact points and relative resistance of each layer

with much larger electrical cross-sections. In addition, the measuring current needs to be run perpendicular to the plane of the film instead of in the plane of the film.

Creating an MTJ sensor that covers a large area is challenging because a single small pinhole defect in the thin but highly resistive tunneling layer can short out the entire MTJ. In contrast, spin valves and GMR stacks are much more tolerant to pinhole defects, the probability of which is a function of the sensor area. Furthermore, the necessity of a top electrical lead on an MTJ increases the distance between the magnetic nanoparticles and the free sensing layer. Consequently, a careful design of top electrode shape is required to detect 10 nm sized particles [17].

For our MagArrays it is desirable to select an MR sensor that is able to cover a large measuring area, because the resulting larger number of sampled sites will reduce the stochastic noise in low-concentration measurements where binding events are widely scattered and sporadic. Furthermore, it is desirable to select an MR sensor with a sufficient saturation field (also called stiffness in magnetic recording), because higher tickling fields can be used which elicit a stronger signature from the superparamagnetic labels. Lastly, an MR sensor with a strongly pinned reference layer has better baseline stability, and is thus more suited for quantitative analytic assays that typically run tens of minutes. Considering these requirements for large area coverage, pinhole tolerance, saturation avoidance, and strong pinning, we have chosen a spin valve with synthetic antiferromagnetic pinning for our MagArray chip.

## 15.3.1 Sensor Shape

The width of a magnetoresistive sensor appears to be one of the main performance-determining parameters. The narrower the magnetoresistive sensor is, the more sensitive it is to magnetic labels, particularly when very small magnetic labels are used [2, 18]. The benefit of reducing sensor width may, however, be limited, because theory predicts that the edge of the spin valve creates an inverse (negative) signal and thus reduces the (positive) signal from the diminishing area of the spin valve [13].

When designing an MR sensor for magnetic biochips, consideration should also be given to the magnetic domain formation and the final resistance of the sensor. The layout of the sensor can influence the way in which the free layer responds to a changing tickling field. Edges create a local demagnetizing field which favors alignment of the magnetization parallel to any edges that the sensor has. This can be used to instill a bias into the magnetization of the free layer, which, for example, would tend to align with the long axis of a linear segment. Similarly, curved or bent segments may hinder a coherent domain rotation and reduce the linearity and reproducibility of domain rotations, which might lead to increased measurement noise.

The other consideration is the sensor's electrical resistance resulting from the design. The electrical resistance should fall into a range that is easily measured at low voltages, because higher measuring voltages place more stress on the very thin passivation that electrically isolates the sensor from the aqueous reagents. Yet, too low a sensor resistance would necessitate a more complex four-wire measurement setup.

### 15.3.2 Sensor Passivation

MagArray chips face a particular challenge with regard to the sensor passivation which electrically isolates the sensors from the aqueous analyte environment. This passivation needs to be durable enough to minimize leakage currents and prevent sensor corrosion, but on the other hand, this passivation needs to be as thin as possible to maximize the sensitivity of the finished chip (see Fig. 15.1).

For our MagArray chips, we have found it useful to adopt a 50 nm thin trilayer oxide–nitride–oxide (ONO) passivation consisting of a single layer of $Si_3N_4$ sandwiched between two layers of $SiO_2$. This type of passivation was reported to be among the most durable passivations for biochips with sensors exposed to aqueous environments, presumably because the nitride layer provides a good diffusion barrier, and because the use of two $SiO_2$ layers results in a symmetric passivation film with lower residual stresses compared to a two-layer film [19]. Most SV sensors experience no corrosion damage, as is evident from a reproducible signal baseline during repeat exposure to water or buffer, and as is further evident from stable long-term sensor resistances.

Although corrosion is not an issue, immersed (wet) sensors can still experience some degree of AC current leakage from parasitic capacitance across the 50 nm passivation. This parasitic capacitance is usually small enough to not result in any significant cross-talk between neighboring sensors, but it can still result in a "water signal", which is a small reversible signal baseline shift that occurs whenever the sensors are transitioned between wet and dry states. This reversible signal shift occurs when an open-well MagArray chip is used, where reagent changes are often marked by a brief dry period as shown in Fig. 15.5. In contrast, a microfluidic MagArray chip would not require any dry periods for complete reagent changes and should not exhibit any baseline shifts during reagent cycling.

### 15.3.3 Magnetic Nanotags

As mentioned earlier, superparamagnetic particles are preferred to avoid clustering and precipitation. A small diameter enhances their rate of diffusion and helps limit the observation volume to surface-bound nanotags only. Nanotags should also have as high a magnetic moment as possible, however, their most important performance-determining characteristic is probably their surface chemistry and stability in suspension. Particle precipitation must not occur at any rate, because it would lead to a continuous rise in the signal baseline that could obscure the equilibration of the binding reactions, especially at low concentrations. Similarly, it is important that the surface chemistry of the particles leads to highly selective and strong binding reactions so that the molecules of interest are labeled exclusively and irreversibly.

The MACS particles (~40 nm diameter) from magnetic cell separation technology provider Miltenyi work acceptably in this regard. Their magnetic properties are far from ideal, but they do appear to be very stable in suspension and highly specific in binding. We have also used custom-made 16 nm diameter $Fe_3O_4$ nanotags in some evaluation experiments.

## 15.4 Prototype MagArray Chip for Assay Development

In the development of the MagArray biochip, an essential challenge lies in optimizing the chip and the assay chemistry concurrently. To aid this assay optimization, we chose to develop the microfluidics separately and carry out experiments with our open-well MagArray chip as shown in Fig. 15.3. A multichannel microfluidic version of this chip is planned for 2007, but as the following pages show, successful assays can be performed without the added complexity of microfluidics.

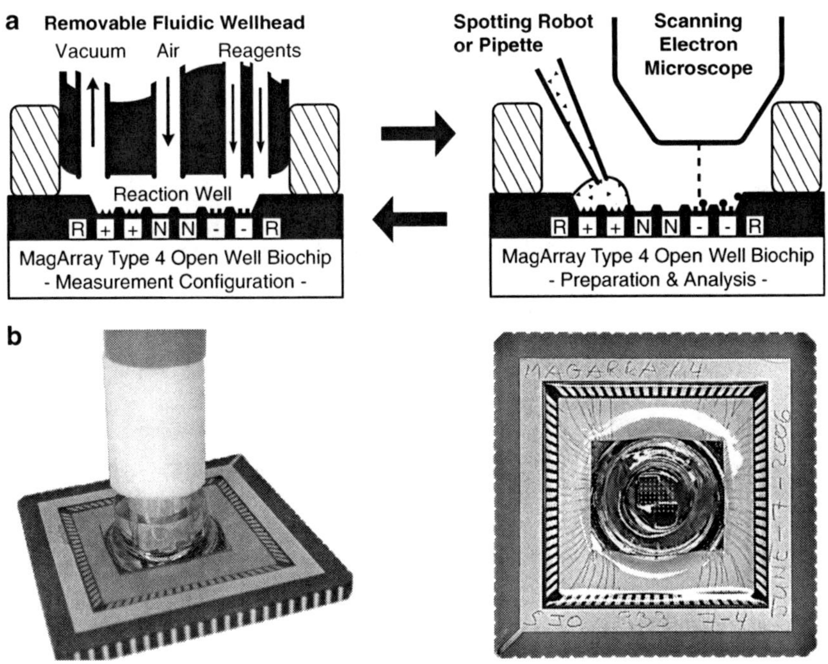


**Fig. 15.3** For rapid assay development, a prototype MagArray biochip with an open reaction well is used. It contains an 8 × 8 array of 64 individually addressable spin valve sensors. The open well makes it possible to interrupt a measurement and access the sensors directly. This simplifies biofunctionalization and inspections under a scanning electron microscope, where the adsorption density and adsorption specificity of the nanoparticles can be verified

The chip contains 64 sensors, each having 32 linear spin valve segments of 1.5 μm width to cover a total measuring area of approximately 90 × 90 μm per sensor. The sensors are electrically insulated with the mentioned 50 nm ONO passivation. After groups of sensors on the chip have been differentially biofunctionalized as shown in Fig. 15.4, common assay reagents, for example, the analyte, washing buffers, and finally magnetic nanotag solutions, are delivered to the entire chip with a fluidic well head as shown in Fig. 15.3. Real-time data are recorded during the relevant steps of the assay as shown in Fig. 15.5. To subsequently remove an assay reagent, clean air is forcefully drawn onto the center of the chip and towards an activated vacuum line, which carries any droplets from the well with it.

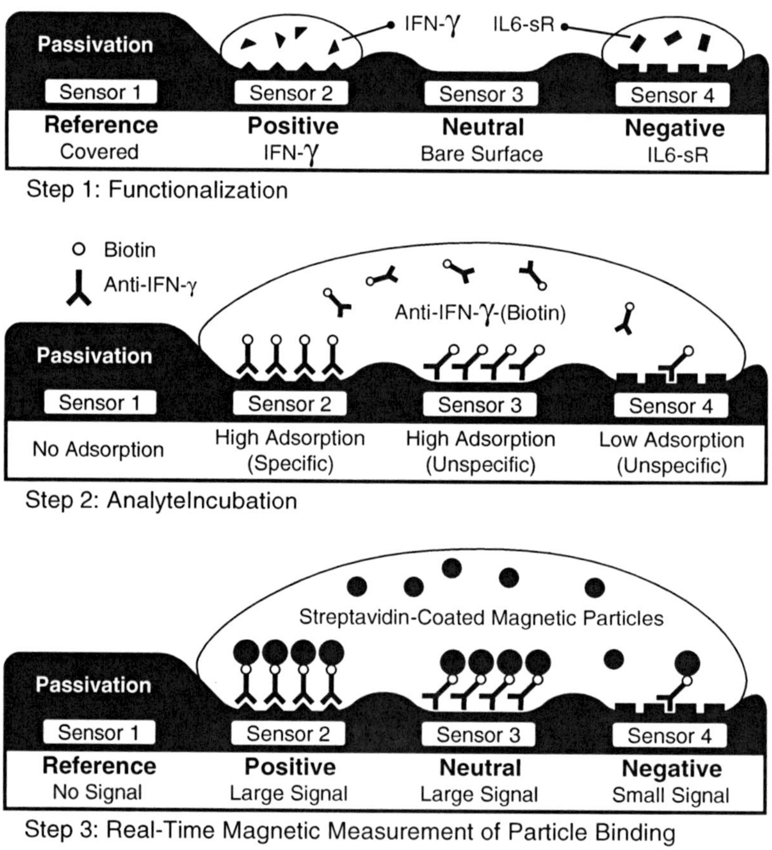


**Fig. 15.4** Schematically shown is a direct proteomics assay that quantifies the concentration of biotinylated anti-IFN-γ in PBS buffer. Quantification occurs during step 3, when magnetic nanoparticle labels are captured by the analyte molecule. This nanoparticle capture is recorded by the spin valve sensors in real-time as shown in Fig. 15.5

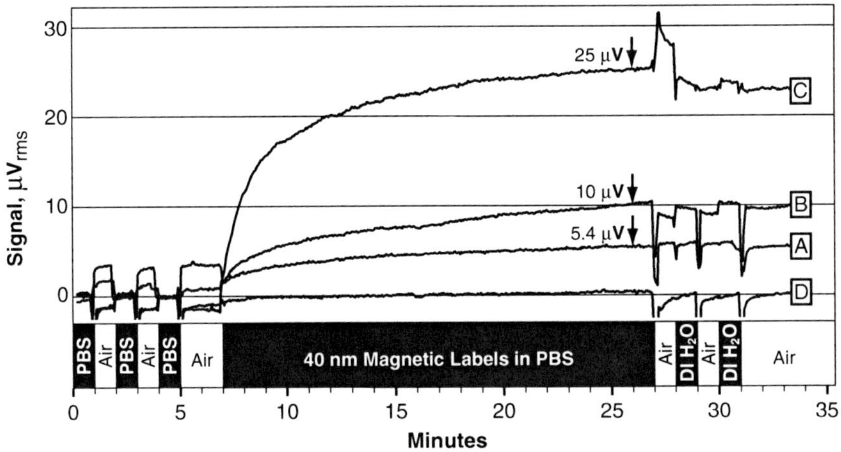


**Fig. 15.5** Real-time magnetic measurements of nanoparticle binding on open-well MagArray chips. The degree of nanoparticle binding on positive sensors is a function of analyte concentration (A = 10 ng/mL; B = 100 ng/mL; C = 1000 ng/mL of biotinylated anti-IFN-γ). No significant nanoparticle binding is observed on a negative sensor D, which was exposed to 1000 ng/mL of biotinylated anti-IFN-γ and located on the same chip as sensor C

## 15.5 Assays with Magnetic Nanotags

High-sensitivity DNA and proteomics assays can both be easily run on MagArrays, as long as the chosen biochemistry can bind and immobilize the nanoparticle labels. In this chapter we demonstrate an early simplified proteomics assay for illustrative purposes. Simplifications include that (a) the analyte is delivered in PBS buffer without a serum component, (b) only one analyte is present, and (c) the analyte is an antibody that permits direct labeling with nanoparticles. However, these simplifications do not enhance the apparent performance of the MagArray, which in our experience has a slightly better sensitivity in sandwich assays where the dilute analyte is an antigen, which is typically smaller and faster diffusing than an antibody.

As illustrated in Fig. 15.4, portions of the sensor array are functionalized in one of four different ways. Reference sensors like Sensor 1 are covered with a physical barrier such as an extra thick passivation that positively prevents the magnetic nanoparticles from entering the reference sensors' observation volume. Such reference sensors will not respond to nanoparticle labels and provide an absolute reference level. Positive sensors like Sensor 2 are saturated with IFN-γ, 100 μg/mL in 1x PBS buffer. This can be done manually by positioning a 0.5 μL droplet on a portion of the sensor array for 30 minutes at 4°C during which the IFN-γ is physically adsorbed onto the sensor passivation. Positive sensors are expected to specifically bind the analyte of interest. Neutral sensors like Sensor 3 are not saturated with a protein during the functionalization step and are thus particularly susceptible to later nonspecific adsorption of the analyte, the nanoparticle labels, or any other passing molecules. Negative sensors like Sensor 4 are saturated with IL6-sR,

100 μg/mL in 1x PBS buffer. Such sensors are expected to be well protected from nonspecific adsorption, particularly with regard to the analyte of interest.

Several chips are functionalized this way and then rinsed with a 1% BSA in 1x PBS buffer solution. Each chip well is then filled with 100 μL of analyte at a chosen concentration, that is, 10/100/1000 ng/mL of biotinylated anti-IFN-γ in 1x PBS buffer. The analyte is incubated in the entire well of the chips for 1.5 hours at 30°C. The chips are then rinsed with 0.1% BSA in TPBS and transferred to the measuring station for subsequent analyte quantification.

For quantification, on each chip at least two differential pairs of sensors are measured. One differential pair, consisting of a positive sensor and a reference sensor, is set up to record the nanoparticle capture on the positive sensors, thereby revealing the amount of specifically adsorbed analyte. Another differential pair, consisting of a negative sensor and a reference sensor, is set up to record the nanoparticle capture on negative sensors, thereby revealing the amount of cross-reactivity between the analyte and a nonmatching functionalization.

Additional differential pairs can be set up as desired, for example, consisting of a neutral sensor and a reference sensor, which would reveal the amount of nonspecific adsorption of the analyte to the bare substrate. Such experiments have shown that the analyte itself will in most cases significantly adsorb onto surfaces that have not already been saturated, or blocked, with another protein. Similarly, such experiments seem to indicate that the MACS nanoparticles that we use in this experiment are highly stable in suspension and will neither precipitate nor adsorb nonspecifically. However, in our experience several other nanoparticles do tend to precipitate, so numerous control assays had to be performed to find a magnetic label of adequately high performance.

Once the appropriate differential sensor pairs have been set up, data recording for all pairs of interest is initiated, and the chip is primed via the fluidic well head several times with 1x PBS buffer as shown in the timeline of Fig. 15.5. This is to ensure that simple contact with the buffer solution does not induce any irreversible baseline drift, which could occur, for example, from sensor corrosion. During the wet/dry transitions of these priming rinses, the baseline shifts back and forth due to parasitic capacitance in the thin passivation as explained earlier, but the shift is reversible and reproducible.

After the chip has been primed a few times with PBS buffer and the stability of the baseline has been confirmed, at time $t$ = 7 minutes, 100 μL of undiluted nanoparticle solution (Miltenyi MACS, 130-048-102, 40 nm diameter, in PBS) are delivered to the reaction well of the chip and incubated for 20 minutes at room temperature. As the streptavidin-coated nanoparticles are captured by the biotin ligand of the analyte, they are immobilized within the observation volume of the sensors, and the signal level rises accordingly as shown in Fig. 15.5. At the end of the 20 minute nanoparticle incubation period, excess particles are flushed from the reaction well with air, and the well is twice filled with deionized water for 1 minute each. This rinsing step with DI water removes any excess salts and facilitates a later confirmation of the nanoparticle coverage in the scanning electron microscope.

The final signal can now be assessed in two ways. For the signal level corresponding to a homogeneous assay without a final rinse, the signal level at the end of the nanoparticle incubation period ($t$ = 26 minute) is recorded. This method of

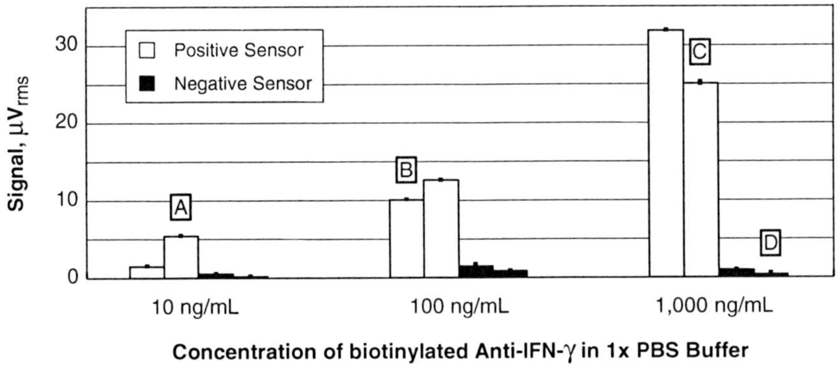


**Fig. 15.6** The signal levels from three different open-well MagArray chips after 19 minutes of magnetic nanoparticle incubation ($t = 26$ in Fig. 15.5). Each chip was exposed to a different concentration of biotinylated anti-IFN-γ in 1x PBS buffer. On each chip, data from two positively and two negatively functionalized sensors were recorded. The real-time data curves from sensors A, B, C, and D are superimposed in Fig. 15.5

quantification is generally more reproducible, and sample results are summarized in Fig. 15.6. However, when one desires to relate the signal level to SEM observations of nanoparticle coverage, then the signal levels after the final rinses with DI water are more relevant ($t = 33$ minutes). This "dry state" method of quantification may need to be adjusted for the baseline shift that occurs during wet/dry transitions and is also less reproducible, presumably because the final rinse with DI water can denature a part of the binding chemistry, which reduces the apparent nanoparticle coverage.

Note that a live signal trace from a microfluidic MagArray is expected to reduce or eliminate the wet/dry transition signal jumps, but such a chip may be less accessible for control quantifications of particle coverage in the SEM.

## 15.6 Sensitivity and Signal-to-Noise Ratio

In Fig. 15.6, the small error bars on the data represent the electronic ripple noise of the sensor signal, which is about 400 nV$_{rms}$. This type of noise is quite small compared to the measured signal levels, which means that the electronic signal-to-noise ratio (SNR) of a MagArray chip can be quite good, for example, 22 dB at 10 ng/mL and 36 dB at 1,000 ng/mL (sensors A and C in Fig. 15.6).

A better figure of merit is the biochemical SNR based on the ratio of positive to negative signals, which ensures that the nonspecific adsorption and cross-reactivity are also taken into account. In Fig. 15.6 the average negative signal is around 1 μV, which indicates a relatively low amount of nonspecific adsorption and a biochemical SNR of approximately 15 dB and 28 dB for sensors A and C in Fig. 15.6, respectively.

It is notable that the greatest source of measurement uncertainty results from the difference of the positive signal levels, which on average is 4.4 μV in this experiment. On our MagArray there are probably two main sources for this relatively large amount of variability, the first of which is the open-well design. The critical first biofunctionalization is applied to the sensors with small droplets that cover multiple sensors, and the resulting protein deposition may be uneven due to evaporation and convection inside the droplet; this is akin to fluorescent arrays where the dyes often form rings, rather than evenly filled circles. In addition, reagent flow through the well is turbulent rather than laminar and may not be perfectly uniform across various sensors. This source of variability can be eliminated in a microfluidic MagArray if air-free laminar flow is used to apply an in situ biofunctionalization.

The second source of variability is likely the surface chemistry itself. The proteins can degrade quickly, and developing an assay protocol that gives consistent results requires a great amount of time and expertise in biochemistry.

The SNR and reproducibility of the data will set the lower bound for the sensitivity of a MagArray. Physically the sensitivity is limited to the smallest amount of magnetic matter that can still be detected. With the low magnetic content MACS nanoparticles, approximately 12 nanoparticles per square micrometer are needed to achieve a 6 dB (2 μV) signal on the MagArray with 1.5 μm wide SV sensors, but this estimate is very approximate due to the clustering and polydisperse size distribution of these particles as seen in Fig. 15.7.

The maximum achievable nanoparticle coverage density, which to a first approximation seems to exhibit a Langmuir isotherm adsorption behavior, will determine the upper limit of detection and the dynamic range of a MagArray. For the MACS

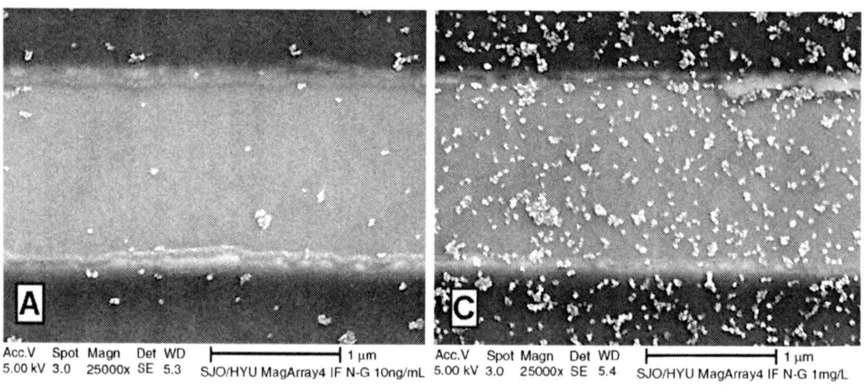


**Fig. 15.7** Scanning electron microscope (SEM) images of spin valve sensors after completion of the assay. Both sensors were positively functionalized with IFN-γ. Sensors A (left) and C (right) were subsequently exposed to 10 and 1000 ng/mL of biotinylated anti-IFN-γ in 1x PBS, respectively. Upon exposure to 40 nm streptavidin-coated magnetic nanoparticles sensors A and C produced signals of 5 and 25 microvolts, respectively (Figs. 15.5 and 15.6). Rinses with DI water ($t = 28$, 30 in Fig. 15.5) removed any salts before SEM imaging

particles, the coverage shown on the right side in Fig. 15.7 is roughly 170 nanoparticles per square micrometer, well below the achievable coverage limit.

## 15.7 DNA Detection and Future Outlook

The MagArray chips described above have also been applied to DNA hybridization detection. We have experimentally observed a molar sensitivity of ~10 pM with 1.5 × 2800 μm sensors, and an absolute sensitivity of a few hundred oligos per 0.2 × 2 μm sensor. Due to the space constraints here the DNA assay results appear elsewhere [20, 21].

Ultimately, the overall performance of a MagArray will largely depend on the stability of the assay chemistry and system-level integration. For example, protein sensitivities of approximately 1 pg/mL appear to be possible with a few improvements, for example, the use of microfluidics, proteomic sandwich assays [22], similar size but higher magnetic moment nanoparticle labels, narrower sensors, and sensors with a higher magnetoresistance ratio.

**Acknowledgments** The authors would like to thank the following collaborators: N. Pourmand and H. Yu for their expertise in biochemistry, S. Sun for expertise in nanoparticle synthesis, and S. Han for electronic instrumentation. Funding was generously provided by the U.S. Department of Defense, the National Institutes of Health, and the Whitaker Foundation for Biomedical Engineering.

## References

1. Li, G.X., Sun, S.H., and Wang, S.X. (2006) Spin valve biosensors: Signal dependence on nanoparticle position. *J. Appl. Phys*. 99, 08P107.
2. Li, G.X., Wang, S.X., and Sun, S.H. (2004) Model and experiment of detecting multiple magnetic nanoparticles as biomolecular labels by spin valve sensors. *IEEE Trans. Magn*. 40, 3000–3002.
3. Kotitz, R., Matz, H., Trahms, L., Koch, H., Weitschies, W., Rheinlander, T., Semmler, W., and Bunte, T. (1997) SQUID based remanence measurements for immunoassays. *IEEE Trans. Appl. Supercond*. 7, 3678–3681.
4. Baibich, M.N., Broto, J.M., Fert, A., Vandau, F.N., Petroff, F., Eitenne, P., Creuzet, G., Friederich, A., and Chazelas, J. (1988) Giant magnetoresistance of (001)Fe/(001)Cr magnetic superlattices. *Phys. Rev. Lett*. 61, 2472–2475.
5. Wang, S.X. and Taratorin, A.M. (1999) *Magnetic Information Storage Technology*. Academic Press, San Diego.
6. Baselt, D.R., Lee, G.U., Natesan, M., Metzger, S.W., Sheehan, P.E., and Colton, R.J. (1998) A biosensor based on magnetoresistance technology. *Biosen. Bioelectron*. 13, 731–739.
7. Edelstein, R.L., Tamanaha, C.R., Sheehan, P.E., Miller, M.M., Baselt, D.R., Whitman, L.J., and Colton, R.J. (2000) The BARC biosensor applied to the detection of biological war-fare agents. *Biosen. Bioelectron*. 14, 805–813.
8. Rife, J.C., Miller, M.M., Sheehan, P.E., Tamanaha, C.R., Tondra, M., and Whitman, L.J. (2003) Design and performance of GMR sensors for the detection of magnetic microbeads in biosensors. *Sens. and Act. A* 107, 209–218.

9. Graham, D.L., Ferreira, H., Bernardo, J., Freitas, P.P., and Cabral, J.M.S. (2002) Single magnetic microsphere placement and detection on-chip using current line designs with integrated spin valve sensors: biotechnological applications. *J. Appl. Phys.* 91, 7786–7788.
10. Schotter, J., Kamp, P.B., Becker, A., Puhler, A., Reiss, G., and Bruckl, H (2004) Comparison of a prototype magnetoresistive biosensor to standard fluorescent DNA detection. *Biosens. Bioelect.* 19, 1149–1156.
11. Li, G.X., Joshi, V., White, R.L., Wang, S.X., Kemp, J.T., Webb, C., Davis, R.W. (2003) Detection of single micron-sized magnetic bead and magnetic nanoparticles using spin valve sensors for biological applications. *J. Appl. Phys.* 93, 7557–7559.
12. Joshi, V., Li, G., Wang, S.X., and Sun, S. (2004) Biochemical stability of components for use in a DNA detection system. *IEEE Trans. Magn.* 40, 3012–3014.
13. Li, G.X., Sun, S.H. Wilson, R.J., White, R.L., Pourmand, N., and Wang, S.X. (2006) Spin valve sensors for ultrasensitive detection of superparamagnetic nanoparticles for biological applications. *Sens. Actuat. A* 126, 98–106.
14. Shen, W.F., Liu, X.Y., Mazumdar, D., and Xiao, G. (2005) In situ detection of single micron-sized magnetic beads using magnetic tunnel junction sensors. *Appl. Phys. Lett.* 86, 253901.
15. Prinz, G.A. (1999) Magnetoelectronics applications. *J. Magn. Magn. Matl.* 200, 57–68.
16. O'Handley, R.C. (2000) *Modern Magnetic Materials: Principles and Applications*. John Wiley, New York.
17. Wang, S.X., Bae, S.Y., Li, G., Sun, S.H., White, R.L., Kemp, J.T., and Webb, C.D. (2005) Towards a magnetic microarray for sensitive diagnostics. *J. Mag. Mag. Mat.* 293, 731–736.
18. Li, G.X. and Wang, S.X. (2003) Analytical and micromagnetic modeling for detection of a single magnetic microbead or nanobead by spin valve sensors. *IEEE Trans. Magn.* 39, 3313–3315.
19. Schmitt, G., Schultze, J.W., Fassbender, F., Buss, G., Luth, H., and Schoning, M.J. (1999) Passivation and corrosion of microelectrode arrays. *Elec. Acta* 44, 3865–3883.
20. Xu, L., Yu, H., Akhras, M.S., Han, S.J., Osterfeld S.J., White, R.L., Pourmand, N., and Wang, S.X. (2008) Giant magnetoresistive biochip for DNA detection and HPV genotyping. Biosens Bioelectorn 24:99–103.
21. Han, S.J., Yu, H., Xu, L., Wilson, R.J., Pourmand, N., and Wang, S.X. (2006) CMOS integrated DNA microarray based on GMR sensors. *Proceedings of International Electron Device Meeting (IEDM)*, Paper #28.2, San Francisco, CA, December 11–13.
22. Osterfeld, S.J., Yu, H., Gaster, R.S., Caramuta, S., Xu, L., Han, S.J., Hall, D.A., Wilson, R.J., Sun, S.H., White, R.L., David, R.W., Pourmand, N., and Wang, S.X. (2008) Multiplex Protein Assays Based on Real-Time Magnetic Nanotag Sensing. Submitted to PNAS.

# Chapter 16
# Bar Coding Platforms for Nucleic Acid and Protein Detection

**Uwe R. Müller**

**Abstract** A variety of novel bar coding systems has been developed as multiplex testing platforms for applications in biological, chemical, and biomedical diagnostics. Instead of identifying a target through capture at a specific locus on an array, target analytes are captured by a bar coded tag, which then uniquely identifies the target, akin to putting a UPC bar code on a product. This requires an appropriate surface functionalization to ensure that the correct target is captured with high efficiency. Moreover the tag, or bar code, has to be readable with minimal error and at high speed, typically by flow analysis. For quantitative assays the target may be labeled separately, or the tag may also serve as the label. A great variety of materials and physicochemical principles has been exploited to generate this plethora of novel bar coding platforms. Their advantages compared to microarray-based assay platforms include in-solution binding kinetics, flexibility in assay design, compatibility with microplate-based assay automation, high sample throughput, and with some assay formats, increased sensitivity.

## 16.1 Introduction

The debut of microarray technology in the late 1980s was driven by the growing need for massive multiplex analysis in many different research and clinical applications, a concept that originated with the Southern blot in 1975 [1], and continued with the development of microtiter plates from 96 wells per plate in the 1970s to 9600 wells some 20 years later [2]. Although microarrays are typically more suited to the multiplexing of assays, microplates in combination with robotic liquid handling provide for multiplexing of samples. The combination of both approaches (i.e., multiplexing of samples and assays) became feasible with

U.R. Müller
Nanosphere, Inc., 4088 Commercial Avenue, Northbrook, IL, 60062, USA

K. Dill et al. (eds.), *Microarrays: Preparation, Microfluidics, Detection Methods, and Biological Applications*,

the introduction of arrays at the bottom of microtiter plates [3], but had significant limitations with regard to array size and readout technology. The next step in multiplexing technology, a microbar code-based assay platform, was aimed at overcoming these disadvantages. This type of assay format provides the optimum in flexibility, automation, and multiplexing, because it combines the advantages of the microplate format with the multiplexing power of array technology in a liquid assay system.

A variety of different bar coding formats and related assay platforms has emerged over the past 20 years aimed at applications that range from high-throughput screening in pharmaceutical drug discovery to clinical diagnostics, biodefense, security, environmental analysis, and process monitoring among others. In addition to their target application they differ significantly in the chemical or physical principles that were exploited to generate and read the bar code. This review focuses only on platforms designed for molecular testing and is organized by the underlying principle employed to generate the bar code.

Bar codes are unique, machine-readable tags, which in the most simple form consist of a linear array of alternating white and black parallel bars of varying widths. The most common bar code system, used worldwide for the identification of goods at the point of sale, is the Uniform Product Code (UPC), coding typically for a 12 digit number to uniquely identify the manufacturer and the specific product (11 data digits and a check digit). Other codes exist for different applications, and with the appropriate increase in the number of bars, it is possible to encode text and all ASCII symbols [4, 5].

The actual size of the bar code is restricted by the code used, whereby the length is of course a function of the amount of information encoded (i.e., the number of characters) whereas the height is restricted more by the application. The smallest UPC tag, for example, cannot be less than ½ in. in height. The reason is obviously the cost involved in the equipment to print and read the bar code, which is to be read at high speed and without error. These are indeed the main reasons that bar codes have become so ubiquitous. Although this may not sound so true the next time you wait in a long checkout line, the 12-character UPC code can be scanned in approximately the time it takes a keyboard operator to make two keystrokes, but with an error rate that is 1000 to 2 million times lower (depending on the coding system used). No wonder then that the bar code has evolved from a patented concept in 1952, and its first market introduction in 1974, to become part of virtually every product that is sold today.

But this evolution is far from its peak; although the currently dominating bar code systems consist of printed codes that are read by laser-based scanners in close proximity, they may soon go the way of the slide rule, being displaced by a faster and much more powerful system, that is, radio frequency identification (RFID) that does not require the visual link between the scanner and the tag. RFID tags are small devices that can transmit an electromagnetic wave of specific frequency to a wireless receiver in response to an incoming signal. The size of these transponders depends on the frequency at which they transmit and whether they are active or passive.

Active transponders contain a battery and can therefore transmit over a longer range, whereas passive transponders convert the energy of the incoming signal and have therefore a very limited reach. Their advantage is that they can be made as small as a fraction of a square millimeter and, unlike bar codes, can be incorporated into products without being obvious or even detectable. Because of their small size they have also found a new use in multiplex diagnostic assays, where they function to uniquely identify a captured target analyte, as dicussed in detail below.

## 16.2 Bar Codes Based on Natural DNA Sequences

When Woodland and Silver filed the first patent on bar codes in 1949 [6] they envisioned not only a linear array of bars, but also one consisting of concentric rings, similar to the growth rings on trees that have been used for age determinations of trees since the 1920s. Although dendrochronology may not have had an impact on Woodland and Silver's invention, the term bar code and the intent of uniquely identifying an item by a simple and generic coding system has undoubtedly left a huge mark on systematic biology, where taxonomists have struggled with a simple and robust method of classifying living organisms since the time of Carolus Linnaeus. Although still hotly debated [7–9], the use of short othologous DNA sequences as species identifiers (i.e., DNA bar codes) to aid in the classification of organisms has been applied successfully to many different organisms in the animal, plant, fungal, and microbial world [10–14]. In fact, a real effort is now underway to assign such a bar code to any of the estimated 10 million species on this earth [15–17].

Different forms of DNA bar coding have also been developed for human diagnostics, including forensic and paternity applications [18–20]. Most important, however, has been the application to cytogenetics, starting with the invention of chromosome banding techniques in the late 1960s and culminating in sophisticated multicolor bar coding of specific gene sequences by fluorescent in situ hybridization to diagnose chromosomal aberrations in different cancers as well as heritable genetic diseases [21–26].

It is important to note that these DNA bar coding techniques rely on natural DNA sequences and are not to be confused with the synthetic DNA codes discussed below. In other words, the genetic material of any organism can be extracted and prepared in a way that gives the appearance of a linear array. This bar code-like arrangement is then easily scanned by eye or with the aid of imaging systems for classification or mutation analysis.

## 16.3 Bar Codes Based on Synthetic Nucleotide or Amino Acid Sequences

All the variety of nature is encoded in the primary, secondary, and higher-order structures of nucleic acids and proteins, and it seems obvious that even a relatively short sequence of nucleotides or amino acids could be used to generate a huge number of unique codes. Yet, the difficulty is not the generation of such codes, but the development of assays and instrumentation for simple, rapid, low-cost, and error-free decoding.

### *16.3.1 Bar Coding with Nucleic Acids*

The ability of PCR to amplify extremely small quantities of DNA (less than 100 molecules per reaction) to levels that suffice for subsequent sequence analysis has led to the proposal of tagging objects with unique amplifiable DNA sequences, either directly [27] or by encapsulating the DNA to protect it from degradation. Although such a system could be extremely sensitive, decoding still requires a significant effort and costly reagents and instrumentation. Yet, the high coding capacity, sensitivity, and accuracy of this method has found applications in expression analysis [28, 29] and in mutation analysis, where the tags are decoded either by hybridization [30, 31] or by mass spectrometry [32].

A more complex encoding strategy is based on DNA secondary and tertiary structure. With computer-aided sequence design DNA can be folded by self-assembly into 2D and 3D patterns, which are then decoded by atomic force microscopy [33]. Given that the stability of these structures is a function of the local environment and can therefore be changed by heat or ionic strength, this approach holds the potential for a dynamic coding system. Although this concept is intriguing, commercial applications appear far off.

A simpler application of folded DNA for nano-bar coding involves the formation of DNA-based nanostructures as carriers for multiple fluorophors in precise ratios [34] and is discussed in Section 4.5 below.

### *16.3.2 Peptide-Based Coding*

Bar coding based on peptide sequences is equally possible, but in the absence of a PCR-like amplification system for peptides, the amounts of material required for a decoding based on peptide sequence or fragment analysis does not support a diagnostic application. A method to overcome these problems was developed at Vysis, Inc. (now Abbott, Downers Grove, IL), which employed fluorescently labeled peptide tags that were engineered to display a specific and predetermined isoelectric point (pI, the position in a pH gradient at which the sum surface charges of the peptide is 0; [35]).

After attachment of these tags to DNA or antibody probes via a cleavable bond, they serve as surrogate targets in an assay system similar to the Biobarcode™ platform described below. In short, target analytes are sandwiched between a solid surface and a detection probe in a multiplex assay. After washing the captured sandwiches to remove unbound probes, the peptide bar codes are released and decoded by isoelectric focusing in a capillary. During this process the peptides are concentrated into a very small volume and aligned within the capillary in a linear array according to their pI. This volume reduction has a significant signal amplification effect, and their relative position can therefore be identified by laser-induced fluorescence with high sensitivity.

The development and use of 11 different focusing entities demonstrated the potential for multiplexing [36]. Although the design of these tags is somewhat more difficult than the design and synthesis of DNA-based bar codes, their advantages include the ability to focus in microchannels with high sensitivity and decoding times of less than 30 seconds for the complete set [37].

## 16.4 Bar Codes Based on Micro- or Nanostructures

In most diagnostic applications the need for multiplexing has to be balanced with the need for high sensitivity. Bar codes based on short DNA or peptide sequences can carry only a limited number of small label molecules, and their detectability is therefore limited by the cost and complexity of the detection system, generating the need for more powerful and more easily detectable tags. Provided that the most sensitive detection technologies are based on the measurement of photons, it is not surprising that the search for such tags has focused on a combination of high extinction coefficient and some means of releasing the captured photon energy with a good quantum yield. Most bar code platforms are designed to only identify the captured target, which then has to be detected and quantified using a separate labeling system in the presence of, and in addition to, detecting and decoding the bar code. In other systems the bar code serves simultaneously as an identification tag as well as a label for the target analyte. These variations combined with different requirements for decoding speed, accuracy, simplicity, and low cost have resulted in a plethora of micro-or nanostructured bar coding strategies.

### *16.4.1 Bar Codes Based on Particle Size*

The simplest approach to generating bar codes based on structure is the use of differently sized particles, such as polystyrene beads with diameters in the micrometer range, or nanoparticles in the submicron range. The factors limiting the number of bar codes that can be generated are obviously the precision with which these particles can be manufactured to a specific size, and the resolution of the decoding system used. The assay sensitivity is determined by the ability of the detection system to measure the amount of captured target in the presence of the signal received from the bar code. In practical applications such beads or nanoparticles are functionalized with specific antigens, nucleic acid probes, or antibodies for the specific capture of DNA, RNA, or protein targets.

The targets may be labeled directly (typically with fluorophors) or in a sandwich structure using a labeled secondary binding moiety. This was first implemented almost 30 years ago using flow cytometry for simultaneous identification of differently sized beads by scatter analysis, and detection of attached targets by their fluorescence [38, 39]. In a more sophisticated form, the combination of microspheres and nanoparticles was used in a process termed "coupled particle light scattering" (COPALIS). However, the level of multiplexing achieved by this technique was modest (simultaneous analysis of 2–3 target analytes), and the assay is not very sensitive (detection limits ~0.5 ng/ml) [40–42].

A newer version of this concept has been introduced recently, using nanostructures of much smaller dimensions. Two sets of particles, one for capture (300 nm diameter) and one for detection (10 nm diameter) were designed to couple in the presence of a target analyte, forming a uniquely sized structure that could be identified by atomic force microscopy [43]. When aggregates of this type are formed with metallic nanoparticles that are less than 100 nm in diameter, the unique photonic characteristics of such particles [44–47] allow the discrimination of differently sized particles and their aggregates by changes in absorption frequency that is even detectable by eye if sufficient target analyte is present [48]. Much more sensitive is the analysis of scatter light which changes in color upon particle aggregation due to surface plasmon resonance, a feature that was recently exploited by Nanosphere Inc. for the detection of zeptomole quantities of target [49].

### *16.4.2 Bar Codes Based on Particle Shape*

In 1821 Louis Braille developed a coding system that relies on six dots in a 2 × 3 pattern, where some of the dots are physically raised from the surface in a sufficiently small area to be easily decoded by the human finger. Each letter in the alphabet is represented by a unique combination of raised dots, which has become the universal written communication system for the blind.

This concept of encoding by surface structure underlies many bar coding systems in the micro- or nanoscale that are commercially available or in development. For example, square particles (100 × 100 × 35 μm) with a pattern of holes (ImageCodes) was developed by 3D Molecular Sciences Ltd (3DM; Cambridge, UK) using a common polymer (SU-8) and microfabrication technology [50]. The complexity of this pattern required decoding in a static mode by microscopy and image analysis, and a version of this technology is now available under the tradename UltraPlex™ by

Smartbead Technologies Ltd [51]. For higher throughput, a smaller particle (20 × 8 × 5 μm) was engineered by 3DM to be decoded by flow cytometry (FloCodes), but this technology is not yet ready for commercialization.

Although it is easy in principle to generate millions of different codes with either system, the challenge of generating particles with surface features that can be decoded accurately when passing by a detection window with high speed is significantly larger, provided that not all particles are in the same orientation.

In addition to generating micron-sized plastic particles with patterns of holes, a variety of other strategies has been pursued to generate bar coding systems based on shapes of particles, using etching, bleaching, metallic deposition, or micromachining. For more detail the reader is referred to the excellent review by Finkel et al. [52]. Of course the application of these particles in multiplex bioassays requires that their surfaces can be adequately functionalized to serve as carriers for biomolecules, and that the signal from captured and labeled target analytes can be detected and quantified with the same speed and in the same mode with which the bar code is decoded. At this point there is a paucity of data with regard to the analytical performance of these detection platforms.

### 16.4.3 Bar Codes Based on Light Reflection Properties

An extremely miniaturized version of the UPC bar code, termed Nanobarcodes™, was first conceived of by Mike Natan and his colleagues at Pennsylvania State University, and was later commercially developed at SurroMed and its spin-out company Nanoplex Technologies, Inc. [53, 54].

Nanobarcodes are metallic rodlike particles of a few hundred nanometers in diameter and up to several microns long. They are produced within narrow channels by sequential electroplating with different metals (e.g., gold, silver, and platinum). By controlling the thickness of each metal layer, the final nanorod has a banding pattern characterized by different light-reflecting properties [55].

The number of bands and metals used in each rod determines the coding capacity, reaching about 30,000 different codes with just ten bands and three different metals per rod. Readout requires an optical microscope for image acquisition in static applications, where the nanorods are deposited on a surface for analysis. Specifically developed image analysis software counts and decodes each rod in the image with an accuracy that exceeds 90% [56]. A much faster system, albeit more complex and expensive, is based on flow cytometry, whereby each rod passes in a fluid stream by a laser beam and pinhole, and the reflected light is analyzed by a photomultiplier tube.

There are multiple ways by which these nanorods could be used in molecular assays. For example, if functionalized with streptavidin they could replace the enzyme label in a standard microplate-based ELISA. Instead of generating a color as signal, the nanobarcodes could be released from all wells, pooled, and analyzed in a single image. The presence of a captured target would then be detected by the presence of that bar code in the image, and the number of bar codes present would allow quantitation of the target.

In more complex applications the nanobarcodes are functionalized with a specific capture probe or primary antibody and serve to specifically capture as well as identify a specific target [57]. Detection and quantitation of a captured target is then achieved through sandwich formation with a fluorescently labeled detection probe or secondary antibody. Because different fluorophors could be used one can easily envision how multiplexing of samples and tests can be achieved simultaneously.

### 16.4.4 Bar Codes Based on Light Diffraction Properties

Using the principles of diffraction gratings Galitonov et al. [58] have recently generated nanostructured bar codes by etching submicronwide parallel lines into the surface of chromium coated glass with electron beam lithography. Such gratings generate diffraction beams when interrogated with a laser at a 90° angle, whereby the angle of the diffracted light is a function of the pitch of the grating. Because different gratings can be superimposed on each other, tens of thousands of different codes can be generated with just three superimposed gratings, assuming a line resolution of 100 nm. Theoretically billions of codes are possible with higher resolution and more superimposed gratings, but the practical limits will be set by instrument cost for error-free decoding at high readout speed. More important, the cost of manufacturing such gratings not on a flat glass slide but on micron-sized particles may also play a significant role in the commercialization of this very novel approach.

### 16.4.5 Bar Codes Based on Spectral Signatures

The most extensively traveled path in the development of micro-bar coding technologies has involved bar codes with a spectral signature, whereby a significant fraction of the electromagnetic spectrum has been used, ranging in frequency from a few hundred MHz (radio band) to about $5 \times 10^8$ MHz (visible range). This was driven in part by readily available technology for the measurement of electromagnetic waves with high resolution, in theory allowing for a huge number of electromagnetic codes. Secondly, these codes can be read on the same (or similar) instrumentation as the high-sensitivity fluorescent labels that are now in widespread use for the detection of nucleic acid or protein analytes. But as discussed below, there are still significant practical hurdles to the generation and readout of more than just a few hundred codes.

#### 16.4.5.1 Fluorescent Beads

The need for multiplexing of DNA probes for use in fluorescence in situ hybridization (FISH) sparked the development of multicolor probes (chromosomal bar coding) in the early 1990s [24]. The limitations imposed by the number of fluorescent labels that could be used simultaneously, and the need to differentiate multiple genetic regions or chromosomal fragments in a single karyotype led to the development of combinatorial color labeling [59] and color ratioing [60]. As a result, many more probes could be used simultaneously than would otherwise have been possible.

This principle was also applied to the generation of color-coded beads. The first commercial platform, the FlowMetrix™ system (Luminex Corporation), consisted of 5.5 micron diameter polystyrene microspheres that were dyed with different ratios of two fluorophors to generate 64 different beads. A third fluorophor was used for quantitation of the captured target analyte, and decoding as well as target measurements were performed by flow cytometry [61]. A refinement of this technology is now available under the tradename xMAP® (Luminex®) that allows the differentiation of 100 beads, using just two colors for coding. Although this level of multiplexing does not compare to the power of microarrays, the advantage of "in solution" binding kinetics, combined with the reliability, high speed, and versatility of this assay system has lead to widespread research and clinical applications in infectious disease and molecular diagnostics [62–65].

A similar technology was introduced by BD Biosciences under the tradename BD® Cytometric Bead Assay (CBA), which uses only a single fluorophor at different intensity levels to generate the coded beads. This limits the multiplexing power (currently six simultaneous analyses) but allows a very small assay volume and simpler instrumentation [66–68].

A different approach to labeling beads was described by Li et al., whereby bar coding is accomplished through assembly of small DNA dendrimers with a precise ratio of two different fluorophors. In the presence of target these DNA nanostructures are then bound to beads for easy detection by microscopy or flow cytometry [34].

An alternative way to analyze beads or tags is through capillary electrophoresis instead of flow cytometry. The concept is similar in that tags pass by an interrogation window in the capillary in single file, and are analyzed by laser-induced fluorescence measurement to decode the tags and quantify the captured target. The advantage of this approach, which is being pursued by the early stage company Singulex (St. Louis, MO), is that only a very small amount (1–5 ul) of sample is required for each assay (Singulex website). More important, by interrogating individual beads in small confinement the background from the surrounding solution and other beads is significantly reduced or eliminated, allowing for detection of very small signals. This has allowed Singulex to reach detection limits in the low pg/ml range, and for some protein analytes even in the fg/ml range [69].

#### 16.4.5.2 Fluorescent Cell Bar Coding

The concept of differentially labeling beads with varying ratios of fluorophors has recently been applied to generate a cell-based multiplexing technique, referred to as fluorescent cell bar coding. In order to study the cellular response of several drugs, test cells were split into different batches, each to be treated with a different drug. Cells were then fixed, and after permeabilization with methanol each batch was treated with a different concentration of an amine-reactive fluorescent dye for covalent attachment to intracellular proteins. After pooling the cells from all batches they were treated with fluorescently labeled antibodies to different target proteins. By appropriately choosing the fluorescent tags on the antibodies the cellular responses to different drug treatment regimens could be measured in a single analysis on a multicolor flow cytometer. When three different fluorophors were used for bar coding, the cells from 96 well microplates could be combined for analysis in a single run [70].

#### 16.4.5.3 Quantum Dots

Quantum Dots (QDs), inorganic semiconductor crystals such as CdSe that can adsorb photon energy and emit it as luminescence, represent a significant improvement over the organic fluorophors. Ranging in diameter between 1–10 nm, they have a much larger extinction coefficient and consequently capture and emit more photons. The detailed process involves multiple electron energy levels and the transitions of electrons and holes between them by which QDs can adsorb energy from multiple wave lengths (series of overlapping adsorption peaks). They emit this energy with high quantum yield at a wavelength that is slightly longer than the lowest adsorption peak.

Thus, the emission spectrum is independent of the excitation frequency (as long as it is above the lowest adsorption band), is very narrow, and is solely a function of the composition and size of the quantum dot. Therefore, the narrower the size distribution of a QD preparation is, the narrower the emission spectrum of that preparation. Using different materials QDs can be engineered to emit light from deep blue (365 nm) to near-infrared (2300 nm), and are commercially available from manufacturers such as the Quantum Dot Corporation (Hayward, CA) or Evident Technologies (Troy, NY). Typically their surface is functionalized with chemical groups to make them water soluble and allow for linkage to nucleic acids, proteins, and other biomolecules.

The availability of differently colored QDs allows for a modest amount of assay multiplexing when used as tags directly, however, the more powerful bar coding strategy combines populations of different QDs in different ratios, akin to the dying of polystyrene beads with different fluorophors. Nie and colleagues have demonstrated that porous silica and polystyrene beads can be doped with quantum dots to generate spectrally different tags. Because their emission peaks are typically narrower than those of organic fluorophors, more colors can be used simultaneously [71–75].

A similar strategy was pioneered by Molecular Probes, producing highly fluorescent 35–40 nm diameter polystyrene nanoparticles. By mixing different fluorophors with overlapping emission and excitation spectra the excitation energy is transferred between the dye molecules (FRET), generating large Stokes shifts; that is, these particles share the same excitation peak but different and tunable emission peaks that are removed from the excitation peak by over 100 nm [75]. These types of particles are now available under the trade name TransFluoSpheres® from Invitrogen. More recently Wang and Tan have color coded silica nanoparticles using the same principles of multicolor FRET [76]. Because these nanoparticles have a diameter that is about 20 times larger than that of typical QDs, their extinction coefficient is also much larger and consequently they are much brighter on a particle basis [71].

Although in theory these strategies may provide for hundreds to millions of different bar codes, a high level of multiplexing assumes that the difficulties associated with the precise manufacturing requirements as well as accurate spectral decoding at high speeds can be overcome. A bead that is colored with multiple different QDs, for example, can be excited with a single laser source at the blue end of the spectrum, but readout still requires either special filters for each color or the reading of the whole spectrum between all colors used. First products based on quantum dot encoded beads (Qbead™; Quantum Dot Corporation) will therefore be less ambitious, providing fewer than 1000 different codes, although the principle has been demonstrated in biological assays [78–80].

#### 16.4.5.4 Bead Array

The difficulties associated with accurate manufacture and decoding of large numbers of spectral bar codes can be circumvented by breaking the decoding process into multiple steps. Illumina (San Diego, CA) has developed a coding technology (BeadArray™) that involves 3 µm beads, each coated with single-stranded oligonucleotides of specific sequence for identification. The beads are captured in tiny wells at the end of a bundle of optical fibers, and the oligo sequences are then decoded by sequential hybridizations to fluorescently labeled complementary oligonucleotides (decoders).

Note that the position of each bead in the array does not change during this process. The number of different beads that can be used simultaneously depends on the decoding strategy. For example, 1520 different beads can be decoded in seven consecutive hybridization reactions that color a bead in one of two colors (two different fluorophors with little spectral overlap) or leave it without a label [81], reducing the complexity of the readout equipment and improving read accuracy. Once the beads are decoded, the array can be used for the capture of labeled target sequences, such as mRNAs or genomic DNA fragments. Moreover, with available automation multiple bead arrays (i.e., fiber bundles) are used to interrogate each well on a microplate simultaneously, providing for multiplexing of samples as well as assays.

#### 16.4.5.5 Striped and Colored Micro Bar Codes

The ultimate in bar coding power would combine the micro striping technology employed by Nanoplex with a spectral bar coding technique for each of the stripes. This was first realized by a research team at Corning [82], making use of Corning's experience in glass fiber technology.

Starting with different rare earth-doped glass bars that were fused together in a furnace and then drawn into a 20 × 100 μm ribbon fiber, bar codes could be produced by etching the fiber with a laser and then sonicating under water to break it into individual 20 × 20 × 100 μm particles. The drawing process is very accurate and does not distort the aspect ratios of the fused glass bars. Thus, each particle contained as many colored stripes as the number of glass bars that were fused into the fiber. The rare earth dopants were selected to have relatively large Stokes shifts and narrow emission bands in the visible range to maximize the number of resolvable emission bands.

Because the spectrum of each stripe instead of the resulting color can be determined, the number of different stripes is at least equal to what could be produced by doping microspheres or Q-dots, making the coding power of striped particles enormous. For example, the coding power of a small stack of six stripes could generate over 200,000 nondegenerate codes, using just nine different colors. If each stripe is spectrally scanned, many more colors (i.e., spectral signatures) can be resolved and hence many more codes are feasible. Additional advantages include the absence of quenching, no photobleaching, a glass surface optimal for easy attachment of biomolecules based on available silane chemistry, and compatibility with organic and inorganic solvents. Although feasibility for use in biological assays was demonstrated, their application for high-level multiplexing is still hampered by the absence of suitable low-cost equipment for high-speed decoding.

#### 16.4.5.6 Encoding Particles with a Raman Signature

When particles adsorb light energy, some of the energy is converted to heat, and a fraction of the photons are scattered elastically (Rayleigh), meaning they do not change in frequency. A very small portion, however, gain or lose energy in the process (nonelastic scattering), leading to characteristic Raman spectra. Although this technique per se is extremely insensitive, the sensitivity can be increased by many logs if the molecules to be analyzed are attached to rough metallic surfaces [83, 84]. This process, termed surface-enhanced Raman scattering or SERS, allows detection of single Raman active dye molecules when they are attached to silver or gold nanoparticles [85].

Because of the narrow spectral lines generated by SERS (30–60 nm spectral width), a spectral scan should be able to discriminate hundreds or even thousands of different dye molecules if they are amplified in this manner. This is the basis for a bar coding concept that involves nanoparticle probes with different SERS signatures. It was first proposed by Vo-Dinh and collaborators in 1998 [86], and implemented a few years later by different investigators with different approaches. Cao et al. developed spectrally different probes by attachment of different Raman dye molecules to gold nanoparticles via modified DNA oligonucleotides, and used these probes for detection of nucleic acid [87, 88] and protein targets [89], captured on regular glass slides.

When used in this format probes were amplified with silver to achieve the SERS effect, but provided detection limits in the femtomolar range for DNA targets. This amplification step can be avoided when dye-labeled silver, gold, or core-shell nanoparticles are aggregated or used in combination with a metallic surface [89–91], and extreme amplification power is provided when these particles are engineered to have sharp edges [92]. The potential for high-level multiplexing or bar coding by this approach, combined with the potential for significant increases in detection limits that surpass fluorescence, suggest that bar code platforms based on organic or inorganic particles and SERS technology may become available in the not too distant future.

#### 16.4.5.7 Encoding Particles with an RF Signature

The advent of combinatorial chemistry has generated the need for high-level bar coding in the pharmaceutical industry to allow the generation and tagging of chemical libraries with hundreds of thousands of different compounds. A successful solution was presented by the development of miniaturized radiofrequency (RF) tags, miniaturized transponders with integrated circuits that can transmit a unique identification code when queried by an RF source of a specific frequency. Because they encode 40 bits of information, over one trillion different codes can be produced [93].

More recently an even smaller version RF tag was introduced by PharmaSeq (Monmouth Junction, NJ) that is activated by visible light. These tags contain a photovoltaic cell to collect power from a laser, 50 bits of memory for storage of the electronic code, an antenna for transmitting the code, and the necessary synchronization and read logic. The integrated circuits are all embedded in a 500 $\mu m^2$ silicon microchip, and smaller tags (250 $\mu m^2$) are in development. When activated, these tags transmit their code three times in less than one millisecond and over a distance of a few millimeters. This enables flow analysis with relatively high speed, whereby the transponder travels by capillary flow (cell sorter) to pass an interrogation window. The light from a 650 nm laser is used to power this system as well as to excite any fluorophors captured by the RF tag [94–95].

It was demonstrated that fluorescently labeled PCR fragments could be captured specifically on appropriately functionalized transponder surfaces in the presence of control targets containing a single nucleotide mismatch. The discrimination obtained (20:1) was sufficient to differentiate tags with captured target DNA from controls without DNA by CCD imaging in a static mode, and presumably this ratio will also suffice for high-speed flow analysis. It will be interesting to see how well this system performs in multiplex analyses, where the discrimination ratios may be smaller, owing to the difficulties in generating many different isothermal capture probes that perform equally well under the same stringency conditions.

#### 16.4.5.8 Encoding Particles with a Chemical Signature

Chemical encoding is used by the pharmaceutical industry as an alternative to the RF tagging because of high and robust coding capacity and with the available additional feature that the code can be modified at each synthesis step to indicate through which stages of the chemical synthesis a particular compound has gone. In the combinatorial "split-pool synthesis method", for example, polystyrene beads in the 100 μm diameter range serve as solid support for the synthesis of an organic compound that is modified in each successive step. A unique mixture of tagging compounds (e.g., aromatic fluoro- or chlorocarbons) is also attached to the bead in each step in concentrations that allow easy identification by gas chromatography when released from a single bead [93, 96, 97].

Because all beads contain the compound of interest, only a single bead has to be decoded from a given synthesis batch, and only those batches that contain a compound of interest need to be decoded. Thus, the decoding time of a few minutes per bead is acceptable for this application, but makes this particular platform unsuitable for molecular analyses where high sensitivity and rapid decoding times are required.

Improvements in both sensitivity and speed are easily achievable, of course, and chemical encoding is becoming an option for biomedical diagnostics. Attaching a mixture of taggants to each bead that generates a spectral signature (Raman, SERS, IR, MS) increases both sensitivity and decoding speed [98–101], and adds to the repertoire of spectral bar codes discussed above.

#### 16.4.5.9 Encoding Particles with a DNA Signature

When the concept of chemical encoding was first introduced, the proposed encoding mechanism was DNA. A unique code was generated on each bead through addition of a specific trinucleotide to a growing primer sequence after each cycle of chemical synthesis. Akin to the "bar coding with DNA" discussed above, the final DNA tag could be amplified by PCR and subsequently decoded by sequence analysis, providing a novel and powerful method for encoded combinatorial chemistry [102].

More recently, the concept of using DNA sequences as bar coding tags for beads has been applied to molecular bioassays with a unique twist. In an assay platform termed Biobarcode™ the bar code oligonucleotides do not only serve as unique identification tags, but also play a key role in the assay process where they are used to amplify the signal. As a consequence this nanoparticle-based assay platform, originally introduced by Nam et al. for the ultrasensitive detection of proteins [103, 104], does not require any enzymes to reach detection limits that are typically only achieved with the aid of enzymatic target or signal amplification schemes.

As depicted in Fig. 16.1 the assay format consists of three steps. The first step is similar to an ELISA, whereby a target analyte is captured by magnetic beads that are functionalized with specific antibodies. In the second step a 30 nm amplifier nanoparticle, which is co-loaded with secondary Ab and the double-stranded bar code oligonucleotides, is attached to the captured protein target. Because only one of the strands is attached covalently, the complementary strand can be released by increasing

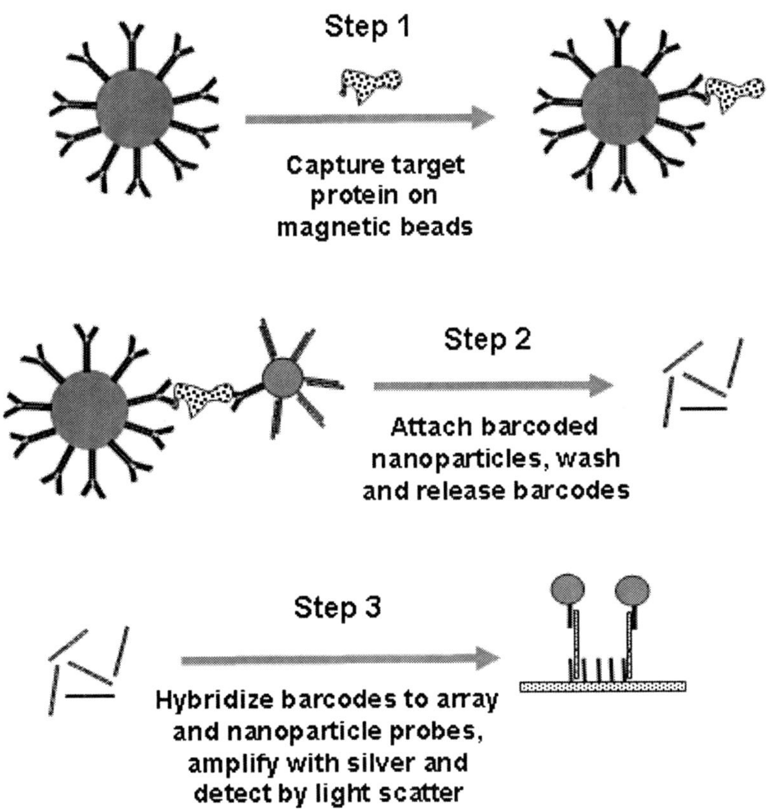


**Fig. 16.1** Schematic of Biobarcode™ assay platform

the stringency of the wash buffer. Many bar codes can be attached to each amplifier nanoparticle, therefore each captured target protein is effectively converted into many surrogate DNA bar code targets. In theory this assay can be multiplexed for multiple target proteins by concomitant use of different amplifier nanoparticles that are each coated with an antibody and a bar code, both being specific for one of the targets to be detected. In the third step the released bar codes are decoded by hybridization to an array. Detection is via a second hybridization event that uses 13 nm nanoparticle probes with silver amplification. This process has been shown to lend extraordinary sensitivity to the detection of nanoparticle probes [105], which contributes significantly to the high sensitivity of this system. When applied to the detection of prostate specific antigen (PSA) [104], a widely used tumor marker, or to ADDL peptides [106] that are implicated in Alzheimer's disease, the reported detection limits were in the attomolar range, several logs more sensitive than standard ELISA assays.

By replacing both the primary antibody on the magnetic bead and the secondary antibody on the amplifier nanoparticle with specific DNA sequences that are complementary to a chosen target nucleic acid, the same platform can also be used for detection of DNA or RNA, but with even higher sensitivities. For example, an anthrax lethal factor specific DNA sequence was detected at 500 zM [107], probably due to the lower dissociation constant of nucleic acid hybridizations compared to that of antibody binding reactions.

A colorimetric version of the bio bar code assay was recently reported by Nam et al. [108], reporting the detection of cytokines in the attomolar range. A commercial development of the bio bar code platform was undertaken by Nanosphere, Inc. (Northbrook, IL), and by modification of the assay format the dose response could be improved 10,000-fold without loss in sensitivity [109]. This platform is now in full development at Nanosphere for protein and nucleic acid targets of clinical utility. Initial products will make use of the superior assay sensitivity compared to standard ELISA systems to lower the threshold for the detection of important disease markers for cancer and cardiac and neurological diseases.

## 16.5 Advantages of Bar Code Versus Microarray-Based Platforms

The plethora of bar code-based platforms for molecular testing currently available or under development would suggest that this assay strategy holds some advantage over microarray technologies. Yet, under close examination the advantages are not as obvious as they might seem at first glance. For example, bar coded particles functionalized with target-specific oligonucleotides or antibodies are expected to have much improved hybridization or binding kinetics, respectively, when contacted with target analyte in solution. Whether this actually translates into a better sensitivity (lower detection limit and/or improved dose response) depends on many additional assay parameters, which could easily void the kinetic advantage.

In assays where target concentrations do not limit the binding reactions, such as in some PCR-based genetic tests, the advantages may well be the flexibility to mix and match bar codes as needed, and to identify target analytes (e.g., PCR products) rapidly and cost-effectively. This is especially relevant for all situations that demand a high throughput of samples. Although bar code-based testing formats are very compatible with microtiter plate-based assay formats, their success depends on the speed, accuracy, and cost with which the decoding and target quantitation can be accomplished. Thus, systems that require image analysis in a static mode are less likely to succeed in this application, whereas flow-based systems may offer significant savings in time and cost compared to microarrays.

Perhaps the most powerful application of bar coding technology involves coded nanostructures that are used also as signal generators to indicate the presence of target.

Fig. 16.2 shows a theoretical comparison of microplate-based ELISAs with microarrays and bar coded nanoparticle-based assays. The assumption is that the same antibody pair is used to develop a sandwich-type assay for protein biomarkers, such as cardiac troponin (a marker for acute myocardial infarction), PSA, or interleukins. The detection limit (LOD) for these types of targets, where excellent antibodies are available, has been reported to range from 30 pg/ml to 50 ng/ml of serum in the best commercially available ELISA assays [110, 111], and LODs on the order of 1 pg/ml have been achieved in research environments [112] Fig. 16.2A).

The same antibody target sandwich can be generated on the surface of a slide in a microarray format, and different label and detection schemes have been developed for this purpose. Assuming similar reaction conditions and binding kinetics, the LOD would be mostly a function of the type of label and readout instrumentation employed, and in principle no significant difference in sensitivity is expected [113].

When nanoparticles are used as labels instead of fluorophors in such assay formats (Fig. 16.2 B), the LODs may be driven down by one or two orders of magnitude due to their much larger extinction coefficients [105, 114, 115]. Further improvements in sensitivity should be achievable when the nanoparticles are released from the capture complex and detected by flow (Fig. 16.2C), as demonstrated in principle by Singulex (see above). In this situation a bar coded particle would be of particular advantage, because the target capture and sandwich formation could be done in the same microtiter well for multiple targets. The presence of specific target analytes is then revealed by detection of the cognate bar coded tag, either by static image analysis or counting in flow mode. Either way individual tags are counted, and inasmuch as the number of tags counted should correlate with the concentration of target in the sample, this method of quantitation is not only more sensitive but far more accurate than a bulk quantitation.

Of course, the sensitivity will be mostly a function of how well the background can be reduced, that is, how well the antibody–target–nanoparticle complex can be washed free of unbound nanoparticles before the release of specifically captured particles. (Note, however, that detection limits are often set by nonspecific signal resulting from cross-reacting antibodies). Because a single particle tag is detectable, in theory a single target could be detected by this assay format, if the antibody

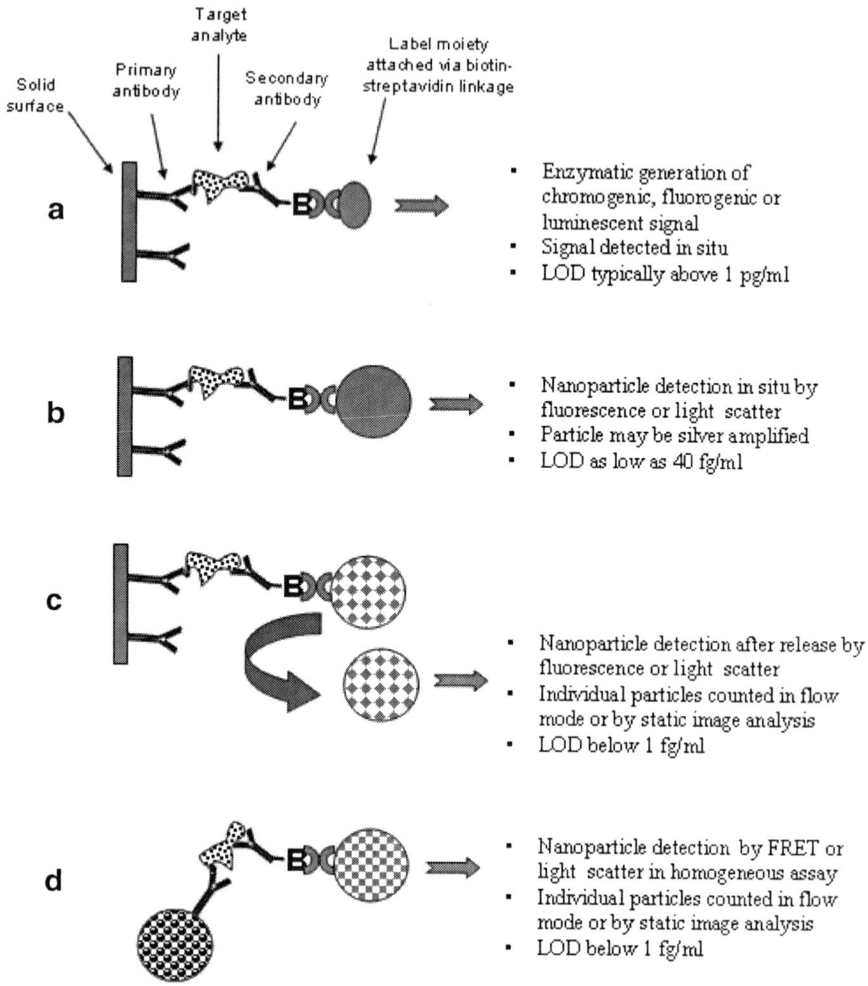


**Fig. 16.2** Illustrative comparison of ELISA assay with microarrays and bar coded nanoparticle-based assay platforms. The solid surface in A through C may be the wall of a microtiter well or the surface of a microarray slide

binding reaction can be driven to capture a single target in the first place. The capture of target molecules in very low concentrations requires a high concentration of antibodies (i.e., antibody coated tags), thus it is nearly impossible to reduce the background of nonspecifically bound tags. One way around this issue is the use of two particles that have to bind to the same target in order to generate a specific signal (Fig. 16.2D). This can be achieved in various ways, such as FRET [116] or resonance shift of metallic nanoparticles [49], as discussed above. This would also enable a homogeneous assay format, which combined with flow analysis may prove to be the most advantageous assay platform for high-sensitivity multiplex analysis [117].

## References


1. Southern, E.M., 1975 Detection of specific sequences among DNA fragments separated by gel electrophoresis. *J Mol Biol* 98: 503–517.
2. Manns, R.L., 2003 Necessity is the mother of invention – The history of microplates. PharmaTech Series Business Briefing: Future Drug Discovery 2003, Business Briefing Ltd, pp. 108–112.
3. Mendoza, L.G., McQuary, P., Mongan, A., Gangadharan, R., Brignac, S., and Eggers, M., 1999 High-throughput microarray-based enzyme-linked immunosorbent assay (ELISA). *Biotechniques* 27: 778–780, 782–786, 788.
4. Taltech website, Introduction to Bar Coding, http://www.taltech.com/TALtech_web/resources/intro_to_bc/bcsymbol.htm
5. Worth Data website, Bar Code Basics, http://www.barcodehq.com/primer.html
6. Woodland, N.J. and Silver, B. 1952. U.S. Patent No. 2,612,994.
7. Ebach, M.C. and Holdrege, C., 2005 DNA barcoding is no substitute for taxonomy. *Nature* 434: 697.
8. Moritz, C. and Cicero, C., 2004 DNA barcoding: Promise and pitfalls. *PLoS Biol* 2: e354.
9. Schindel, D.E. and Miller, S.E., 2005 DNA barcoding a useful tool for taxonomists. *Nature* 435: 417.
10. Campbell, J., Francesconi, S., Boyd, J., Worth, L., and Moshier, T., 1999 Aug Environmental air sampling to detect biological warfare agents. *Mil Med* 164: 541–542.
11. Hajibabaei, M., Janzen, D.H., Burns, J.M., Hallwachs, W., and Hebert, P.D., 2006 DNA barcodes distinguish species of tropical Lepidoptera. *Proc Natl Acad Sci U S A* 103: 968–971.
12. Kress, W.J., Wurdack, K.J., Zimmer, E.A., Weigt, L.A., and Janzen, D.H., 2005 Use of DNA barcodes to identify flowering plants. *Proc Natl Acad Sci U S A* 102: 8369–8374.
13. Lorenz, J.G., Jackson, W.E., Beck, J.C., and Hanner, R., 2005 The problems and promise of DNA barcodes for species diagnosis of primate biomaterials. *Philos Trans R Soc Lond B Biol Sci* 360: 1869–1877.
14. Summerbell, R.C., Levesque, C.A., Seifert, K.A., Bovers, M., Fell, J.W., Diaz, M.R., Boekhout, T., de Hoog, G.S., Stalpers, J., and Crous, P.W., 2005 Microcoding: The second step in DNA barcoding. *Philos Trans R Soc Lond B Biol Sci* 360: 1897–1903.
15. Blaxter, M., Mann, J., Chapman, T., Thomas, F., Whitton, C., Floyd, R., and Abebe, E., 2005 Defining operational taxonomic units using DNA barcode data. *Philos Trans R Soc Lond B Biol Sci* 360: 1935–1943.
16. DeSalle, R., Egan, M.G., and Siddall, M., 2005 The unholy trinity: Taxonomy, species delimitation and DNA barcoding. *Philos Trans R Soc Lond B Biol Sci* 360: 1905–1916.
17. Savolainen, V., Cowan, R.S., Vogler, A.P., Roderick, G.K., and Lane, R., 2005 Towards writing the encyclopedia of life: An introduction to DNA barcoding. *Philos Trans R Soc Lond B Biol Sci* 360: 1805–1811.
18. Leclair, B. and Scholl, T., 2005 Application of automation and information systems to forensic genetic specimen processing. *Expert Rev Mol Diagn* 5: 241–250.
19. Saji, F., 1993 Application of DNA fingerprinting to obstetrics and gynecology. Nippon Sanka Fujinka Gakkai Zasshi 45: 815–821.
20. Sullivan, K.M., Hopgood, R., and Gill, P., 1992 Identification of human remains by amplification and automated sequencing of mitochondrial DNA. *Int J Legal Med* 105: 83–86.

21. Caspersson, T., Farber, S., Foley, G.E., Kudynowski, J., Modest, E.J., Simonsson, E., Wagh, U., and Zech, L., 1968 Chemical differentiation along metaphase chromosomes. *Exp Cell Res* 49: 219–222.
22. Chaudhuri, J.P., Vogel, W., Voiculescu, I., and Wolf, U., 1971 A simplified method of demonstrating Giemsa-band pattern in human chromosomes. *Humangenetik* 14: 83–84.
23. Florijn, R.J., Bonden, L.A., Vrolijk, H., Wiegant, J., Vaandrager, J.W., Baas, F., den Dunnen, J.T., Tanke, H.J., van Ommen, G.J., and Raap, A.K., 1995 High-resolution DNA Fiber-FISH for genomic DNA mapping and colour bar-coding of large genes. *Hum Mol Genet* 4: 831–836.
24. Lengauer, C., Speicher, M.R., Popp, S., Jauch, A., Taniwaki, M., Nagaraja, R., Riethman, H.C., Donis-Keller, H., D'Urso, M., Schlessinger, D., and et al.., 1993 Chromosomal bar codes produced by multicolor fluorescence in situ hybridization with multiple YAC clones and whole chromosome painting probes. *Hum Mol Genet* 2: 505–512.
25. Muller, S., Eder, V., and Wienberg, J., 2004 A nonredundant multicolor bar code as a screening tool for rearrangements in neoplasia. *Genes Chromosomes Cancer* 39: 59–70.
26. Muller, S. and Wienberg, J., 2001 "Bar-coding" primate chromosomes: Molecular cytogenetic screening for the ancestral hominoid karyotype. *Hum Genet* 109: 85–94.
27. Cox, J.P., 2001 Bar coding objects with DNA. *Analyst* 126: 545–547.
28. Qiu, F., Guo, L., Wen, T.J., Liu, F., Ashlock, D.A., and Schnable, P. S., 2003 DNA sequence-based "bar codes" for tracking the origins of expressed sequence tags from a maize cDNA library constructed using multiple mRNA sources. *Plant Physiol* 133: 475–481.
29. Velculescu, V.E., Zhang, L., Vogelstein, B., and Kinzler, K.W., 1995 Serial analysis of gene expression. *Science* 270: 484–487.
30. Cai, H., White, P.S., Torney, D., Deshpande, A., Wang, Z., Keller, R.A., Marrone, B., and Nolan, J.P., 2000 Flow cytometry-based minisequencing: A new platform for high-throughput single-nucleotide polymorphism scoring. *Genomics* 66: 135–143.
31. Fan, J.B., Chen, X., Halushka, M.K., Berno, A., Huang, X., Ryder, T., Lipshutz, R.J., Lockhart, D.J., and Chakravarti, A., 2000 Parallel genotyping of human SNPs using generic high-density oligonucleotide tag arrays. *Genome Res* 10: 853–860.
32. Haff, L.A. and Smirnov, I.P., 1997 Multiplex genotyping of PCR products with MassTag-labeled primers. *Nucleic Acids Res* 25: 3749–3750.
33. Yan, H., LaBean, T.H., Feng, L., and Reif, J.H., 2003 Directed nucleation assembly of DNA tile complexes for barcode-patterned lattices. *Proc Natl Acad Sci U S A* 100: 8103–8108.
34. Li, Y., Cu, Y.T.H., and Luo, D., 2005. Multiplexed detection of pathogen DNA with DNA-based fluorescence nanobarcodes. *Nat. Biotechnol* 23: 885–889.
35. Müller, U.R. 1998. U.S. Patent No. 5,824,478.
36. Cruickshank, K.A., Olvera, J., and Müller, U.R., 1998 Simultaneous multiple analyte detection using fluorescent peptides and capillary isoelectric focusing. *J Chromatogr A* 817: 41–47.
37. Hofmann, O., Che, D., Cruickshank, K.A., and Müller, U.R., 1999 Adaptation of capillary isoelectric focusing to microchannels on a glass chip. *Anal Chem* 71: 678–686.
38. Horan, P.K. and Wheeless, L.L., Jr, 1977 Quantitative single cell analysis and sorting. *Science* 198: 149–157.
39. McHugh, T.M., Miner, R.C., Logan, L.H., and Stites, D.P., 1988 Simultaneous detection of antibodies to cytomegalovirus and herpes simplex virus by using flow cytometry and a microsphere-based fluorescence immunoassay. *J Clin Microbiol* 26: 1957–1961.
40. Baetens, D.G. and Van Renterghem, L.M., 2001 Coupled particle light scattering: a new technique for serodiagnosis of Epstein-Barr virus infection. *J Med Virol* 64: 519–525.
41. Benecky, M.J., Post, D.R., Schmitt, S.M., and Kochar, M.S., 1997 Detection of hepatitis B surface antigen in whole blood by coupled particle light scattering (Copalis). *Clin Chem* 43: 1764–1770.
42. Benecky, M.J., McKinney, K.L., Peterson, K.M., and Kamerud, J.Q., 1998 Simultaneous detection of multiple analytes using Copalis technology: A reduction to practice. *Clin Chem* 44: 2052–2054.

43. Pris, A.D. and Porter, M.D., 2004 Nanoparticle coding: Size-based assays using atomic force microscopy. *Langmuir* 20: 6969–6973.
44. Link, S. and El-Sayed, M.A., 2003 Optical properties and ultrafast dynamics of metallic nanocrystals. *Annu Rev Phys Chem* 54: 331–366.
45. Yguerabide, J. and Yguerabide, E.E., 1998a Light-scattering submicroscopic particles as highly fluorescent analogs and their use as tracer labels in clinical and biological applications. II. Experimental Characterization. *Anal Biochem* 262: 157–176.
46. Yguerabide, J. and Yguerabide, E.E., 1998b Light-scattering submicroscopic particles as highly fluorescent analogs and their use as tracer labels in clinical and biological applications. I. Theory. *Anal Biochem* 262: 137–156.
47. Yguerabide, J. and Yguerabide, E.E., 2001 Resonance light scattering particles as ultrasensitive labels for detection of analytes in a wide range of applications. *J Cell Biochem Suppl* 37: 71–81.
48. Elghanian, R., Storhoff, J.J., Mucic, R.C., Letsinger, R.L., and Mirkin, C.A., 1997 Selective colorimetric detection of polynucleotides based on the distance-dependent optical properties of gold nanoparticles. *Science* 277: 1078–1081.
49. Storhoff, J.J., Lucas, A.D., Garimella, V., Bao, Y.P., and Müller, U.R., 2004a Homogeneous detection of unamplified genomic DNA sequences based on colorimetric scatter of gold nanoparticle probes. *Nat Biotechnol* 22: 883–887.
50. Evans, M., Sewter, C., and Hill, E., 2003 An encoded particle array tool for multiplex bioassays. *Assay Drug Dev Technol* 1: 199–207.
51. Smartbead Technologies website, UltraPlexTM Barcodes Molecules. http://www.smartbead.com
52. Finkel, N.H., Lou, X., Wang, C., and He, L., 2004 Barcoding the microworld. *Anal Chem* 76: 352A–359A.
53. Penn, S.G., He, L., and Natan, M.J., 2003 Nanoparticles for bioanalysis. *Curr Opin Chem Biol* 7: 609–615.
54. Zhou, H., Roy, S., Schulman, H., and Natan, M.J., 2001 Solution and chip arrays in protein profiling. *Trends Biotechnol* 19: S34–39.
55. Nicewarner-Pena, S.R., Freeman, R.G., Reiss, B.D., He, L., Pena, D.J., Walton, I.D., Cromer, R., Keating, C.D., and Natan, M.J., 2001 Submicrometer metallic barcodes. *Science* 294: 137–141.
56. Walton, I.D., Norton, S.M., Balasingham, A., He, L., Oviso, D.F., Jr, Gupta, D., Raju, P.A., Natan, M.J., and Freeman, R.G., 2002 Particles for multiplexed analysis in solution: detection and identification of striped metallic particles using optical microscopy. *Anal Chem* 74: 2240–2247.
57. Freeman, R.G., Raju, P.A., Norton, S.M., Walton, I.D., Smith, P.C., He, L., Natan, M.J., Sha, M.Y., and Penn, S.G., 2005 Use of nanobarcodes particles in bioassays. *Methods Mol Biol* 303: 73–83.
58. Galitonov, G.S., Birtwell, S.W., and Zheludev, N.I., 2006 High capacity tagging using nanostructured diffraction barcodes. *Optics Express* 14: 1382–1387.
59. Speicher, M.R., Gwyn Ballard, S., and Ward, D.C., 1996 Apr Karyotyping human chromosomes by combinatorial multi-fluor FISH. *Nat Genet* 12: 368–375.
60. Morrison, L.E. and Legator, M.S., 1997 Two-color ratio-coding of chromosome targets in fluorescence in situ hybridization: Quantitative analysis and reproducibility. *Cytometry* 27: 314–326.
61. Fulton, R.J., McDade, R.L., Smith, P.L., Kienker, L.J., and Kettman, J.R., Jr, 1997 Advanced multiplexed analysis with the FlowMetrix system. *Clin Chem* 43: 1749–1756.
62. Dias, D., Van Doren, J., Schlottmann, S., Kelly, S., Puchalski, D., Ruiz, W., Boerckel, P., Kessler, J., Antonello, J.M., Green, T., Brown, M., Smith, J., Chirmule, N., Barr, E., Jansen, K.U., and Esser, M.T., 2005 Optimization and validation of a multiplexed luminex assay to quantify antibodies to neutralizing epitopes on human papillomaviruses 6, 11, 16, and 18. *Clin Diagn Lab Immunol* 12: 959–969.
63. Dunbar, S.A., 2006 Applications of Luminex xMAP technology for rapid, high-throughput multiplexed nucleic acid detection. *Clin Chim Acta* 363: 71–82.

64. Hansson, O., Zetterberg, H., Buchhave, P., Londos, E., Blennow, K., and Minthon, L., 2006 Association between CSF biomarkers and incipient Alzheimer's disease in patients with mild cognitive impairment: A follow-up study. *Lancet Neurol* 5: 228–234.
65. Lash, G.E., Scaife, P.J., Innes, B.A., Otun, H.A., Robson, S.C., Searle, R.F., and Bulmer, J.N., 2006 Comparison of three multiplex cytokine analysis systems: Luminex, SearchLight and FAST. *Quant J Immunol Meth* 309: 205–208.
66. Chen, R., Lowe, L., Wilson, J.D., Crowther, E., Tzeggai, K., Bishop, J.E., and Varro, R., 1999 Simultaneous quantification of six human cytokines in a single sample using microparticle-based flow cytometric technology. *Clin Chem* 45: 1693–1694.
67. Morgan, E., Varro, R., Sepulveda, H., Ember, J.A., Apgar, J., Wilson, J., Lowe, L., Chen, R., Shivraj, L., Agadir, A., Campos, R., Ernst, D., and Gaur, A., 2004 Cytometric bead array: A multiplexed assay platform with applications in various areas of biology. *Clin Immunol* 110: 252–266.
68. Tarnok, A., Hambsch, J., Chen, R., and Varro, R., 2003 Cytometric bead array to measure six cytokines in twenty-five microliters of serum. *Clin Chem* 49: 1000–1002.
69. Goix, P, 2006. Single molecule "flow immunoassay" detection: Repurposing existing marker for clinical validation. CHI Clinical Biomarker Summit Presentation, San Diego.
70. Krutzig, P.O. and Nolan, G.P., 2006 Fluorescent cell barcoding in flow cytometry allows high-throughput drug screening and signal profiling. *Nature Meth* 3: 361–368.
71. Chan, W.C., Maxwell, D.J., Gao, X., Bailey, R.E., Han, M., and Nie, S., 2002 Luminescent quantum dots for multiplexed biological detection and imaging. *Curr Opin Biotechnol* 13: 40–46.
72. Gao, X., Chan, W.C., and Nie, S., 2002 Quantum-dot nanocrystals for ultrasensitive biological labeling and multicolor optical encoding. *J Biomed Opt* 7: 532–537.
73. Gao, X. and Nie, S., 2005 Quantum dot-encoded beads. *Methods Mol Biol* 303: 61–71.
74. Han, M., Gao, X., Su, J.Z., and Nie, S., 2001 Quantum-dot-tagged microbeads for multiplexed optical coding of biomolecules. *Nat Biotechnol* 19: 631–635.
75. Bhalgat, M.K., Haugland, R.P., Pollack, J.S., Swan, S., and Haugland, R.P., 1998 Green- and red-fluorescent nanospheres for the detection of cell surface receptors by flow cytometry. *J Immunol Methods* 219: 57–68.
76. Wang, L. and Tan, W., 2006 Multicolor FRET silica nanoparticles by single wavelength excitation. *Nano Lett* 6: 84–88.
77. Lian, W., Litherland, S.A., Badrane, H., Tan, W., Wu, D., Baker, H.V., Gulig, P.A., Lim, D.V., and Jin, S., 2004 Ultrasensitive detection of biomolecules with fluorescent dye-doped nanoparticles. *Anal Biochem* 334: 135–144.
78. Mattheakis, L.C., Dias, J.M., Choi, Y.J., Gong, J., Bruchez, M.P., Liu, J., and Wang, E., 2004 Optical coding of mammalian cells using semiconductor quantum dots. *Anal Biochem* 327: 200–208.
79. Xu, H., Sha, M.Y., Wong, E.Y., Uphoff, J., Xu, Y., Treadway, J.A., Truong, A., O'Brien, E., Asquith, S., Stubbins, M., Spurr, N.K., Lai, E.H., and Mahoney, W., 2003 Multiplexed SNP genotyping using the Qbead system: A quantum dot-encoded microsphere-based assay. *Nucleic Acids Res* 31: e43.
80. Eastman, P.S., Ruan, W., Doctolero, M., Nuttall, R., de Feo, G., Park, J.S., Chu, J.S.F., Cooke, P., Gray, J.W., Li, S., and Chen, F.F., 2006 Qdot nanobarcodes for multiplexed gene expression analysis. *Nano Lett* 6: 1059–1064.
81. Gunderson, K.L., Kruglyak, S., Graige, M.S., Garcia, F., Kermani, B. G., Zhao, C., Che, D., Dickinson, T., Wickham, E., Bierle, J., Doucet, D., Milewski, M., Yang, R., Siegmund, C., Haas, J., Zhou, L., Oliphant, A., Fan, J.B., Barnard, S., and Chee, M.S., 2004 Decoding randomly ordered DNA arrays. *Genome Res* 14: 870–877.
82. Dejneka, M.J., Streltsov, A., Pal, S., Frutos, A.G., Powell, C.L., Yost, K., Yuen, P.K., Müller, U., and Lahiri, J., 2003 Rare earth-doped glass microbarcodes. *Proc Natl Acad Sci U S A* 100: 389–393.
83. Garcia-Vidal, F.J. and Pendry, J.B., 1996 Collective theory for surface enhanced Raman scattering. *Phys Rev Lett* 77: 1163–6.
84. Helmenstine, A., Uziel, M., and Vo-Dinh, T., 1993 Measurement of DNA adducts using surface-enhanced Raman spectroscopy. *J Toxicol Environ Health* 40: 195–202.

85. Nie, S. and Emory, S.R., 1997 Probing single molecules and single nanoparticles by surface-enhanced Raman scattering. *Science* 275: 1102–1106.
86. Isola, N.R., Stokes, D.L., and Vo-Dinh, T., 1998 Surface-enhanced Raman gene probe for HIV detection. *Anal Chem* 70: 1352–1356.
87. Cao, Y.C., Jin, R., and Mirkin, C.A., 2002 Nanoparticles with Raman spectroscopic fingerprints for DNA and RNA detection. *Science* 297: 1536–1540.
88. Cao, Y.C., Jin, R., Nam, J.M., Thaxton, C.S., and Mirkin, C.A., 2003 Raman dye-labeled nanoparticle probes for proteins. *J Am Chem Soc* 125: 14676–14677.
89. Doering, W.E. and Nie, S., 2003 Spectroscopic tags using dye-embedded nanoparticles and surface-enhanced Raman scattering. *Anal Chem* 75: 6171–6176.
90. Faulds, K., Smith, W.E., and Graham, D., 2004 Jan 15 Evaluation of surface-enhanced resonance Raman scattering for quantitative DNA analysis. *Anal Chem* 76: 412–417.
91. Graham, D., Mallinder, B.J., Whitcombe, D., Watson, N.D., and Smith, W.E., 2002 Simple multiplex genotyping by surface-enhanced resonance Raman scattering. *Anal Chem* 74: 1069–1074.
92. Lu, Y., Liu, G.L., Kim, J., Mejia, Y.X., and Lee, L.P., 2005 Nanophotonic crescent moon structures with sharp edge for ultrasensitive biomolecular detection by local electromagnetic field enhancement effect. *Nano Lett* 5: 119–124.
93. Kraemer, T., Antonenko, V.V., Mortezaei, R., Kulikov, N.V. 2002 Encoding technologies. In *Handbook of Combinatorial Chemistry: Drugs, Catalysts, Materials*, 170–189. Wiley-VCH, Weinheim.
94. Cain, J.T., Clark, W.W., Schaefer, L.A., Ulinski, D.J., Mickle, M.H., and Mandecki, W.M., 2001 Energy harvesting for DNA gene sifting and sorting. *Int J Parallel Distrib Syst Netw* 4: 140–149.
95. Mandecki, W, Pappas, M, Kogan, N, Wang, Z, Zamlynny, B. 2002 Light-powered micro-transponders for high multiple-level analyses of nucleic acids. ACS Symposium Series 815, 57–69.
96. Czarnik, A.W., 1997 Encoding strategies in combinatorial chemistry. *Proc Natl Acad Sci U S A* 94: 12738–12739.
97. Ohlmeyer, M.H., Swanson, R.N., Dillard, L.W., Reader, J.C., Asouline, G., Kobayashi, R., Wigler, M., and Still, W.C., 1993 Complex synthetic chemical libraries indexed with molecular tags. *Proc Natl Acad Sci U S A* 90: 10922–10926.
98. Chun, S., Xu, J., Cheng, J., Ding, L., Winograd, N., and Fenniri, H., 2006 Spectroscopically encoded resins for high throughput imaging time-of-flight secondary ion mass spectrometry. *J Comb Chem* 8: 18–25.
99. Fenniri, H., Chun, S., Ding, L., Zyrianov, Y., and Hallenga, K., 2003 Preparation, physical properties, on-bead binding assay and spectroscopic reliability of 25 barcoded polystyrene-poly (ethylene glycol) graft copolymers. *J Am Chem Soc* 125: 10546–10560.
100. Fenniri, H., Terreau, O., Chun, S., Oh, S.J., Finney, W.F., and Morris, M.D., 2006 Classification of spectroscopically encoded resins by Raman mapping and infrared hyperspectral imaging. *J Comb Chem* 8: 192–198.
101. Su, X., Zhang, J., Sun, L., Koo, T.W., Chan, S., Sundararajan, N., Yamakawa, M., and Berlin, A.A., 2005 Composite organic-inorganic nanoparticles (COINs) with chemically encoded optical signatures. *Nano Lett* 5: 49–54.
102. Brenner, S. and Lerner, R.A., 1992 Encoded combinatorial chemistry. *Proc Natl Acad Sci U S A* 89: 5381–5383.
103. Nam, J.M., Park, S.J., and Mirkin, C.A., 2002 Bio-barcodes based on oligonucleotide-modified nanoparticles. *J Am Chem Soc* 124: 3820–3821.
104. Nam, J.M., Thaxton, C.S., and Mirkin, C.A., 2003 Nanoparticle-based bio-bar codes for the ultrasensitive detection of proteins. *Science* 301: 1884–1886.
105. Storhoff, J.J., Marla, S.S., Bao, P., Hagenow, S., Mehta, H., Lucas, A., Garimella, V., Patno, T.J., Buckingham, W., Cork, W.H., and Müller, U.R., 2004b Gold nanoparticle-based detec-

tion of genomic DNA targets on microarrays using a novel optical detection system. *Biosens Bioelectron* 19: 875–883.

106. Georganopoulou, D.G., Chang, L., Nam, J.M., Thaxton, C.S., Mufson, E.J., Klein, W.L., and Mirkin, C.A., 2005 Nanoparticle-based detection in cerebral spinal fluid of a soluble pathogenic biomarker for Alzheimer's disease. *Proc Natl Acad Sci U S A* 102: 2273–2276.
107. Nam, J.M., Stoeva, S.I., and Mirkin, C.A., 2004 Bio-bar-code-based DNA detection with PCR-like sensitivity, *J Am Chem Soc* 126: 5932–5933.
108. Nam, J.-M., Wise, A.R., and Groves, J.T., 2005 Colorimetric bio-barcode amplification assay for cytokines. *Anal Chem* 77: 6985–6988.
109. Bao, Y.P., Wei, T.F., Lefebvre, P.A., An, H., He, L., Kunkel, G.T., and Müller, U.R., 2006 Detection of protein analytes via nanoparticle-based bio bar code technology. *Anal Chem* 78: 2055–2059.
110. Brongersma, M.L., 2003 Nanoscale photonics: Nanoshells: Gifts in a gold wrapper. *Nat Mater* 2: 296–297.
111. Gerhardt, W., Ljungdahl, L., Collinson, P.O., Lovis, C., Mach, F., Sylven, C., Rasmanis, G., Leinberger, R., Zerback, R., Muller-Bardorff, M., and Katus, H.A., 1997 An improved rapid troponin T test with a decreased detection limit: A multicentre study of the analytical and clinical performance in suspected myocardial damage. *Scand J Clin Lab Invest* 57: 549–557.
112. Siddiqui, J. and Remick, D.G., 2003 Improved sensitivity of colorimetric compared to chemiluminescence ELISAs for cytokine assays. *J Immunoassay Immunochem* 24: 273–283.
113. Saviranta, P., Okon, R., Brinker, A., Warashina, M., Eppinger, J., and Geierstanger, B.H., 2004 Evaluating sandwich immunoassays in microarray format in terms of the ambient analyte regime. *Clin Chem* 50: 1907–1920.
114. Bao, P., Frutos, A.G., Greef, C., Lahiri, J., Müller, U., Peterson, T.C., Warden, L., and Xie, X., 2002 High-sensitivity detection of DNA hybridization on microarrays using resonance light scattering. *Anal Chem* 74: 1792–1797.
115. Wei, J., Mu, Y., Song, D., Fang, X., Liu, X., Bu, L., Zhang, H., Zhang, G., Ding, J., Wang, W., Jin, Q., and Luo, G., 2003 A novel sandwich immunosensing method for measuring cardiac troponin I in sera. *Anal Biochem* 321: 209–216.
116. Ihara, T., Tanaka, S., Chikaura, Y., and Jyo, A., 2004 Preparation of DNA-modified nanoparticles and preliminary study for colorimetric SNP analysis using their selective aggregations. *Nucleic Acids Res* 32: e105.
117. Storhoff, J.J., Lucas, A., Müller, U.R., Bao, Y.P., Senical, M., and Garimella, V., 2006 U.S. Patent Appl. No. 20050250094.

# Chapter 17
# Electrochemical Nanoparticle-Based Sensors

**Joseph Wang**

**Abstract** Electrochemical devices are extremely useful for delivering analytical information in a fast, simple, and low-cost fashion, and are thus uniquely qualified for meeting the demands of point-of-care diagnostics. In particular, nanoparticles offer elegant ways for interfacing biomolecular recognition events with electronic signal transduction, for dramatically amplifying the resulting electrical response, and for designing novel coding strategies. Nanoparticles, such as colloidal gold or inorganic nanocrystals, offer considerable promise as quantitation tags for biological assays owing to their unique amplification and coding capabilities.

## 17.1 Introduction

### *17.1.1 Particle-Based Bioassays*

The emergence of nanotechnology is opening new horizons for the application of nanoparticles in bioanalytical chemistry [1, 2]. Metal nanoparticles linked to biomolecules have received considerable interest in the rapidly growing field of nanobiotechnolgy [3]. Nanoparticles, such as colloidal gold or inorganic nanocrystals, offer great promise as versatile quantitation tags for biological assays owing to their unique amplification and coding capabilities [4, 5]. The enormous amplification afforded by nanoparticle tracers provides the basis for ultrasensitive assays of proteins and nucleic acids [6] and opens up the possibility of detecting disease diagnosis markers present at ultralow levels during early stages of the disease progression. In addition, the novel size-dependent optical and metal-dependent electrical properties of inorganic nanocrystals render

J. Wang
Department of Nano Engineering, UCSD,
La Jolla, 92093, USA

K. Dill et al. (eds.), *Microarrays: Preparation, Microfluidics, Detection Methods, and Biological Applications*,

them ideal for multiplexed coding of biomolecules [7]. Such nanoparticle tags hold considerable promise for highly sensitive bioaffinity and for novel biosensor protocols that employ electronic [8], optical [9], or microgravimetric [10] signal transduction.

This chapter highlights recent developments based on the use of bioconjugated nanoparticles for electrochemical transduction of biomolecular recognition events. Particular attention is given to new signal amplification and coding strategies based on metal and semiconductor nanoparticle quantitation tags for electrochemical bioaffinity assays of nucleic acids and proteins. Such developments provide a pathway for diverse and exciting opportunities.

### *17.1.2 Electrochemical Bioaffinity Assays*

Electroanalytical techniques are concerned with the interplay between electricity and chemistry, namely the measurements of electrical quantities, such as current, potential, or charge, and their relationship to chemical parameters [11]. Controlled-potential (potentiostatic) techniques deal with the study of charge transfer processes at the electrode/solution interface, and are based on dynamic situations. Here, the electrode potential is being used to derive an electron-transfer reaction and the resulting current is measured. The role of the potential is analogous to that of the wavelength in optical measurements. Such a controllable parameter can be viewed as "electron pressure", which forces the chemical species to gain or lose an electron (reduction or oxidation, respectively).

The past two decades have seen major advances in electroanalytical chemistry, including the development of ultramicroelectrodes, the design of modified electrodes, the coupling of biological components or nanoscale materials with electrical transducers, the microfabrication of molecular devices, and the introduction of "smart" sensors and sensor arrays. Electrochemical devices are extremely useful for delivering analytical information in a fast, simple, and low-cost fashion, and are thus uniquely qualified for meeting the demands of decentralized testing. The required instrumentation is simple and can be miniaturized with low power requirements. Readers are referred to a recent book and a review for comprehensive information on electrochemical systems [4, 11].

Electrochemical devices have received considerable recent attention in the development of bioaffinity sensors [12–14]. Such affinity electrochemical biosensors and bioassays exploit selective binding of certain biomolecules (e.g., antibodies, oligonucleotides, glycans) toward the target analyte for triggering useful electrical signals. Electrochemical devices offer elegant routes for interfacing (at the molecular level) the biorecognition binding event and the signal-transduction element. The electrochemical transduction of binding events is commonly detected using enzyme or redox labels. The use of nanoparticle tags in electrochemical detection is relatively new, and offers attractive opportunities for electronic transduction of biomolecular interactions, for the biosensing of proteins, nucleic acids and glycans, and for biodiagnostics, in general.

## 17.2 Nanoparticle-Based Electrochemical Bioaffinity Sensors and Assays

Inspired by the novel use of nanoparticles in optical bioassays [5, 9, 15], recent studies have focused on developing analogous particle-based electrochemical routes for the detection of proteins and nucleic acids. The majority of this work has focused on metal nanoparticles and inorganic (quantum dot) nanocrystals. Additional work included nanowires, polymeric carrier spheres, or magnetic beads. These nanoscale materials offer attractive avenues for interfacing biomolecular recognition events with electrochemical signal transduction, for dramatically amplifying the resulting response, and for designing powerful coding strategies.

Most of these protocols have relied on a highly sensitive stripping-voltammetric measurement of the captured metal tag. The remarkable sensitivity of such electrochemical stripping measurements is attributed to the "built-in" accumulation step, during which the target metals are electroplated onto the working electrode [16]. The preconcentration step is followed by the stripping (measurement) step, which involves the dissolution (stripping) of the deposit. Because the metals are preconcentrated into the electrode by factors of 100–1000, detection limits are lowered by two to three orders of magnitude compared to solution-phase voltammetric measurements. Four to six metals can thus be measured simultaneously in various matrices at concentration levels down to $10^{-11}$ M, utilizing relatively inexpensive and portable instrumentation. Such ultrasensitive electrochemical detection of metal tracers has been accomplished in connection with a variety of new and novel DNA- or protein-linked particle nanostructure networks. The successful realization of these nanoparticle-based signal amplification strategies requires proper attention to nonspecific adsorption issues (see discussion in Section 17.2.4).

In a typical bioassay the DNA probe, antibody, or aptamer is immobilized onto the surface of the working electrode, on the walls of polymeric microwells, or onto functionalized magnetic beads in connection with different anchor chemistries. This can be accomplished in connection with adsorption into microtiter plate wells [17, 18], through the use of a high-density mixed monolayer of alkanethiols on the gold surface [19], or via streptavidin-coated magnetic beads [20]. Electropolymerization represents another attractive route for localizing the probes on small electrode surfaces, as desired for the fabrication of high-density arrays.

The preparation of DNA-functionalized gold nanoparticles or CdS nanocrystals often involves the use of thiolated-terminated oligonucleotides. Antibodies can be conjugated to nanoparticle tracers through common bifunctional linkers (e.g., carbodiimide), coupled to terminal groups on the functionalized nanocrystal.

### 17.2.1 Gold and Silver Metal Tags for Electrochemical Detection of DNA and Proteins

Several groups have developed powerful nanoparticle-based electrochemical bioaffinity assays based on capturing gold [17–20] or silver [21] nanoparticles to the bound target, followed by dissolution and anodic-stripping voltammetric measurement of the solubilized metal tracer. These protocols facilitated the detection of DNA and proteins down to the picomolar and subnanomolar levels. For example, Limoges's group [17] demonstrated an electrochemical metalloimmunoassay based on stripping voltammetric detection of a colloidal gold label (Fig. 17.1). The same group demonstrated the utility of the gold nanoparticle-based stripping protocol for the hybridization detection of the 406-base human cytomegalovirus DNA sequence [18].

Further sensitivity enhancement can be achieved by catalytic enlargement of the gold tag in connection with nanoparticle-promoted precipitation of gold [20] or silver [21–23]. Combining such catalytic enlargement of the metal-particle tracers, with the effective built-in amplification of electrochemical stripping analysis led to subpicomolar detection limits [22]. The silver enhancement involves the chemical reduction of silver ions by hydroquinone to silver metal on the surface of the gold nanoparticles. The silver reduction time must be controlled as a tradeoff between a larger signal enhancement and a larger nonspecific background. A significant reduction of the silver staining background signals was obtained by using an indium–tin oxide (ITO) electrode possessing low silver-enhancing properties or by modifying the gold transducer with a polyelectrolyte multilayer [23]. A simplified gold-nanoparticle based procedure was reported [24], relying on the pulse-voltammetric monitoring of the gold–oxide wave at ~1.20 V at disposable graphite pencil electrode. A detection limit of 0.78 fmol was reported for PCR amplicons

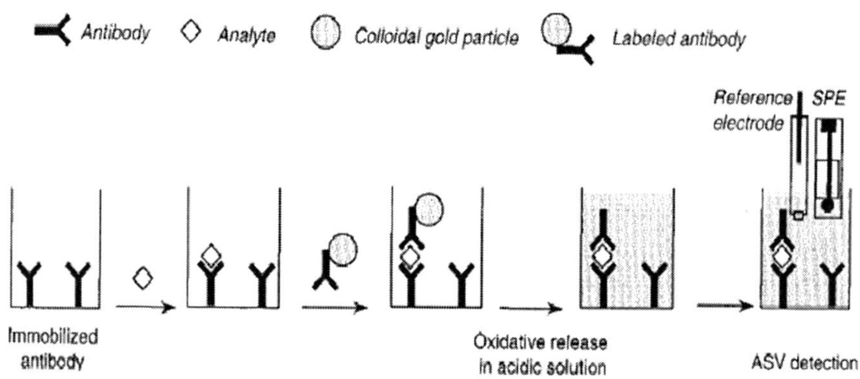


**Fig. 17.1** Sandwich electrochemical immunoassay based on the use gold nanoparticle tags and electrochemical stripping detection of the dissolved tag (based on [17])

bound to the pencil electrode in connection with hybridization to oligonucleotide–nanoparticle conjugates.

The procedures described above have been based on the use of one nanoparticle tag per one binding event. It is possible to further enhance the sensitivity by capturing multiple nanoparticles per binding event. For example, we have demonstrated an electrochemical triple-amplification hybridization assay, combining polymeric carrier-sphere amplifying units (loaded with numerous gold nanoparticles tracers) with the built-in preconcentration of the electrochemical stripping detection and catalytic enlargement of the multiple gold-particle tags [25].

The gold-tagged spheres were prepared by binding biotinylated metal nanoparticles to streptavidin-coated polystyrene spheres. This multiple-amplification route offered a substantial enhancement of the sensitivity and a lowering of the detection limits by approximately three orders of magnitude. Such enlargement of numerous gold nanoparticles tags (on a supporting sphere carrier) represents the fourth generation of amplification (Fig. 17.2), starting with the early use of "first-generation" single gold nanoparticle tags [17, 18, 20] and a gradual increase of the amount of captured gold per binding event.

It is also possible to use gold nanoparticles as carriers of redox markers for amplified biodetection [26]. Gold nanoparticles covered with 6-ferrocenylhexaenthiol were used for this purpose in connection with a sandwich DNA hybridization assay. Due to the elasticity of the DNA strands, the ferrocene/Au-nanoparticle con-

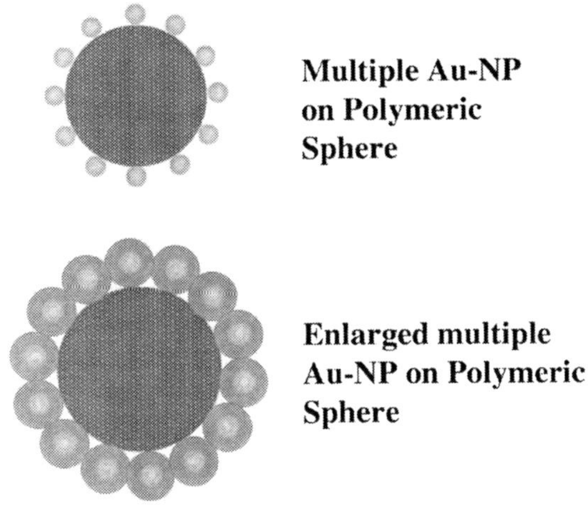


**Fig. 17.2** Different generations of amplificati on platforms for bioelectronic detection based on gold nanoparticle tracers: (A) a single nanoparticle tag; (B) catalytic enlargement of the nanoparticle tag; (C) polymer carrier bead loaded with numerous gold nanoparticle tags; (D) catalytic enlargement of multiple tags on the carrier bead

jugates were positioned in closed proximity to the underlying electrode to allow a facile electron-transfer reaction. An extremely low detection limit of 10 amol was observed, along with linearity up to 150 nM. Applicability to PCR products related to the hepatitis B virus was reported.

It is possible also to detect nucleic acid hybridization based on preparing the metal marker along the DNA backbone [27]. Such a procedure relies on DNA-template induced generation of conducting nanowires as a mode of capturing the metal tracer. The use of DNA as a metallization template has evoked substantial research activity directed to the construction of functional circuits. This approach was applied to growing silver clusters on DNA templates. The use of such DNA-templated growth of metal wires for detecting DNA hybridization consists of the vectorial electrostatic "collection" of silver ions along the captured DNA target, followed by the hydroquinone-induced reductive formation of silver aggregates along the DNA skeleton, along with dissolution and stripping detection of the nanoscale silver cluster.

### *17.2.2 Inorganic Nanocrystal Tags: Toward Electrical Coding*

Semiconductor (quantum dot) nanocrystals have received considerable interest for optical DNA detection due to their unique (tunable-electronic) properties [7]. Recent efforts have demonstrated the utility of such inorganic crystals for improved electrochemical detection of nucleic acids, proteins, and glycans [28–30].

We reported on the detection of DNA hybridization in connection with cadmium-sulfide nanoparticle tags and stripping voltammetric measurements of the cadmium [31]. A nanoparticle-promoted cadmium precipitation was used to enlarge the nanoparticle tag and amplify the stripping DNA hybridization signal. In addition to measurements of the dissolved cadmium ion we demonstrated solid-state measurements following a magnetic collection of the magnetic-bead/DNA-hybrid/CdS-tracer assembly onto a screen-printed electrode transducer. Such a procedure combined the amplification features of nanoparticle/polynucleotides assemblies and highly sensitive chronopotentiometric stripping detection of cadmium, with an effective magnetic isolation of the duplex. The low detection limit (100 fmol) was coupled with good reproducibility (RSD = 6%). A substantially enhanced signal was obtained by encapsulating multiple CdS nanocrystals into the host bead or by loading onto carbon-nanotube carriers [32].

Inorganic nanocrystals offer an electrodiverse population of electronic tags that can be used to detect profiles of multiple target biomolecules. Such metal-dependent electrical properties of inorganic nanocrystals thus render them ideal for simultaneous analysis of different protein or DNA targets [28, 29]. The different electronic signals from multiple metal-sulfide nanoparticles (e.g., CdS, PbS, ZnS, CuS) can thus be readily resolved, hence allowing the detection of multitude binding events in a single run, using a single sensing electrode. Each binding event yielded a distinct voltammetric peak, whose size and position reflected the level and identity, respectively, of the corresponding antigen or DNA target (e.g., Fig. 17.3). The encoding nanoparticles have thus been used to differentiate the signals of four proteins or DNA targets in connection with a sandwich immunoassay and hybridization assay, respectively, along with stripping voltammetry of the corresponding metals. Conducting massively parallel assays (in microwells of microtiter plates or using multichannel microchips, with each microwell or channel carrying out multiple measurements) could thus lead to a high-throughput analysis of proteins or nucleic acids, and offer concentration profiles of multiple biomarkers in relevant clinical samples.

Recent activity in our laboratory has led to large particle-based libraries for electrical coding, based on the judicious design of encoded "identification" beads [33] or striped metal nanowires [34]. By incorporating different predetermined levels (or lengths) of multiple metal markers, such spheres or rods can lead to a large number of recognizable voltammetric signatures, and hence to a reliable identification of a large number of biomolecules. For example, multimetal cylindrical particles can be prepared by template-directed electrochemical synthesis, by plating indium, zinc, bismuth, and copper onto a porous membrane template. Capping the rod with a gold end facilitates its functionalization with a thiolated oligonucleotide probe. Each nanowire thus yields a characteristic multipeak voltammogram, whose peak potentials and current intensities reflect the identity of the corresponding DNA target.

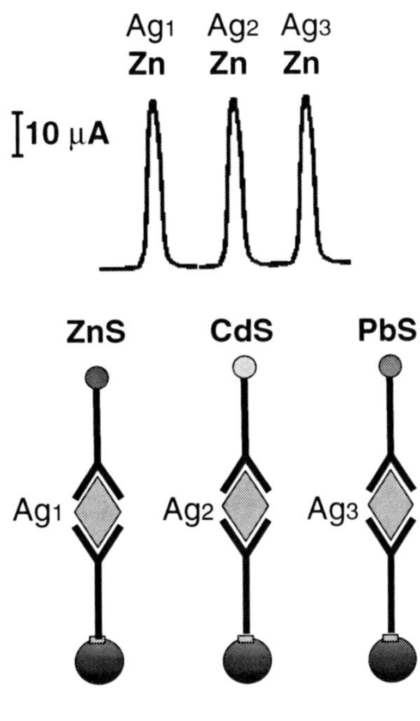


**Fig. 17.3** Use of different quantum-dot tracers for electrical detection of multiple protein targets. Top: stripping voltammogram for a solution containing dissolved ZnS, CdS, and PbS nanoparticle tracers, corresponding to the three protein targets ($Ag_1$–$Ag_2$; based on [29])

Thousands of usable codes could be generated in connection with five to six different potentials and four to five different current intensities. In addition to powerful bioassays, such "identification beads" hold great promise for the identification of counterfeit products and related authenticity testing. The template-directed electrochemical route can also be used for preparing micrometer-long metal nanowire tags for ultransensitive DNA detection ([34]). The linear relationship between the charge passed during the preparation and the size of the resulting nanowire allows tuning of the sensitivity of the electrical DNA assay. For example, nanowires prepared by indium plating into the pores of a host membrane offered a substantially lower detection limit (250 zmol) compared to analogous bioassays based on spherical nanoparticle tags. Indium offers an attractive stripping-voltammetric behavior and is not normally present in biological samples or reagents. Solid-state chronopotentiometric measurements of the indium nanowires can be accomplished using a magnetic collection of the DNA-linked particle assembly onto the screen-printed working electrode.

The amplification and coding features of inorganic nanocrystals have been shown to be extremely useful for monitoring aptamer–protein interactions. Recently we described a highly sensitive and selective simultaneous electrochemical displacement assay of several proteins using a self-assembled monolayer of several thiolated aptamers conjugated to proteins carrying different inorganic nanocrystals [19]. Stripping-voltammetric detection of the nondisplaced nanocrystal tracers resulted in a remarkably low detection limit (down to the attomole level), that is significantly lower than those of existing aptamer biosensors (Fig. 17.4). Unlike two-step sandwich assays used in early QD-based electronic hybridization or immunoassays [27, 29], this aptamer biosensing scheme relies on a single-step displacement protocol. Aptamers hold considerable promise for sensitive displacement assays because the tagged protein has a significantly lower affinity to the aptamer compared to the unmodified analyte.

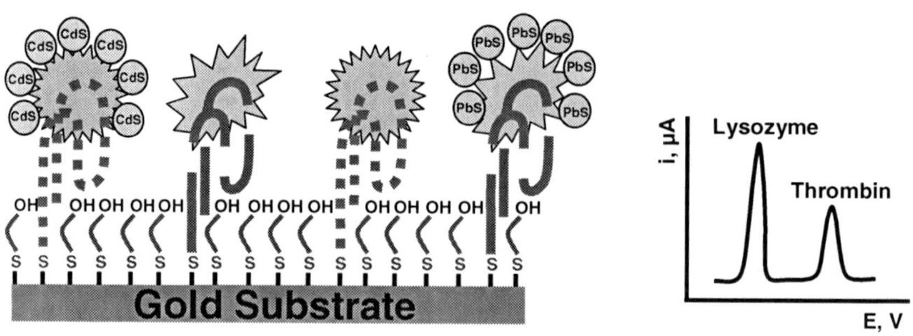


**Fig. 17.4** Aptamer/quantum-dot (QD) based dual-analyte biosensor, involving displacement of the tagged proteins by the target analytes. The protocol involves the co-immobilization of several thiolated aptamers, along with binding of the corresponding QD-tagged proteins on a gold surface, addition of the protein sample, and monitoring the displacement through electrochemical detection of the remaining nanocrystals (based on [19])

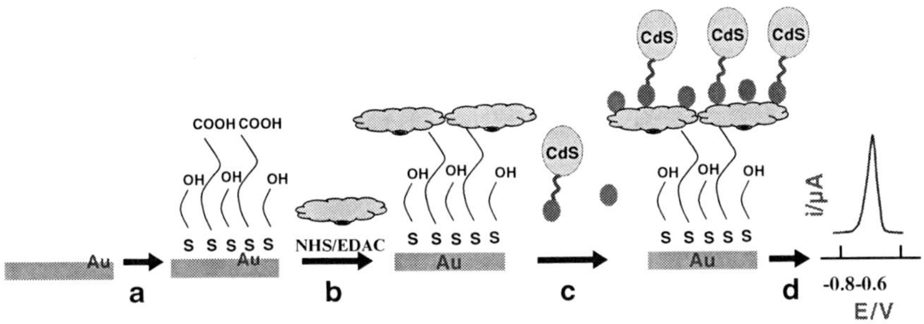


**Fig. 17.5** Operation of the nanoparticle-based bioelectronic sensor for glycans involving competition of the tagged sugar with the target analy tes for the binding sites of the immobilized lectin: (a) mixed self-assembled monolayer on the gold substrate; (b) covalent immobilization of the lectin; (c) addition of the tagged and untagged sugars; (d) dissolution of the captured nanocrystals followed by their stripping-voltammetric detection (based on [35])

Very recently we reported on a nanoparticle-based electrochemical biosensing of sugars based on their interaction with surface-functionalized lectins [35]. As illustrated in Fig. 17.5, the new sugar biosensor involves the immobilization of the lectin, the carbohydrate recognition element, onto the gold surface (in connection with a mixed self-assembly monolayer and EDAC/NHS coupling, (a) and (b), competition between a nanocrystal (CdS)-labeled sugar and the target sugar for the carbohydrate binding sites on lectins (c), and monitoring the extent of competition through highly sensitive electrochemical stripping detection of the captured nanocrystal (d). These developments will allow decentralized testing for disease-related sugar markers, glycan profiles, and lectin–sugar interactions to be performed more rapidly, sensitively, inexpensively, and reliably.

Metal nanoparticles have also been shown to be useful for electronic coding of single nucleotide polymorphisms (SNP; [36]). This protocol involves the hybridization of monobase-modified gold nanoparticles with the mismatched bases. The binding event leads to changes in the gold oxide peak and holds great promise for coding all mutational changes. Analogous SNP electronic coding protocols, based on different inorganic nanocrystals, have been developed in our laboratory [37]. The protocol involves the addition of CdS, PbS, ZnS, and CuS crystals linked to cytidine, guanosine, adenosine, and thymidine mononucleotides, respectively. Each mutation captures a different nanocrystal–mononucleotide conjugate via base pairing, to give a distinct electronic fingerprint.

### *17.2.3 Use of Magnetic Beads*

Several of the procedures described above [20, 22] have combined the signal amplification of electrochemical stripping voltammetry with an effective discrimination against nontarget biomolecules. In addition to efficient isolation of the duplex, magnetic beads can be useful open for triggering and controlling the electrochemical detection of DNA and proteins [38, 39].

For example, an attractive magnetic triggering of the electrical DNA detection was accomplished via a "magnetic" collection of the magnetic-bead/DNA-hybrid/metal-tracer assembly onto a screen-printed electrode transducer that allowed direct electrical contact of the silver precipitate [40]. This bioassay involved the hybridization of a target oligonucleotide to probe-coated magnetic beads, followed by binding of the streptavidin-coated gold nanoparticles to the captured target, catalytic silver precipitation on the gold-particle tags, a magnetic collection of the DNA-linked particle assembly and solid-state chronopotentiometric stripping detection. Such a magnetic collection avenue greatly simplifies the electrochemical detection of metal tracers as it eliminates the acid dissolution step.

Magnetic beads have also been used for triggering the electron-transfer reactions of DNA [40]. Changing the position of the magnet (under the thick film electrode) was thus used for switching on/off the DNA oxidation (through attraction and removal of DNA functionalized-magnetic spheres). The process was reversed and repeated upon switching the position of the magnet, with and without oxidation signals in the presence and absence of the magnetic field, respectively. Such magnetic triggering of the DNA oxidation holds great promise for DNA arrays (based on closely spaced electrodes and guanine-free inosine-substituted probes). Willner and coworkers described an amplified detection of viral DNA and of single-base mismatches using oligonucleotide-functionalized magnetic spheres and an electrochemoluminescence (ECL) detection [41]. The magnetic attraction of the labeled magnetic spheres and their rotation on the electrode surface was used to amplify the ECL signal.

Magnetic spheres can also be used as tags for DNA hybridization detection in connection with stripping voltammetric measurements of their iron content [42]. A related protocol, developed in the same study, involved probes labeled with gold-coated iron core-shell nanoparticles. In both cases, the captured iron-containing particles were dissolved following the hybridization, and the released iron was quantified by adsorptive stripping voltammetry in the presence of the ligand 1-nitroso-2-naphthol and a bromate catalyst. Core-shell copper–gold nanoparticle tracers were also shown to be useful for combining the attractive electrochemical behavior of the copper core with the attractive surface modification properties of the gold shell [43].

### *17.2.4 Ultrasensitive Biodetection Based on Multiple Amplification Schemes*

We already discussed several amplification processes such as catalytic enlargement of the metal tracer and its electrolytic accumulation onto the electrode surface. Such protocols have been based on the use of one reporter per one binding event. It is possible to further enhance the sensitivity by using multiple tracers per

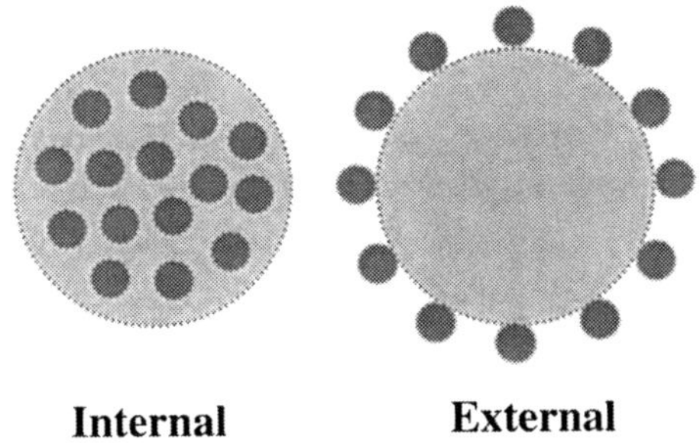


**Fig. 17.6** Polymeric "carrier" beads amplifying units based on loading numerous redox tags externally (on their outer surface) or internally (via encapsulation)

binding event. This can be achieved using polymeric microbeads loaded with multiple redox tracers externally (on their surface) or internally (via encapsulation; Fig. 17.6). Coupled with additional amplification units and processes, such bead-based multiamplification protocols meet the high sensitivity demands of electrochemical affinity biosensors. For example, a substantial sensitivity enhancement has been observed in sandwich bioassays involving capturing of polymeric spheres loaded with numerous gold nanoparticles tracers [25], using the triple-amplification scheme described in Section 17.2.1.

Internal encapsulation of electroactive tracers within polymeric carrier beads offers an attractive alternative to their external loading. For example, a remarkably sensitive electrochemical detection of DNA hybridization was reported based on polystyrene beads impregnated with a ferrocene marker [44]. The resulting "electroactive beads" were capable of carrying a huge number of molecules of the ferrocene tracer and thus offer a remarkable amplification of single hybridization events. This allowed electrochemical detection of the DNA target down to the $5.1 \times 10^{-21}$mol level (~31,000 molecules) using a 20 min hybridization time and "release" of the marker in an organic medium.

The dramatic signal amplification advantage is coupled with an effective discrimination against a large excess ($10^7$-fold) of nontarget nucleic acids. SEM images indicated that the 10 μm electroactive beads are cross-linked to the smaller (~0.8 μm) magnetic spheres through the DNA hybrid. Other marker encapsulation strategies hold great promise for electrochemical bioassays. Particularly attractive are nanoencapsulated microcrystalline particles, prepared by the layer-by-layer (LBL) assembly technique, that offer large marker/biomolecule ratios and greatly amplified bioassays [45].

A critical requirement for the successful realization of ultrasensitive nanoparticle-based electrochemical bioassays is the ability to minimize nonspecific binding of coexisting biomolecules. Proper attention should be given to the surface chemistry and to the washing step as desired for minimizing nonspecific adsorption of the nanoparticle amplifiers. Surface blocking steps should thus be employed to avoid amplification of background signals (associated with nonspecific adsorption of the nanoparticle amplifiers).

### 17.2.5 Nanoparticle-Induced Conductivity Detection

The formation of conductive domains as a result of biomolecular interactions provides an alternative and attractive route for electrochemical transduction of biorecognition events. Nanoparticle-induced changes in the conductivity across a microelectrode gap were exploited by Mirkin's team for highly sensitive and selective detection of DNA hybridization [46]. This protocol involved capturing the nanoparticle-tagged DNA targets by probes immobilized within the gap between the two closely spaced microelectrodes, and a subsequent silver precipitation (Fig. 17.7). This resulted in a conductive metal layer bridging the gap, and led to a measurable conductivity signal. Such hybridization-induced conductivity signals, associated with resistance changes across the electrode gap, offered high sensitivity with detection limit down to the 0.5 picomolar level. Control of the salt concentration allowed an excellent mismatch discrimination without thermal stringency.

Similarly, Velev and Kaler exploited the catalytic features of nanoparticles for analogous conductivity immunoassays of proteins in connection with antibody-functionalized latex spheres placed between two closely spaced electrodes [47]. A sandwich immunoassay thus led to the binding of a secondary gold-labeled antibody, followed by catalytic precipitation of a silver layer "bridging" the two electrodes. Such generation of conductive bridging paths enabled the detection of human IgG down to the $2 \times 10^{-13}$ M level.

## 17.3 Conclusions and Future Directions

Recent years have witnessed the development of a variety of nanomaterial-based bioelectronic devices exhibiting novel functions. The use of nanomaterials in such sensing devices has taken off rapidly and will surely continue to expand. Nanomaterials offer new and unique opportunities for designing powerful electrochemical bioassays and biosensors. The ability to tailor the composition, size, and shape of nanoscale materials is expected to lead to entirely new types of electrochemical biosensors. The examples described above demonstrate the broad potential of bioconjugated nanoparticles for electrochemical transduction of biomolecular recognition events.

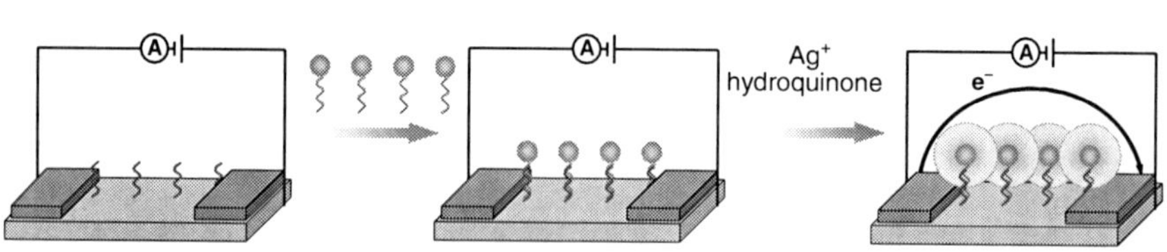


**Fig. 17.7** Conductivity detection of nanoparticle-based microelectrode arrays. The capture of the nanoparticle-tagged DNA targets by probes confined to the gap, and a subsequent silver enlargement, electrically short the gap and lead to a measurable conductivity signal (based on [46])

The use of nanoparticles in electrochemical bioassay has taken off rapidly and will surely continue to expand. Metal and semiconductor nanoparticles, 1-D nanotubes, and nanowires have rapidly become attractive labels for bioaffinity assays, offering unique signal amplification and multiplexing capabilities. The coupling of different nanoparticle-based amplification platforms and amplification processes dramatically enhances the intensity of the analytical signal and leads to ultrasensitive bioassays of proteins and nucleic acids. The enormous amplification afforded by such nanoparticle-based amplification schemes opens up the possibility of detecting disease markers, infectious agents, or biothreat agents that cannot be measured by conventional methods. These highly sensitive biodetection schemes could provide an early detection of diseases or warning of a terrorist attack. Such ultrasensitive bioelectroanalytical protocols could not be achieved by standard electrochemical methods.

A critical requirement for the successful realization of ultrasensitive nanoparticle-based electrochemical bioassays is the ability to minimize nonspecific binding of coexisting nontarget biomolecules. Proper attention should be given to the surface chemistry and to the washing step as desired for minimizing nonspecific adsorption of the nanoparticle amplifiers. The combination of high sensitivity, specificity, and multitarget detection capabilities permits nanoparticle-based electronic bioaffinity assays to rival the most advanced optical protocols. Such nanomaterials-based bioelectronic assays and devices are expected to have a major impact upon clinical diagnostics, environmental monitoring, security surveillance, or for ensuring our food safety. Future innovative research is expected to lead to advanced particle-based electrochemical biodetection strategies, that coupled with other major technological advances will result in cost-effective, easy-to-use, handheld portable devices and chip-based array formats for broad range of target biomolecules.

**Acknowledgment** Financial support from the National Science Foundation (Grant Number CHE 0506529) and NIH (R01A 1056047-01 and R01 EP 0002189) is gratefully acknowledged.

## References

1. Niemeyer, C.M. (2001). Nanoparticles, proteins, and nucleic acids: Biotechnology meets materials science, *Angew. Chem. Int. Ed.* 40, 4128.
2. Alivisatos, P. (2004). The use of nanocrystals in biological detection, *Nature Biotechnology*, 22, 47.
3. Niemeyer, C.M. and Mirkin, C.A. (2004). *Nanobiotechnology*, Wiley-VCH, Weinheim.
4. Katz, E. and Willner, I. (2004). Integrated nanoparticle-biomolecule hybrid systems: Synthesis, properties, and applications, *Angew Chemie Int. Ed.* 43, 6042.
5. Rosi, N.L., Mirkin, C.A. (2005). Nanostructures in biodiagnostics, *Chem. Rev.*, 105, 1547.
6. Wang, J. (2005a). Nanomaterial-based amplified transduction of biomoleular interactions. *Small*, 1, 1036.
7. Han, M., et al. (2001) Quantum-dot-tagged microbeads for multiplexed optical coding of biomolecules. *Nat Biotechnol.*, 19, 631.
8. Wang J. (2003a). Nanoparticle-based electrical DNA assays, *Anal. Chim. Acta*, 500, 247.

9. Storhoff, J.J., Elghanian, R., Mucic, R.C., Mirkin, C.A., and Letsinger, R.L. (1998). One-pot colorimetric differentiation of polynucleotides with single base imperfections using gold nanoparticle probes, *J. Am. Chem. Soc.* 120, 1959.
10. Willner, I., Patolsky, F., Weizmann, Y., and Willner, B. (2002). Amplified detection of single-base mismatches in DNA using micro gravimetric quartz-crystal-microbalance transduction, *Talanta*, 56, 847.
11. Wang, J. (2006). *Analytical Electrochemistry* (3rd Edition), Wiley, New York.
12. Mikkelsen, S.R. (1996). Electrochemical biosensors for DNA sequence detection, *Electroanalysis,* 8, 15.
13. Palecek, E. and Fojta, M. (2001). Detecting DNA hybridization and damage, *Anal. Chem.* 73, 75A.
14. Skladal P. (1997). Advances in electrochemical immunosensors, *Electroanalysis*, 9, 737.
15. Taton, T.A., Mirkin, C.A., and Letsinger, R.L. (2000). Scanometric DNA array detection with nanoparticle probes, *Science* 289, 1757.
16. Wang, J. (1985). *Stripping Analysis*, VCH, New York.
17. Dequaire, M., Degrand, C., and Limoges, B. (2000). An electrochemical metalloimmunoassay based on a colloidal gold label, *Anal. Chem.* 72, 5521.
18. Authier, L., Grossiord, C., Berssier, P., and Limoges, B. (2001). Gold nanoparticle-based quantitative electrochemical detection of amplified human Cytomegalovirus DNA using disposable microband electrodes, *Anal. Chem.* 73, 4450.
19. Hansen, J., Wang, J., Kawde, A., Xiang, Y., Gothelf, K.V., and Collins, G. (2006). Quantum-dot/aptamer-based ultrasensitive multi-analyte electrochemical biosensor. *J. Am. Chem. Soc.,* 128, 2228.
20. Wang, J., Xu, D., Kawde, A., and Polsky, R. (2001a). Metal nanoparticle-based electrochemical stripping potentiometric detection of DNA hybridization, *Anal. Chem.* 73, 5576.
21. Cai, H., Xu, Y., Zhu, N., He, P., and Fang, Y. (2002). An electrochemical DNA hybridization detection assay based on a silver nanoparticle label, *Analyst*, 127, 803.
22. Wang, J., Polsky, R., and Danke, X. (2001b). Silver-enhanced colloidal gold electrical detection of DNA hybridization, *Langmuir,* 17, 5739.
23. Lee, T.M.H., Li, L.L., and Hsing, I.M. (2003). Enhanced electrochemical detection of DNA hybridization based on electrode-surface modification. *Langmuir*, 19, 4338.
24. Ozsoz, M., Erdem, A., Kerman, K., Okzan, D., Tugrul, B., Topcuoglo, N., Ekren, H., and Taylan, M. (2003). Electrochemical genosensor based on colloidal gold nanoparticles for the detection of factor V eiden mutation using disposable pencil graphite electrodes, *Anal. Chem.* 75, 2181.
25. Kawde, A. and Wang, J. (2004). Amplified electrical transduction of DNA hybridization based on polymeric beads loaded with multiple gold nanoparticles tags, *Electroanalysis*, 16, 101.
26. Wang, J., Li, J., Baca, A., Hu, J., Zhou, F., Yan, W., and Pang, D.W. (2003b). Amplified voltammetric detection of DNA hybridization via oxidation of ferrocene caps on gold nanoparticle/streptavidin conjugates, *Anal. Chem.* 75, 3941.
27. Wang, J., Rincon, O., Polsky, R., and Dominguez, E. (2003c). Electrochemical detection of DNA hybridization based on DNA-templated assembly of silver cluster, *Electrochem. Commun.*, 5, 83.
28. Wang, J., Liu, G., and Merkoçi, A. (2003d). Electrochemical coding technology for simultaneous detection of multiple DNA targets, *J. Am. Chem. Soc.*, 125, 3214.
29. Liu, G., Wang, J., Kim, J., Jan, M., and Collins, G. (2004). Electrochemical coding for multiplexed immunoassays of proteins. *Anal Chem.*, 76, 7126.
30. Escosura-Muniz, A. de la Ambrosie, A., and Merckici, A. (2008). Electrochemical analysis with nanoparticle-based biosystems, *Trends Anal. Chem.* 27, 568.
31. Wang, J., Liu, G. and Polsky, R. (2002a). Electrochemical stripping detection of DNA hybridization based on CdS nanoparticle tags, *Electrochemistry Commun.,* 4, 819.
32. Wang, J., Liu, G., Jan, R., and Zhu, Q. (2003e). Electrochemical detection of DNA hybridization based on carbon-nanotubes loaded with CdS tags, *Electrochemistry Commun.*, 5, 1000.

33. Wang, J., Liu, G., and Rivas, G. (2003f). Encoded beads for electrochemical identification, *Anal. Chem.*, 75, 4661.
34. Wang, J., Liu, G., and Zhou, J. (2003g). Indium microrod tags for electrical detection of DNA hybridization, *Anal. Chem.*, 75, 6218.
35. Dai, Z., Kawde, A., Xiang, Y., La Belle, J., Gerlach, J., Bhavanandan, V.P., Joshi, J., and Wang, J. (2006). Nanoparticle-based bioelectronic sensing of glycan-lectin interactions, *J. Am. Chem. Soc.*, 128, 10018.
36. Kerman, K., Saito, M., Morita, Y., Takamura, Y., Ozsoz, M., and Tamiya, E. (2004). Electrochemical coding of single-nucleotide polymorphisms by monobase-modified gold nanoparticles, *Anal. Chem.* 76, 1877.
37. Wang, J., Lee, T., and Liu, G. (2005b), Nanocrystal-based bioelectronics coding of SNP, *J. Am. Chem. Soc.* 127, 38.
38. Wang, J., Xu, D., and Polsky, R.(2002b). Magnetically-induced solid-state electrochemical detection of DNA hybridization, *J. Am. Chem. Soc.*, 124, 4208.
39. Palecek, E., Fojta, M., and Jelen, F. (2002). New approaches in the development of DNA sensors: Hybridization and electrochemical detection of DNA and RNA at two different surfaces, *Bioelectrochemistry* 56, 85.
40. Wang, J. and Kawde, A. (2002c). Magnetic-field stimulated DNA oxidation, *Electrochemistry Commun.*, 4, 349.
41. Patolsky, F., Weizmann, Y., Katz, E., and Willner, I. (2003). Magnetically amplified DNA assays (MADA): Sensing of viral DNA and single-base mismatches by using nucleic acid modified magnetic particles: *Angew Chemie Int. Ed.* 42, 2372.
42. Wang, J., Liu, G.D., and Merkoci, A. (2003i). Particle-based detection of DNA hybridization using electrochemical stripping measurements of an iron tracer, *Anal. Chim. Acta*, 482, 149.
43. Cai, H., Zhu, N., Jiang, Y., He, P. and Fang, Y.Z. (2003), Cu-Au alloy nanoparticle as oligo nucleotides lables for electrochemical stripping detection of DNA hybridization. *Bioelectronis*, 18, 1311.
44. Wang, J., Polsky, P., Merkoci, A., and Turner, K. (2003h). Electroactive beads for ultrasensitive DNA detection, *Langmuir*, 19, 989.
45. Trau, D., Yang, W., Seydack, M., Carusu, F., and Renneberg, R. (2002). Nanoencapsulated microcrystalline particles for superamplified biochemical assays, *Anal. Chem.* 74, 5480.
46. Park, S., Taton, T.A., and Mirkin, C.A. (2002). Array-based electrical detection of DNA with nanoparticle probes, *Science*, 295, 1503.
47. Velev, O.D. and Kaler, E.W. (1999). In situ assembly of colloidal particles into miniaturized biosensors, *Langmuir*, 15, 3693.

# Index

Printed in the United States
136882LV00005B/72/P